TABLES OF POWERS AND ROOTS

No.	Squares	Cubes	Square Roots	Cube Roots	No.	Squares	Cubes	Square Roots	Cube Roots
1	1	1	1.000	1.000	51	2 601	132,651	7.141	3.708
2	4	8	1.414	1.259	52	2 704	140,608	7.211	3.732
3	9	27	1.732	1.442	53	2 809	148,877	7.280	3.756
4	16	64	2.000	1.587	54	2 916	157,464	7.348	3.779
5	25	125	2.236	1.709	55	3 025	166,375	7.416	3.802
6	36	216	2.449	1.817	56	3 136	175,616	7.483	3.825
7	49	343	2.645	1.912	57	3 249	185,193	7.549	3.848
8	64	512	2.828	2.000	58	3 364	195,112	7.615	3.870
9	81	729	3.000	2.080	59	3 481	205,379	7.681	3.892
10	100	1 000	3.162	2.154	60	3 600	216,000	7.745	3.914
11	121	1 331	3.316	2.223	61	3 721	226,981	7.810	3.936
12	144	1 728	3.464	2.289	62	3 844	238,328	7.874	3.957
13	169	2 197	3.605	2.351	63	3 969	250,047	7.937	3.979
14	196	2 744	3.741	2.410	64	4 096	262,144	8.000	4.000
15	225	3 375	3.872	2.466	65	4 225	274,625	8.062	4.020
16	256	4 096	4.000	2.519	66	4 356	287,496	8.124	4.041
17	289	4 913	4.123	2.571	67	4 489	300,763	8.185	4.061
18	324	5 832	4.242	2.620	68	4 624	314,432	8.246	4.081
19	361	6 859	4.358	2.668	69	4 761	328,509	8.306	4.101
20	400	8 000	4.472	2.714	70	4 900	343,000	8.366	4.121
21	441	9 261	4.582	2.758	71	5 041	357,911	8.426	4.140
22	484	10,648	4.690	2.802	72	5 184	373,248	8.485	4.160
23	529	12,167	4.795	2.843	73	5 329	389,017	8.544	4.179
24	576	13,824	4.898	2.884	74	5 476	405,224	8.602	4.198
25	625	15,625	5.000	2.924	75	5 625	421,875	8.660	4.217
26	676	17,576	5.099	2.962	76	5 776	438,976	8.717	4.235
27	729	19,683	5.196	3.000	77	5 929	456,533	8.774	4.254
28	784	21,952	5.291	3.036	78	6 084	474,552	8.831	4.272
29	841	24,389	5.385	3.072	79	6 241	493,039	8.888	4.290
30	900	27,000	5.477	3.107	80	6 400	512,000	8.944	4.308
31	961	29,791	5.567	3.141	81	6 561	531,441	9.000	4.326
32	1 024	32,768	5.656	3.174	82	6 724	551,368	9.055	4.344
33	1 089	35,937	5.744	3.207	83	6 889	571,787	9.110	4.362
34	1 156	39,304	5.830	3.239	84	7 056	592,704	9.165	4.379
35	1 225	42,875	5.916	3.271	85	7 225	614,125	9.219	4.396
36	1 296	46,656	6.000	3.301	86	7 396	636,056	9.273	4.414
37	1 369	50,653	6.082	3.332	87	7 569	658,503	9.327	4.431
38	1 444	54,872	6.164	3.361	88	7 744	681,472	9.380	4.447
39	1 521	59,319	6.244	3.391	89	7 921	704,969	9.433	4.464
40	1 600	64,000	6.324	3.419	90	8 100	729,000	9.486	4.481
41	1 681	68,921	6.403	3.448	91	8 281	753,571	9.539	4.497
42	1 764	74,088	6.480	3.476	92	8 464	778,688	9.591	4.514
43	1 849	79,507	6.557	3.503	93	8 649	804,357	9.643	4.530
44	1 936	85,184	6.633	3.530	94	8 836	830,584	9.695	4.546
45	2 025	91,125	6.708	3.556	95	9 025	857,375	9.746	4.562
46	2 116	97,336	6.782	3.583	96	9 216	884,736	9.797	4.578
47	2 209	103,823	6.855	3.608	97	9 409	912,673	9.848	4.594
48	2 304	110,592	6.928	3.634	98	9 604	941,192	9.899	4.610
49	2 401	117,649	7.000	3.659	99	9 801	970,299	9.949	4.626
50	2 500	125,000	7.071	3.684	100	10,000	1,000,000	10.000	4.641

COLLEGE
ALGEBRA

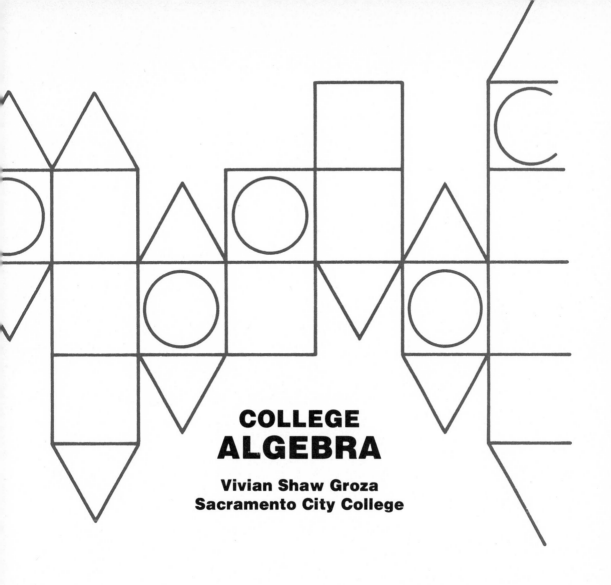

COLLEGE
ALGEBRA

Vivian Shaw Groza
Sacramento City College

1980 Saunders College

Philadelphia

College Algebra

Copyright © 1980 by Saunders College/Holt, Rinehart and Winston.
All rights reserved.

Material on pages 188, 193–194, 196, 235, 236, 237, 238, 239, 240, 241,
242, 349, 350 and 351 are reprinted from *Precalculus Mathematics* by
Vivian Shaw Groza and Susanne Shelley. Copyright © 1972 by Holt,
Rinehart and Winston, Inc. Reprinted by permission of Holt, Rinehart and Winston.

Library of Congress Cataloging in Publication Data

Groza, Vivian Shaw.
  College algebra.

  Includes index.
  1. Algebra. I. Title.
QA154.2.G75   512′.9   79-25365
ISBN 0-03-040376-6

Printed in the United States of America
0 1 2 3   032   9 8 7 6 5 4 3 2 1

PREFACE

I have written this text for the student who has completed the equivalent of one and a half to two years of high school algebra and now wants a more advanced course in preparation for subjects such as calculus and statistics.

Having taught algebra for more than 25 years, I am convinced that the student wants a book that he can easily read and understand. While I have often heard remarks that students rarely read the discussion, it has been my experience that students try to read it and will read it if they possibly can. On the other hand, students complain most about the examples and exercises—especially about the lack of continuity between them. The student wants the subject to make sense as a whole: how did this concept originate? what does it lead to? how is it involved in the development of other ideas? how is it used in applications?

To be consistent with the level of mathematical maturity the student has gained, a course called "College Algebra" should develop an appreciation and understanding of algebra as a logically developed mathematical system. However, I feel that this development should be moderate in scope and should avoid undue rigor and excessive symbolism.

Therefore, in writing this text, I have used the following guidelines:
1. Clear, simple language that is easy to read and understand.
2. Thorough, mathematically sound development of concepts.
3. Well-developed motivation for concepts.
4. Generous number of examples, appropriate to the discussion.
5. Fully explained examples keyed to the exercises.
6. Great variety of applied problems related to concepts of the unit.
7. Numerous exercises, with graduated levels of difficulty.

Each chapter contains a Chapter Review keyed to the chapter sections. If a student has difficulty with a particular problem, he can easily locate the section of the chapter that he needs to restudy. The answers to Chapter Reviews and to the odd-numbered exercises are in the Appendix.

There is a separate Instructor's Manual that contains answers to even-numbered exercises. This manual also has four Chapter Tests for each chapter as well as the answers to these tests. The four tests for each chapter are parallel, and any one of them could be used as a Sample Test or as a Pretest for the chapter.

I am very grateful to the following persons who have assisted in preparing the manuscript for production: Barbara Schott, Acquisition Editor for Holt, Rinehart and Winston, for her excellent services; and Sally Ann Sweeney, for her excellent technical review. I also want to thank the following persons for their constructive criticisms and helpful suggestions, which were used in the preparation of the final manuscript: George Anderson, Central Piedmont Community College; Howard Bird, St. Cloud State University; Paul D. Blanchard, University of Wisconsin at Eau Claire; Charles M. Brennan, Bloomsburg State College; W. Homer Carlisle, Southwest Texas State University; Glynn Creamer, Eastern Kentucky University; Leroy Estergard, Brevard Community College; Joan Golliday, Santa Fe Community College; J. D. Hepworth, Weber State College; Kendell Hyde, Weber State College; and Joe C. Prater, University of Southern Colorado.

V. S. G.
Sacramento, California

CONTENTS

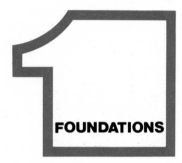

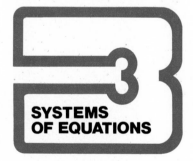

FUNCTIONS AND RELATIONS

SYSTEMS OF EQUATIONS

MATRICES, DETERMINANTS, APPLICATIONS TO SYSTEMS

EXPONENTS AND RADICALS

CONTENTS

1 FOUNDATIONS

Both algebra and arithmetic deal with numbers, relations between numbers, and operations on numbers. Algebra, however, is more general than arithmetic. A few hundred years ago, algebra was known as the "Greater Art" and arithmetic as the "Lesser Art." Algebra was developed to explain number properties and to provide methods for solving problems that have numbers for solutions.

The Pythagoreans, a society of Greeks that lived around 500 B.C., had as their motto "Number rules the Universe." Today it would be even easier to believe this. Numbers and applications of algebra appear in almost all aspects of our everyday life; in our use of time, money, and measurements, in business, economics, art, architecture, nature, music, biology, engineering, electronics, psychology, and on and on.

Algebra is a mathematical system and, as such, similar to plane geometry; it has a logical structure. First, undefined terms are stated. For algebra, these consist of a set of numbers, relations, and operations. Definitions are then formed and assumptions, also called axioms, or postulates, are made about the undefined terms and definitions. A deductive reasoning process is used to obtain more statements about the subject matter. These statements, called theorems, are said to be proved statements. Most of the subject matter of a mathematical system consists of theorems.

In Chapter 1 we examine the set of real numbers, the structure of the real number system, and the set of complex numbers.

In a strict mathematical sense, for the development shown in this text, the set of real numbers is an undefined set. However we are motivated by a knowledge of these numbers that we obtained from our study of arithmetic. So we begin by reviewing special sets and subsets of real numbers and some of their properties. Then the real number system is exhibited as a logical structure, called a field. After this the set of complex numbers is introduced.

1.1 SETS OF REAL NUMBERS

Algebra is concerned with numbers, relations between numbers, and operations on numbers. Symbols are used to specify the numbers, relations, and operations. Ideas from the theory of sets are used to explain the subject matter of algebra.

A **set** is a well-defined collection of objects. This means that it is possible to determine if a given object belongs to the set or not. The objects in the set are also called **members** or **elements** of the set.

If a set has no members, it is called the **empty set,** or **null set**, and is designated in symbols as $\emptyset$ or as { }.

If a set contains all the members that are used in a particular discussion, it is called the **universal set** for the discussion.

The universal set for algebra is the set of complex numbers which consists of real numbers and imaginary numbers. Imaginary numbers are discussed in Section 1.5. For the first four sections of this chapter, the set of real numbers is selected as the universal set.

A **real number** is any number that has either a finite or infinite decimal representation. In other words, a real number can be written as a decimal, using the Hindu-Arabic decimal notation with which we are familiar.

Examples of real numbers:

$$\left.\begin{array}{l} 5 = 5.0 \\ -5 = -5.0 \\ \frac{17}{4} = 4.25 \end{array}\right\} \quad \text{terminating decimals}$$

$$\left.\begin{array}{l} -4\frac{1}{3} = -4.33333\overline{3}\ldots \\ \frac{71}{33} = 2.1515\overline{15}\ldots \end{array}\right\} \quad \text{infinite repeating decimals*}$$

$$\left.\begin{array}{l} \sqrt{2} = 1.41421\ldots \\ \pi = 3.14159\ldots \end{array}\right\} \quad \text{infinite, nonrepeating decimals}$$

*A bar is drawn over a digit or block of digits to indicate that the digit or block of digits repeat.

1.1 SETS OF REAL NUMBERS

There is a **one-to-one correspondence** between the set of real numbers and the set of points on a number line as shown in Figure 1.1.

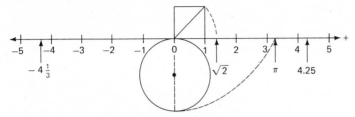

FIG. 1.1. The real number line.

This means there is exactly one real number for each point on the number line and exactly one point on the number line for each real number.

Both letters and numerals, such as those shown in the previous examples, are used as the names of numbers.

A letter (*x, y,* or *z* are commonly used) is called a **variable** if it names a number that belongs to a given set of numbers but whose value is not specified.

A letter is called a **constant** if it names exactly one number.

The expressions $3x^2 + 4x + 2$ and $x^2 - 7x + 6$ are special cases of the general expression $ax^2 + bx + c$. In each of these, the letter x is the variable while the numerals 3, 4, 2, 1, -7, and 6, and the letters $a, b,$ and c are constants. For a special problem, the values of $a, b,$ and c are specified whereas no value is specified for x. However it is understood that x belongs to the universal set of numbers under discussion.

Symbols are also used to indicate the operations of algebra.

For **addition,** the sum of two numbers x and y is written as $x + y$, and is read "x plus y."

For **subtraction,** the difference between x and y when y is subtracted from x is written as $x - y$, and is read "x minus y."

For **multiplication,** the product of 3 and 5 may be written as $3 \cdot 5$, as 3(5), or as (3)(5). No multiplication symbol is needed when one of the two numbers being multiplied is designated by a letter. For example, $3x$ indicates the product of 3 and x and xy indicates the product of x and y.

For **division,** the quotient of two numbers is written usually as $\frac{x}{y}$, and is read "x over y" or "x divided by y."

Set *A* and set *B* are equal, $A = B$, if and only if every member of A is a member of B and every member of B is a member of A.

In other words, two equal sets have exactly the same members.

The set of real numbers R has some important subsets. We say that set A is a **subset** of set B (in symbols, $A \subset B$) if and only if every member of A is also a member of B.

If A is a subset of B and A contains every member of B, then A is said to be an **improper subset** of B. In this case, $A = B$. In every other case, A is called a **proper subset** of B.

The empty set is a subset of every set.

The set of **digits** D is a subset of the set of real numbers. In symbols, $D \subset R$.

$D = \{0, 1, 2, 3, 4, 5, 6, 7, 8, 9\}$

The listing method is used above to describe the set D; that is, its members are written within braces and a comma is written between the members.

The set of **natural numbers** N, also called the **counting numbers**, is a subset of the set of real numbers; $N \subset R$.

$N = \{1, 2, 3, 4, 5, 6, \ldots\}$

The three dots, ..., mean that the list is unending. The set N is an infinite set of numbers whereas the set of digits D is a finite set.

A nonempty set is **finite** if it is possible to assign a natural number as the number of objects in the set.

The **empty set** is considered to be a **finite set.**

An **infinite set** is a set that is not finite. Any list of its members is unending.

The set of **whole numbers** W is the set consisting of the natural numbers and the number zero, 0. Note that $W \subset R$.

$W = \{0, 1, 2, 3, 4, 5, \ldots\}$

The set of natural numbers has some important subsets. To describe these sets, the concepts of multiple and factor are first defined.

DEFINITION OF FACTOR AND MULTIPLE

Let $n = pq$ where n, p, and q are natural numbers. Then

p is a **factor** of n, and
q is a **factor** of n,
n is a **multiple** of p, and
n is a **multiple** of q.

The set of **even numbers** E is the set of natural numbers that are multiples of 2.

$E = \{2, 4, 6, 8, 10, \ldots\}$

The set of **odd numbers** $\mathcal{O}$ is the set of natural numbers that are not even.

$\mathcal{O} = \{1, 3, 5, 7, 9, \ldots\}$

There is another method for describing sets called the **rule, or set-builder method.** Using this method,

$E = \{x \mid x = 2n,$ where n is a natural number$\}$

This is read "E is the set of all numbers x such that x equals the product of 2 and a natural number n."

The vertical bar is read "such that," the expression to the left of the bar identifies a member of the set being defined, and the expression to the right of the bar states a rule or property that determines membership in the set.

Similarly, we can write

$\mathcal{O} = \{x \mid x = 2n - 1,$ where n is a natural number$\}$

A **prime,** or **prime number,** is a natural number that has exactly two factors, the number itself and the number 1.

Using the listing method to describe the set of primes P,

$P = \{2, 3, 5, 7, 11, 13, 17, 19, \ldots\}$

A **composite,** or **composite number,** is a natural number that is different from 1 and is not a prime. Letting $K =$ the set of composites,

$K = \{4, 6, 8, 9, 10, 12, 14, 15, \ldots\}$

The set of primes and the set of composites are both infinite sets. Note that the number 1 is neither a prime nor a composite.

The set of **integers** J, another subset of the real numbers, consists of the natural numbers, the number zero, and the negatives of the natural numbers.

$J = \{\ldots, -5, -4, -3, -2, -1, 0, 1, 2, 3, 4, 5, \ldots\}$

The natural numbers are also called the **positive integers** and the negatives of the natural numbers are called the **negative integers.**

The set of integers can also be described as the union of the set of whole numbers W with the set of the negatives of the natural numbers M.

The **union of set A with set B,** written as $A \cup B$ and read "A union B," is the set of elements that are in A or in B or in both A and B. The statement $J = W \cup M$ is read "J(the set of integers) is the union of W(the set of whole numbers) with M(the set of the negatives of the natural numbers)."

The set of **rational numbers** Q, another subset of the reals, consists of all possible quotients of integers. In symbols,

$$Q = \{x \mid x = \frac{n}{d}, \text{ where } n \text{ and } d \text{ are integers and } d \neq 0\}$$

This is read "Q is the set of numbers x such that x can be written in the form $\frac{n}{d}$, where n and d are integers and d does not equal zero."

Note that 5 is a rational number because 5 can be written as $\frac{5}{1}$. Similarly, the following numbers are rational:

$$-4 = \frac{-4}{1}, \qquad 0 = \frac{0}{1}, \qquad 3.5 = \frac{7}{2}, \qquad 1.66\overline{6}\ldots = \frac{5}{3}$$

In fact, every terminating decimal and every infinite repeating decimal is a rational number and can be written as a quotient of integers.

The set of **irrational numbers** H, another subset of the reals, consists of those numbers that have an infinite decimal representation with no repeating blocks of digits. These numbers are not rational and cannot be written as a quotient of integers.

$$2.1515\overline{15}... = \tfrac{71}{33} \quad \text{is rational}$$
$$1.73205... = \sqrt{3} \quad \text{is irrational}$$
$$-2.23606... = -\sqrt{5} \text{ is irrational}$$
$$3.14159... = \pi \quad \text{is irrational}$$

The set of real numbers is the union of the set of rational numbers and the set of irrational numbers. In symbols, $R = Q \cup H$.

The **intersection of set A and set B,** written as $A \cap B$ and read "A intersection B," is the set consisting of elements that are in A and are also in B.

Note that the intersection of the set of integers with the set of natural numbers is the set of natural numbers, $J \cap N = N$. Note also that the intersection of the set of irrationals with the set of rationals is the empty set, $H \cap Q = \emptyset$.

When the intersection of two sets is the empty set, the sets are said to be **disjoint.**

The set of real numbers can be considered as the union of three disjoint sets; the set consisting of zero only, the set of **positive real numbers** (those to the right of zero on the number line), and the set of **negative real numbers** (those to the left of zero on the number line).

The positive and negative real numbers are also called **signed numbers** since a negative number is written using a minus sign, and a positive number may be written using a plus sign. Note that zero has no sign; it is neither positive nor negative.

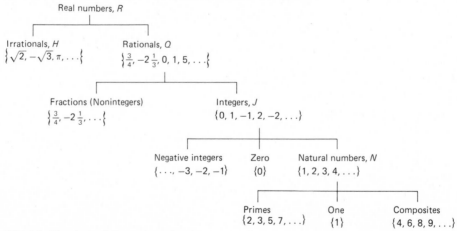

FIG. 1.2. **Structure of the real numbers.**

Figure 1.2 is useful for understanding the structure of the set of real numbers.

EXERCISES 1.1

(1–4) For each given set S of real numbers, list those elements that are (a) natural numbers, (b) integers, (c) rationals, and (d) irrationals.

EXAMPLE 1

$S = \left\{ -8, -\sqrt{5}, \frac{-10}{3}, 0, \frac{1}{5}, \sqrt{6}, 6, \frac{21}{2} \right\}$

Solution
(a) natural numbers, $\{1, 6\}$
(b) integers, $\{-8, 0, 1, 6\}$
(c) rationals, $\left\{ -8, \frac{-10}{3}, 0, \frac{1}{5}, 1, 6, \frac{21}{2} \right\}$
(d) irrationals, $\{-\sqrt{5}, \sqrt{6}\}$

1. $S = \left\{ -3, -\sqrt{3}, \frac{-1}{3}, 0, 1, 3, \sqrt{3}, \frac{19}{3} \right\}$

2. $S = \left\{ -5, -\sqrt{5}, \frac{-1}{5}, 0, \frac{1}{5}, \sqrt{5}, 5, \frac{51}{5} \right\}$

3. $S = \left\{ -\pi, -\sqrt{9}, \frac{-90}{9}, \frac{9}{9}, \sqrt{\frac{1}{9}}, \sqrt{90} \right\}$

4. $S = \{ -3, 3.1, -3.14, 3.14\overline{14}, \ldots, 3.14159, \ldots, 4.0 \}$

(5–28) List the elements in each set described. State whether each set is finite or infinite.

EXAMPLE 2 List and describe: the integers greater than -3
Solution $\{-2, -1, 0, 1, 2, \ldots\}$, infinite

EXAMPLE 3 List and describe: the odd natural numbers between 30 and 40
Solution $\{31, 33, 35, 37, 39\}$, finite

EXAMPLE 4 List and describe: the prime factors of 90
Solution $\{2, 3, 5\}$, finite

EXAMPLE 5 List and describe: the rational numbers having the form $\frac{x}{3}$ where x is an integer

Solution $\left\{ \ldots, -1, \frac{-2}{3}, \frac{-1}{3}, 0, \frac{1}{3}, \frac{2}{3}, 1, \frac{4}{3}, \ldots \right\}$, infinite

EXAMPLE 6 List and describe: the irrational numbers that are integers
Solution $\emptyset$, finite

5. the natural numbers less than 8
6. the natural numbers greater than 12
7. the integers less than -3
8. the negative integers greater than -7
9. the integers between -4 and 4

10. the integers between -10 and -5

11. the odd natural numbers between 10 and 20

12. the even natural numbers between 15 and 24

13. the natural numbers that are factors of 105

14. the natural numbers that are factors of 84

15. the prime factors of 105

16. the prime factors of 84

17. the composite factors of 105

18. the even factors of 15

19. the natural numbers that are multiples of 12

20. the natural numbers that are multiples of 9

21. the whole numbers less than 0

22. the whole numbers less than 7

23. the rational numbers having the form $\dfrac{x}{2}$ where x is an integer

24. the rational numbers having the form $\dfrac{1}{x}$ where x is an integer

25. the irrational numbers having the form $\sqrt{n}$ where n is a natural number

26. the irrational numbers having the form $\dfrac{x}{y}$ where x and y are integers

27. the real numbers that are integers

28. the real numbers that have the form $k\pi$ where k is an integer

(29–40) State whether each given real number is rational or irrational. If a number is rational, state if it is integral (an integer) or not.

(*Note.* A bar on top of a block of digits means this block repeats. If there is no bar, then there is no repeating digit or block of digits.)

EXAMPLE 7 State the kind of real number: 8.31427...

Solution Irrational, because there is no repeating block of digits

EXAMPLE 8 State the kind of real number: $-4.363\overline{36}$...

Solution Rational and nonintegral, because there is a repeating block of digits

EXAMPLE 9 State the kind of real number: 28.43

Solution Rational and nonintegral, since it is a terminating decimal with at least one nonzero decimal digit

EXAMPLE 10 State the kind of real number: 2843

Solution Rational and integral

29. 4040

30. 27.958

31. 2.791218...

32. 27958

33. $8.424\overline{242}$...

34. 2.7958...

35. -5.123456

36. -5.010010001...

37. $-2.7828...$ **38.** $-5010010001...$

39. -151515 **40.** $-5.010\overline{101}...$

(41–50) Find (a) $A \cup B$ and (b) $A \cap B$.

EXAMPLE 11

A = the set of odd factors of 66

B = the set of prime factors of 66

> Solution
>
> The factors of 66 are $\{1, 2, 3, 6, 11, 22, 33, 66\}$.
>
> The set of odd factors, $A = \{1, 3, 11, 33\}$.
>
> The set of prime factors, $B = \{2, 3, 11\}$.
>
> (a) $A \cup B = \{1, 2, 3, 11, 33\}$; (b) $A \cap B = \{3, 11\}$

EXAMPLE 12

$A = \left\{ x \mid x \text{ and } \dfrac{12}{x} \text{ are positive integers} \right\}$

$B = \left\{ x \mid x \text{ and } \dfrac{18}{x} \text{ are positive integers} \right\}$

> Solution
>
> $A = \{1, 2, 3, 4, 6, 12\}$
>
> $B = \{1, 2, 3, 6, 9, 18\}$
>
> (a) $A \cup B = \{1, 2, 3, 4, 6, 9, 12, 18\}$
>
> (b) $A \cap B = \{1, 2, 3, 6\}$

41. A = the set of prime factors of 70

     B = the set of even factors of 70

42. A = the set of odd factors of 70

     B = the set of composite factors of 70

43. $A = \left\{ x \mid x \text{ and } \dfrac{10}{x} \text{ are positive integers} \right\}$

     $B = \left\{ x \mid x \text{ and } \dfrac{25}{x} \text{ are positive integers} \right\}$

44. $A = \left\{ x \mid x \text{ and } \dfrac{x}{6} \text{ are positive integers} \right\}$

     $B = \left\{ x \mid x \text{ and } \dfrac{x}{9} \text{ are positive integers} \right\}$

45. $A = \{x \mid x \text{ is an integer greater than 5}\}$

     $B = \{x \mid x \text{ is an integer greater than 2}\}$

46. $A = \{x \mid x \text{ is an integer greater than 1}\}$

     $B = \{x \mid x \text{ is an integer less than 6}\}$

47. A is the set of rational numbers.

     B is the set of irrational numbers.

48. A is the set of whole numbers.

     B is the set of natural numbers.

49. A is the set of positive integers.

     B is the set of negative integers.

50. A is the set of prime numbers.

     B is the set of composite numbers.

1.2 THE REAL NUMBER SYSTEM

THE EQUALITY RELATION

The statements $2 + 3 = 5$, $2 + 3 = 7$, $x + 3 = 7$, $x + 3 = x$, and $x + y = y + x$ are examples of equations illustrating the equality relation. An equation may be true or it may be false, or it may be an **open equation** if it contains one or more variables. An open equation may become true for all values of the variable(s), for no value(s), or it may become true for some value(s) of the variable(s) and false for others.

If a and b designate real numbers, the equality $a = b$, when true, means that a and b name the same number.

There are some equations that are considered to be true due to the form of the statement only. For example, $5 = 5$, $2 + 3 = 2 + 3$, and in general, $a = a$. This illustrates one of the basic equality axioms. Recall that an axiom is a statement that is assumed to be true.

EQUALITY AXIOMS

For all real numbers a, b, and c:

1. Reflexive axiom $a = a$
2. Symmetric axiom   If $a = b$, then $b = a$.
3. Transitive axiom   If $a = b$ and $b = c$, then $a = c$.
4. Substitution axiom   If $a = b$, then a may replace b in any statement without changing the truth or falsity of the statement.

FIELD AXIOMS

In this development of the real number system, the equality relation and the operations of addition and multiplication are undefined terms (also called primitive terms or primitives). Besides the equality axioms, there are also axioms that describe the operations of addition and multiplication. These axioms, stated below, are called field axioms. Any mathematical system for which the field axioms are valid is called a **field.** Thus the real number system is a field.

FIELD AXIOMS

For all real numbers a, b, and c:

1.	Closure axiom, addition	$a + b$ is a real number
2.	Closure axiom, multiplication	ab is a real number
3.	Commutative axiom, addition	$a + b = b + a$
4.	Commutative axiom, multiplication	$ab = ba$
5.	Associative axiom, addition	$(a + b) + c = a + (b + c)$
6.	Associative axiom, multiplication	$(ab)c = a(bc)$
7.	Distributive axiom	$a(b + c) = ab + ac$ and $(b + c)a = ba + ca$

8. Identity axiom, addition. There exists exactly one real number, 0, called the **additive identity,** such that
$$0 + a = a \quad \text{and} \quad a + 0 = a$$

9. Identity axiom, multiplication. There exists exactly one real number, 1, called the **multiplicative identity,** such that
$$1a = a \quad \text{and} \quad a1 = a$$

10. Inverse axiom, addition. There exists exactly one real number $-a$, called the **opposite** of a (or **additive inverse** of a) such that
$$a + (-a) = 0 \quad \text{and} \quad (-a) + a = 0$$

11. Inverse axiom, multiplication. If $a \neq 0$, then there exists exactly one real number $\dfrac{1}{a}$, called the **reciprocal** of a (or **multiplicative inverse** of a) such that
$$a \cdot \frac{1}{a} = 1 \quad \text{and} \quad \frac{1}{a} \cdot a = 1$$

SUBTRACTION AND DIVISION

Unlike addition and multiplication, the operations of subtraction and division are defined.

DEFINITION OF SUBTRACTION, $a - b$

If a and b are any real numbers, then the difference of a and b, written $a - b$ and read "a minus b," is defined as follows:
$$a - b = a + (-b)$$

The definition of subtraction states that the operation of subtracting b from a is done by adding the opposite of b to a. As examples,
$$7 - 3 = 7 + (-3) = 4$$
$$3 - 7 = 3 + (-7) = -4$$
$$7 - (-3) = 7 + (-(-3)) = 7 + 3 = 10$$

We can show that the definition of subtraction stated here is equivalent to the meaning of subtraction given in arithmetic; that is, $a - b$ is the real number which when added to the subtrahend b produces the number a.

THEOREM

$(a - b) + b = a.$

Proof

Statements	Reasons
1. $(a - b) + b = (a - b) + b$	Reflexive axiom
2. $a - b = a + (-b)$	Definition of subtraction
3. $(a - b) + b = (a + (-b)) + b$	Substitution axiom
4. $(a + (-b)) + b = a + (-b + b)$	Associative axiom, addition
5. $(a - b) + b = a + (-b + b)$	Substitution axiom
6. $-b + b = 0$	Inverse axiom, addition
7. $(a - b) + b = a + 0$	Substitution axiom
8. $a + 0 = a$	Identity axiom, addition
9. $(a - b) + b = a$	Substitution axiom

The proof shown above is called a **formal proof** because there is a "statements" column and a "reasons" column with a reason supplied for each statement. A reason in a mathematical proof must be a definition, an axiom, or a previously proved theorem.

Note how often the substitution axiom was used in this proof. This is typical of many algebraic proofs. Since it becomes tedious to write "substitution axiom" so often, this reason is usually omitted from a proof, and the equality statement for the substitution is omitted. The reason given for the result of the substitution is the reason for the equality statement. The understanding is that the reader is aware of the substitution and can supply the missing statement and reason. The shortened proof that follows illustrates this practice for the proof previously given.

Shortened Proof

1.	$(a - b) + b = (a - b) + b$	Reflexive axiom
2.	$= (a + (-b)) + b$	Definition of subtraction
3.	$= a + (-b + b)$	Associative axiom, addition
4.	$= a + 0$	Inverse axiom, addition
5.	$= a$	Identity axiom, addition

In an informal proof, those statements, or reasons, that the writer of the proof feels are obvious to the reader may be omitted. Also the double column format may be replaced by a set of statements and reasons in paragraph form. It is assumed that the reader can supply the complete formal proof.

DEFINITION OF DIVISION, $\frac{a}{b}$

If a and b are real numbers and $b \neq 0$, then the quotient of a divided by b, written $\frac{a}{b}$ and read "a over b" or "a divided by b," is defined as follows:

$$\frac{a}{b} = a \cdot \frac{1}{b}$$

The definition of division states that the operation of dividing a by b is done by multiplying a by the reciprocal of b. As examples,

$$\frac{10}{2} = 10 \cdot \frac{1}{2} = 5$$

$$\frac{3}{12} = 3 \cdot \frac{1}{12} = \frac{1}{4}$$

$$\frac{0}{4} = 0 \cdot \frac{1}{4} = 0$$

Expressions such as $\frac{5}{0}$ and $\frac{0}{0}$ are said to be undefined or "meaningless." They do not represent real numbers. In general, division by zero is undefined.

In arithmetic, $\frac{a}{b}$ (or equivalenty $a \div b$) is defined as the unique number which when multiplied by b produces a. We can show that the definition of division stated here preserves this meaning; that is, $\frac{a}{b} \cdot b = a$ for $b \neq 0$.

THEOREM

$$\frac{a}{b} \cdot b = a \text{ for } b \neq 0.$$

Proof

Statements	*Reasons*
1. $\frac{a}{b} \cdot b = \frac{a}{b} \cdot b$	Reflexive axion
2. $\quad = \left(a \cdot \frac{1}{b} \right) b$	Definition of division
3. $\quad = a \left(\frac{1}{b} \cdot b \right)$	Associative axiom, multiplication
4. $\quad = a \cdot 1$	Inverse axiom, multiplication
5. $\quad = a$	Identity axiom, multiplication

To fully understand a statement, it is important to be aware of not only what the statement says but also what it does not say.

Addition and multiplication are said to be commutative and associative due to the content of the commutative and associative axioms. However subtraction and division are neither commutative nor associative. As examples,

$$10 - 3 \neq 3 - 10$$
$$12 \div 4 \neq 4 \div 12$$
$$(10 - 5) - 2 \neq 10 - (5 - 2)$$
$$(18 \div 6) \div 3 \neq 18 \div (6 \div 3)$$

On the other hand, since $(a + b) + c = a + (b + c)$ and $(ab)c = a(bc)$, the way in which the terms of a sum or the factors of a product are grouped does not affect the final result, so parentheses are not needed. These expressions can then be written more simply as $a + b + c$ and as abc, respectively.

DEFINITION

$$a + b + c = (a + b) + c$$
$$abc = (ab)c$$

The distributive axiom is another axiom which is very often used and abused. This axiom, $a(b + c) = ab + ac$ and $(b + c)a = ba + ca$, is important because it relates the operations of addition and multiplication. We say that multiplication can be distributed over addition. However, addition cannot be distributed over multiplication nor can multiplication be distributed over multiplication. As examples,

$$2 + (5 \cdot 3) \neq (2 + 5)(2 + 3)$$
$$2(5 \cdot 3) \neq (2 \cdot 5)(2 \cdot 3)$$

EXERCISES 1.2

(1–14) Each of the following is an immediate consequence of one or more of the equality axioms. State the axiom or axioms.

EXAMPLE 1
$5(x + 3) = 5(x + 3)$

Solution   Reflexive axiom, because the left and right sides of the equation are identical.

EXAMPLE 2   If $10 = x + 4$, then $x + 4 = 10$.

Solution   Symmetric axiom, because the second equation can be obtained from the first by interchanging left and right sides.

EXAMPLE 3   If $y = x + 1$ and $x + y = 7$, then $x + (x + 1) = 7$.

Solution   Substitution axiom, because $x + (x + 1) = 7$ can be obtained from $x + y = 7$ by replacing y by its equal $x + 1$.

EXAMPLE 4   If $x = y$ and $y = 5$, then $x = 5$.

Solution   Transitive axiom; if x and 5 are each equal to the same number y, they are equal to each other.
Alternate solution   Substitution axiom, by replacing y in $x = y$ by its equal 5.

(*Note.* The transitive axiom is a special case of the substitution axiom. Whenever the transitive axiom can be used as a reason, the substitution axiom is also a valid reason. However, the converse is not the case. To illustrate, note that, in Example 3, the transitive axiom is not a valid reason.)

1. $x = x$

2. $y - 2 = y - 2$

3. If $3 = x$, then $x = 3$.

4. If $x + 2 = y$, then $y = x + 2$.

5. If $x - 1 = y$ and $y = 8$, then $x - 1 = 8$.

6. If $(x + 2) - 2 = x + (2 - 2)$ and $x + (2 - 2) = x$, then $(x + 2) - 2 = x$.

7. If $xy = 6$ and $y = x - 1$, then $x(x - 1) = 6$.

8. If $3\left(\frac{x}{3}\right) = 7$ and $3\left(\frac{x}{3}\right) = x$, then $x = 7$.

9. If $x^2 - 25 = 0$ and $x^2 - 25 = (x - 5)(x + 5)$, then $(x - 5)(x + 5) = 0$.

10. If $x \neq 6$ and $y = x$, then $y \neq 6$.

11. If $0 = 3x - 6$, then $3x - 6 = 0$.

12. $2x - 10 = 2x - 10$

13. $(x - 5)(x + 5) = (x - 5)(x + 5)$

14. If $a(b + c) = ab + ac$, then $ab + ac = a(b + c)$.

(15–34) Each of the following is an immediate consequence of one of the field axioms. State the axiom.

EXAMPLE 5 $\sqrt{2} + 6$ is a real number.

> Solution   Closure axiom, addition, because $\sqrt{2}$ and 6 are real numbers.

EXAMPLE 6 $\sqrt{3}\,\sqrt{2}$ is a real number.

> Solution   Closure axiom, multiplication, because $\sqrt{3}$ and $\sqrt{2}$ are real numbers.

EXAMPLE 7 $0.5 + x = x + 0.5$ for x a real number.

> Solution   Commutative axiom, addition, because the order of the terms on the right side is the reverse of that on the left.

EXAMPLE 8
$\sqrt{5} \cdot 3 = 3 \cdot \sqrt{5}$

> Solution   Commutative axiom, multiplication, because the order of the factors on the right side is the reverse of that on the left.

EXAMPLE 9
$(89 + 75) + 25 = 89 + (75 + 25)$

> Solution   Associative axiom, addition, because the order of the terms is the same on each side but the grouping is different.

EXAMPLE 10 $7(2y) = 14y$ for y a real number.

> Solution   Associative axiom, multiplication because the order of the factors is the same on each side but the grouping is different. Note that $7(2y) = (7 \cdot 2)y = 14y$.

EXAMPLE 11

$6(x + 5) = 6x + 30$

Solution Distributive axiom. Note that a product has been changed to a sum. Note also that $6x + 6 \cdot 5 = 6x + 30$.

EXAMPLE 12

$0 + \pi = \pi$

Solution Identity axiom, addition. The sum of zero and any real number is that real number.

EXAMPLE 13

$1\pi = \pi$

Solution Identity axiom, multiplication. The product of one and any real number is that number.

EXAMPLE 14

$-\sqrt{3} + \sqrt{3} = 0$

Solution Inverse axiom, addition. The sum of any number and its opposite is zero.

EXAMPLE 15

$(0.2)\left(\frac{1}{0.2}\right) = 1$

Solution Inverse axiom, multiplication. The product of any number and its reciprocal is one.

15. $8 + x = x + 8$ **16.** $1x = x$

17. $-5 + 5 = 0$ **18.** $15x = 3(5x)$

19. $6 \cdot \frac{1}{6} = 1$ **20.** $(n + 1) + 2 = n + 3$

21. $x2 = 2x$ **22.** $3 + (x - 1) = (x - 1) + 3$

23. $-4 \cdot 1 = -4$ **24.** $\dfrac{1}{\frac{1}{2}} \cdot \frac{1}{2} = 1$

25. $2(3y) = 6y$ **26.** $-(-3) + (-3) = 0$

27. $(y + 2) + 5 = y + 7$ **28.** $x^2 + x = x^2 + x + 0$

29. $-2x + 0 = -2x$ **30.** $(3 + 5)x = 3x + 5x$

31. $3(n + 4) = 3n + 12$ **32.** $a(b + c) = (b + c)a$

33. $23.76(9800)$ is a real number   **34.** $5 + \sqrt{6}$ is a real number

(35–50) Each of the following can be readily justified by two or more field or equality axioms. State the field axioms that apply.

EXAMPLE 16

$4(86 \cdot 25) = 86(25 \cdot 4)$

Solution

$$4(86 \cdot 25) = (86 \cdot 25)4 \quad \text{commutative, multiplication}$$
$$= 86(25 \cdot 4) \quad \text{associative, multiplication}$$

EXAMPLE 17

$(6x + 2) + (-2) = 6x$

Solution

$$(6x + 2) + (-2) = 6x + (2 + (-2)) \quad \text{associative, addition}$$
$$= 6x + 0 \quad \text{inverse axiom, addition}$$
$$= 6x \quad \text{identity axiom, addition}$$

35. $(65 + 78) + 35 = (65 + 35) + 78$

36. $4x + x = 5x$

37. $25(69 \cdot 4) = 69(25 \cdot 4)$

38. $5x\left(\frac{1}{5}\right) = x$

39. $(2x + 5) + (-5) = 2x$

40. $6y + 2x = 2(x + 3y)$

41. $\frac{1}{3} \cdot 3x = x$

42. $8(97 \cdot 125) = 97(8 \cdot 125)$

43. $4(5 + x) = 4x + 20$

44. $6\left(x + \frac{1}{6}\right) = 6x + 1$

45. $3x + 5x = 8x$

46. $y(x + 2) + 4(x + 2) = (x + 2)(y + 4)$

47. $2 + 3(x + 1) = 3x + 5$

48. $3x = (-4 + 3x) + 4$

49. $24\left(\frac{1}{4} + \frac{1}{6}\right) = 4 + 6$

50. $(8 + x) + 2 = x + 10$

(51–60) Supply the reasons for the statements in the following proofs. Use as reasons the axioms and definitions of this section.

51. THEOREM

$ab + ac = a(b + c)$

Statements	Reasons
1. $a(b + c) = ab + ac$	_____
2. $ab + ac = a(b + c)$	_____

52. THEOREM

$ab + ac = (b + c)a$

Statements	Reasons
1. $a(b + c) = ab + ac$	_____
2. $ab + ac = a(b + c)$	_____
3. $a(b + c) = (b + c)a$	_____
4. $ab + ac = (b + c)a$	_____

53. THEOREM

$a(b + c + d) = ab + ac + ad$

Statements	Reasons
1. $a(b + c + d) = ab + a(c + d)$	_____
2. $a(c + d) = ac + ad$	_____
3. $a(b + c + d) = ab + ac + ad$	_____

54. THEOREM

$(a + b) + (-b) = a$

Statements   Reasons
1. $(a + b) + (-b) = a + (b + (-b))$ _____
2. $b + (-b) = 0$ _____
3. $(a + b) + (-b) = a + 0$ _____
4. $a + 0 = a$ _____
5. $(a + b) + (-b) = a$ _____

55. THEOREM

$\dfrac{1}{a}(ab) = b$ for $a \neq 0$

Statements   Reasons

1. $\dfrac{1}{a}(ab) = \left(\dfrac{1}{a} \cdot a\right)b$ _____

2. $\dfrac{1}{a} \cdot a = 1$ _____

3. $\dfrac{1}{a}(ab) = 1b$ _____

4. $1b = b$ _____

5. $\dfrac{1}{a}(ab) = b$ _____

56. THEOREM

$ab\left(\dfrac{1}{a} + \dfrac{1}{b}\right) = a + b$ for $a \neq 0$ and $b \neq 0$

Statements   Reasons

1. $ab\left(\dfrac{1}{a} + \dfrac{1}{b}\right) = ab\left(\dfrac{1}{a}\right) + ab\left(\dfrac{1}{b}\right)$ _____

2. $\qquad = ba\left(\dfrac{1}{a}\right) + ab\left(\dfrac{1}{b}\right)$ _____

3. $\qquad = b\left(a \cdot \dfrac{1}{a}\right) + a\left(b \cdot \dfrac{1}{b}\right)$ _____

4. $\qquad = b1 + a1$ _____

5. $\qquad = b + a$ _____

6. $\qquad = a + b$ _____

57. THEOREM

$0 - a = -a$

Statements   Reasons
1. $0 - a = 0 + (-a)$ _____
2. $0 + (-a) = -a$ _____
3. $0 - a = -a$ _____

58. THEOREM

$$b\left(\frac{a}{b}\right) = a \text{ for } b \neq 0$$

Statements   *Reasons*

1. $b\left(\frac{a}{b}\right) = b\left(\frac{a}{b}\right)$ _____

2. $\frac{a}{b} = \frac{1}{b} \cdot a$ _____

3. $b\left(\frac{a}{b}\right) = b\left(\frac{1}{b} \cdot a\right)$ _____

4. $\quad = \left(b \cdot \frac{1}{b}\right)a$ _____

5. $\quad = 1a$ _____

6. $\quad = a$ _____

59. THEOREM

$$-(-a) = a$$

Statements   *Reasons*

1. $-(-a) = -(-a) + 0$ _____

2. $-a + a = 0$ _____

3. $-(-a) = -(-a) + (-a + a)$ _____

4. $\quad = (-(-a) + (-a)) + a$ _____

5. $\quad = 0 + a$ _____

6. $\quad = a$ _____

60. THEOREM

$$\frac{1}{\frac{1}{a}} = a \text{ for } a \neq 0$$

Statements   *Reasons*

1. $\frac{1}{\frac{1}{a}} = \frac{1}{\frac{1}{a}} \cdot 1$ _____

2. $\frac{1}{a} \cdot a = 1$ _____

3. $\frac{1}{\frac{1}{a}} = \frac{1}{\frac{1}{a}}\left(\frac{1}{a} \cdot a\right)$ _____

4. $\quad = \left(\frac{1}{\frac{1}{a}} \cdot \frac{1}{a}\right)a$ _____

5. $\quad = 1a$ _____

6. $\quad = a$ _____

1.3 PROPERTIES OF THE REAL NUMBER SYSTEM

This section is primarily concerned with properties that indicate how operations are performed on real numbers. These properties are derived as theorems. In these theorems, a, b, c, and d designate real numbers.

THEOREM 1 OPPOSITE OF AN OPPOSITE

$$-(-a) = a$$

Proof

1. $-(-a) = -(-a) + 0$		Identity axiom, addition
2. $= -(-a) + (-a + a)$		Inverse axiom, addition
3. $= [-(-a) + (-a)] + a$		Associative axiom, addition
4. $= 0 + a$		Inverse axiom, addition
5. $= a$		Identity axiom, addition

Notice that if $a = +5$, then $-a = -(+5) = -5$, and $-a$ is a negative number. If $a = -5$, then $-a = -(-5) = 5$, and $-a$ is a positive number.

Thus $-a$, which is best read as "the opposite of a," sometimes is positive and sometimes is negative. It is misleading to refer to $-a$ as "negative a."

THEOREM 2 RECIPROCAL OF A RECIPROCAL

For $a \neq 0$,

$$\frac{1}{\frac{1}{a}} = a$$

Proof

1. $\dfrac{1}{\frac{1}{a}} = \dfrac{1}{\frac{1}{a}} \cdot 1$		Identity axiom, multiplication
2. $= \dfrac{1}{\frac{1}{a}} \left(\dfrac{1}{a} \cdot a \right)$		Inverse axiom, multiplication
3. $= \left(\dfrac{1}{\frac{1}{a}} \cdot \dfrac{1}{a} \right) a$		Associative axiom, multiplication

4. $= 1a$ Inverse axiom, multiplication
5. $= a$ Identity axiom, multiplication

Notice for $a = 2$, that $\dfrac{1}{a} = \dfrac{1}{2}$ and $\dfrac{1}{\frac{1}{a}} = \dfrac{1}{\frac{1}{2}} = 2$. For $a = \dfrac{1}{3}$,

then $\dfrac{1}{a} = \dfrac{1}{\frac{1}{3}} = 3$ and $\dfrac{1}{\frac{1}{a}} = \dfrac{1}{3}$.

Theorems 3 through 6 provide information about the equality relation. These theorems are useful for deriving other theorems and for solving equations.

THEOREM 3   ADDITION THEOREM OF EQUALITY

If $a = b$, then $a + c = b + c$.

Proof

1. $a + c = a + c$ Reflexive axiom
2. $a = b$ Given
3. $a + c = b + c$ Substitution axiom

THEOREM 4   MULTIPLICATION THEOREM OF EQUALITY

If $a = b$, then $ac = bc$.

Proof

1. $ac = ac$ Reflexive axiom
2. $a = b$ Given
3. $ac = bc$ Substitution axiom

Theorem 3 informs us that if the same number is added to each of two equal numbers, then the resulting sums are equal. Theorem 4 states that when two equal numbers are each multiplied by the same number, then the resulting products are equal.

Note the use of "Given" as a reason in these two proofs. A supposition provided in the "If" statement may be used in the proof as an assumed fact.

THEOREM 5   CANCELLATION THEOREM FOR ADDITION

If $a + c = b + c$, then $a = b$.

Proof

1. $a + c = b + c$ Given
2. $(a + c) + (-c) = (b + c) + (-c)$ Addition theorem of equality

3. $a + (c + (-c)) = b + (c + (-c))$ Associative axiom, addition
4. $a + 0 = b + 0$ Inverse axiom, addition
5. $a = b$ Identity axiom, addition

THEOREM 6   CANCELLATION THEOREM FOR MULTIPLICATION

If $ac = bc$ and $c \neq 0$, then $a = b$.

Proof

1. $ac = bc$ and $c \neq 0$ Given
2. $ac\left(\dfrac{1}{c}\right) = bc\left(\dfrac{1}{c}\right)$ Multiplication theorem of equality
3. $a\left(c \cdot \dfrac{1}{c}\right) = b\left(c \cdot \dfrac{1}{c}\right)$ Associative axiom, multiplication
4. $a1 = b1$ Inverse axiom, multiplication
5. $a = b$ Identity axiom, multiplication

In essence, Theorem 5 informs us that if the same number is subtracted from two equal numbers, then the resulting differences are equal. Similarly, Theorem 6 states that if two equal numbers are each divided by the same nonzero number, then the resulting quotients are equal.

The next theorem informs us of a very special property of the number 0; namely, the product of zero and any real number is zero.

THEOREM 7   ZERO FACTOR THEOREM

$a \cdot 0 = 0$ and $0 \cdot a = 0$

Proof

1. $0 + 0 = 0$ Identity axiom, addition
2. $(0 + 0)a = 0 \cdot a$ Multiplication theorem of equality
3. $0 \cdot a + 0 \cdot a = 0 \cdot a$ Distributive axiom
4. $0 \cdot a + 0 \cdot a = 0 \cdot a + 0$ Identity axiom, addition
5. $0 \cdot a = 0$ Addition cancellation theorem
6. $a \cdot 0 = 0 \cdot a$ Commutative axiom, multiplication
7. $a \cdot 0 = 0$ Substitution axiom

The next five theorems provide information regarding additive inverses. These theorems are useful for explaining how operations are performed using signed numbers.

These proofs are omitted here. Outlines of the proofs are provided in the Exercises.

THEOREM 8

$-a = -1 \cdot a$ and $-1 \cdot a = -a$

THEOREM 9

$$-(a + b) = (-a) + (-b) \quad \text{and} \quad (-a) + (-b) = -(a + b)$$

THEOREM 10

$$-(a - b) = -a + b = b - a$$

THEOREM 11

$$(-a)b = -ab$$

THEOREM 12

$$a(-b) = -ab$$

THEOREM 13

$$(-a)(-b) = ab$$

THEOREM 14

$$a(b - c) = ab - ac$$

Theorems 9 and 10 are useful in removing parentheses and in the addition of signed numbers. Note that if a and b are positive real numbers, then Theorem 9, in the form $(-a) + (-b) = -(a + b)$, states the rule for adding two negative numbers. For example, $-2 + (-7) = -(2 + 7) = -9$. The rules for the addition of a positive number and a negative number are contained in Theorem 10. For example, $-8 + 3 = -(8 - 3) = -5$ and $-3 + 8 = 8 - 3 = 5$.

Theorems 11, 12, and 13 are applied in the multiplication of signed numbers. Theorems 11 and 12 inform us, for a and b positive, that the product of a positive number and a negative number is negative. For example, $(-5)4 = -20$ and $5(-4) = -20$. Theorem 13, for a and b positive, states that the product of two negative numbers is a positive number. For example, $(-5)(-4) = 20$. Theorem 14 shows that multiplication is distributive over subtraction. For example,

$$5(7 - 2) = 5 \cdot 7 - 5 \cdot 2 = 35 - 10 = 25$$
$$4(x - 3) = 4x - 4 \cdot 3 = 4x - 12$$

Theorems 15 through 19 describe properties of quotients of real numbers. They also provide information indicating how operations are performed on rational numbers. Informal proofs are shown for Theorems 15 and 16 and proofs of the remaining theorems in this chapter are left as exercises.

THEOREM 15 EQUALITY TEST FOR FRACTIONS

For $b \neq 0$ and $d \neq 0$,

$$\frac{a}{b} = \frac{c}{d} \quad \text{if and only if} \quad ad = bc$$

(*Note.* A statement of the form "*A* if and only if *B*" means that both "If *A*, then *B*" and "If *B*, then *A*" are valid. Thus the proof of a theorem having this form must verify both of these statements.)

Part 1 Informal proof. If $\frac{a}{b} = \frac{c}{d}$, then $ad = bc$.

Multiplying each side of $\frac{a}{b} = \frac{c}{d}$ by bd and using the definition of division,

$$\left(a \cdot \frac{1}{b}\right)bd = \left(c \cdot \frac{1}{d}\right)bd$$

Using the commutative and associative axioms of multiplication to rearrange the factors, we obtain

$$ad\left(b \cdot \frac{1}{b}\right) = bc\left(d \cdot \frac{1}{d}\right)$$

Thus $ad = bc$.

Part 2 Informal proof. If $ad = bc$, then $\frac{a}{b} = \frac{c}{d}$.

Multiplying each side of $ad = bc$ by $\frac{1}{b} \cdot \frac{1}{d}$ since $b \neq 0$ and $d \neq 0$ and rearranging factors,

$$a\left(\frac{1}{b}\right)d\left(\frac{1}{d}\right) = c\left(\frac{1}{d}\right)b\left(\frac{1}{b}\right)$$

Using the inverse and identity axioms of multiplication and the definition of division,

$$\frac{a}{b} = \frac{c}{d}$$

THEOREM 16 FUNDAMENTAL THEOREM OF FRACTIONS

$$\frac{ad}{bd} = \frac{a}{b} \quad \text{for } b \neq 0 \quad \text{and} \quad d \neq 0$$

Theorem 16 is readily proved by using Theorem 15, the equality test for fractions. We first obtain $(ad)b = (bd)a$. Rearranging factors by the commutative and associative axioms of multiplication, $abd = abd$, which is true by the reflexive axiom.

Theorem 16 is especially useful for simplifying fractions (reducing to lowest terms) and for building up fractions for addition and subtraction.

THEOREM 17

$$\frac{a}{1} = a \quad \text{and} \quad \frac{a}{a} = 1 \quad \text{for } a \neq 0$$

THEOREM 18   PRODUCT OF TWO RECIPROCALS

$$\frac{1}{b} \cdot \frac{1}{d} = \frac{1}{bd} \quad \text{for } b \neq 0 \quad \text{and} \quad d \neq 0$$

THEOREM 19   SUM OF QUOTIENTS

$$\frac{a}{c} + \frac{b}{c} = \frac{a + b}{c} \quad \text{for } c \neq 0$$

THEOREM 20   PRODUCT OF QUOTIENTS

$$\frac{a}{b} \cdot \frac{c}{d} = \frac{ac}{bd} \quad \text{for } b \neq 0 \quad \text{and} \quad d \neq 0$$

THEOREM 21

$$\frac{a}{-b} = \frac{-a}{b} = -\frac{a}{b} \quad \text{for } b \neq 0$$

THEOREM 22

$$\frac{-a}{-b} = \frac{a}{b} \quad \text{for } b \neq 0$$

Theorem 21 informs us that a negative fraction can be written in three forms. As an example, $\frac{5}{-8} = \frac{-5}{8} = -\frac{5}{8}$. The preferred form is $\frac{-5}{8}$ with the minus sign in the numerator. In general, $\frac{-a}{b}$ is the preferred form. Theorems 21 and 22 provide the rules for the division of signed numbers:

1. The quotient of a positive number and a negative number is a negative number.
2. The quotient of two negative numbers is a positive number.

As examples, $\frac{20}{-5} = -4$, $\frac{-20}{5} = -4$, and $\frac{-20}{-5} = 4$.

Algebra has two more operations besides those of addition, subtraction, multiplication, and division; namely, raising to a power and root extraction.

RAISING TO A POWER

For any real number b and for any natural number n, the expression b^n, which is read "b to the n," is called the **nth power of b.** The real number b is called the **base** and the natural number n is called the **exponent.**

DEFINITION OF b^n

For any real number b and for any natural number n,
$$b^1 = b \quad \text{and} \quad b^n = b \cdot b^{n-1} \quad \text{for } n > 1$$

The formal definition of b^n given here has the familiar meaning of n factors of b. For example,

$5^4 = 5 \cdot 5^3 = 5(5 \cdot 5^2) = 5 \cdot 5 \cdot 5 \cdot 5 = 625$

$x^2 = x \cdot x^1 = xx$

$x^3 = x \cdot x^2 = xxx$

The second power of x, namely x^2, is also called "x square" or "the square of x."

The third power of x, namely x^3, is also called "x cube" or "the cube of x."

ROOT EXTRACTION

DEFINITION OF NTH ROOT

For any real number a and for any natural number n, x is an nth root of a and only if $x^n = a$.

As examples,

5 and -5 are square roots of 25 because $5^2 = 25$ and $(-5)^2 = 25$,

$\frac{3}{4}$ and $\frac{-3}{4}$ are square roots of $\frac{9}{16}$ because $\left(\frac{3}{4}\right)^2 = \frac{9}{16}$ and $\left(\frac{-3}{4}\right)^2 = \frac{9}{16}$,

2 is a cube root of 8 because $2^3 = 8$.

If $x^2 = 2$, then there is no rational number x for which $x^2 = 2$. However there is a positive real number whose square is 2, denoted symbolically as $\sqrt{2}$. Furthermore $\sqrt{2}$ can be represented geometrically as the measure of the diagonal of a unit square. Similarly $\sqrt{3}$, $\sqrt{5}$, and $\sqrt{6}$ denote the positive real numbers whose squares are 3, 5, and 6, respectively; that is, $(\sqrt{3})^2 = 3$, $(\sqrt{5})^2 = 5$, and $(\sqrt{6})^2 = 6$.

If $a > 0$, then $\sqrt{a}$ denotes the positive real number whose square is a, and $(\sqrt{a})^2 = a$. Similarly $\sqrt[3]{a}$ denotes the positive real number whose cube is a and $(\sqrt[3]{a})^3 = a$.

In general, we have the following theorem on the existence of a positive nth root of each positive real number.

THEOREM   EXISTENCE OF POSITIVE ROOTS*

If a is a positive real number and n is any natural number, then there is exactly one positive real number x for which $x^n = a$. This unique positive nth root is called the **principal nth root** of a and is written $\sqrt[n]{a}$.

The expression $\sqrt[n]{a}$ is called a **radical,** $\sqrt{}$ is called the **radical sign,** n is called the **index,** and a is called the **radicand.**

DEFINITIONS

For $a > 0$, and for n a natural number,

1. $\sqrt[n]{a}$ is the unique positive real number such that $(\sqrt[n]{a})^n = a$
2. $\sqrt{a} = \sqrt[2]{a}$
3. $\sqrt[n]{0} = 0$

As examples,
$\sqrt{36} = 6$ because $6^2 = 36$ and 6 is positive.
$\sqrt[3]{8} = 2$ because $2^3 = 8$ and 2 is positive.
$\sqrt{3}$ is positive and $(\sqrt{3})^2 = 3$.
$\sqrt[3]{5}$ is positive and $(\sqrt[3]{5})^3 = 5$.
Every positive real number a also has a negative square root written as $-\sqrt{a}$. For example, $-\sqrt{25} = -5$ and $-\sqrt{3}$ is the negative of the positive number whose square is 3.
When more than one operation is involved in a problem, grouping symbols, such as parentheses ( ), brackets [ ], braces { }, or the bar —, are used to clarify the meaning and to insure the expression names exactly one real number. To avoid using any more symbols than are necessary, the following rules specify the order in which operations are to be done.

Rules for Order of Operations

Do all operations within grouping symbols, and within the innermost set of grouping symbols first, following the next three rules in order.

1. Do all power raisings and/or root extractions in order from left to right.
2. Do all multiplications and/or divisions in order from left to right.
3. Do all additions and/or subtractions in order from left to right.

(*Note.* The order of operations as stated above can be changed only if such a change can be justified by an axiom or a proved theorem.)
When a proposed solution of an equation is checked by substitution, it is extremely important that the operations are performed strictly according to the above rules in order to obtain a valid check.

*The proof of this theorem is beyond the scope of this text.

EXERCISES 1.3

(1–74)   Do the indicated operations.

EXAMPLE 1
$0 - 6 - (-2)$

Solution $-6 - (-2) = -6 + 2 = -4$

EXAMPLE 2
$-6(-2 + 3)$

Solution $-6(1) = -6$

EXAMPLE 3
$\left(\dfrac{-4}{5}\right)\left(\dfrac{-15}{8}\right)$

Solution $\dfrac{(-4)(-15)}{5 \cdot 8} = \dfrac{3(20)}{2(20)} = \dfrac{3}{2}$

EXAMPLE 4
$-4[5 - 2(5 - 7)]$

Solution $-4[5 - 2(-2)] =$
$-4(5 + 4) = -4(9) = -36$

EXAMPLE 5
$(-5)^3$

Solution $(-5)(-5)(-5) = -125$

EXAMPLE 6
$\sqrt{(-5)^2 - 4(-6)}$

Solution $\sqrt{25 + 24} = \sqrt{49} = 7$

EXAMPLE 7
$\dfrac{1000 - 216}{\sqrt[3]{1000} - \sqrt[3]{216}}$

Solution $\dfrac{784}{10 - 6} = \dfrac{784}{4} = 196$

EXAMPLE 8
$(7 - 10)^2 - \{(9 - 6)[4 - 2(-3)^2]\}$

Solution $(-3)^2 - \{(3)[4 - 2(9)]\} =$
$9 - [(3)(-14)] = 9 - (-42) = 51$

1. $-12(-7)$ **2.** $-25(8)$

3. $6(-6)$ **4.** $-9(-9)$

5. $-2(-5)(-7)$ **6.** $-2(-2)(-2)$

7. $125(-2)(-4)$ **8.** $(-5)(10)(-20)$

9. $-76(0)(49)(-5)$ **10.** $(0 - 8)(10 - 0)$

11. $2 - 7 + 0$ **12.** $-17 + 8 + 0$

13. $-5 - 7$

14. $-6 - (-9)$

15. $8 - (-10)$

16. $0 - (-6)$

17. $5 - 8 - 4$

18. $-2 + 5 - 7$

19. $-2(3 + 2 - 7)$

20. $-(-2 - 8 + 4)$

21. $5 - (2 + 8)(2 - 8)$

22. $1 - (7 - 2)(-6)$

23. $8 - 4(3 - 5)$

24. $2 - 3(4 + 2)$

25. $(8 - 4)(3 - 5)$

26. $(2 - 3)(4 + 2)$

27. $(-5 + 2)[(-5)^2 + (-5)(2)]$

28. $[10 - (-5 + 5)][10 - (-5 - 5)]$

29. $\dfrac{5}{12} + \dfrac{1}{12}$

30. $\dfrac{8}{3} - \dfrac{2}{3}$

31. $\dfrac{1}{-5} \cdot \dfrac{1}{6}$

32. $\dfrac{1}{-2} \cdot \dfrac{1}{-4}$

33. $\left(-\dfrac{1}{3}\right)\left(-\dfrac{2}{3}\right)$

34. $\left(\dfrac{-3}{4}\right)\left(\dfrac{2}{9}\right)$

35. $\left(\dfrac{5}{6}\right)\left(\dfrac{-3}{10}\right)(2)$

36. $\left(\dfrac{-2}{3}\right)\left(\dfrac{-6}{5}\right)\left(\dfrac{-5}{4}\right)$

37. $-12\left(\dfrac{3}{8} + \dfrac{5}{8}\right)$

38. $18\left(\dfrac{1}{4} - \dfrac{3}{4}\right)$

39. $\dfrac{-80}{-1}$

40. $\dfrac{-700}{-700}$

41. $-\dfrac{-2.5}{-2.5}$

42. $\dfrac{-1000}{8}$

43. $\dfrac{5000}{-40}$

44. $-\dfrac{-60{,}000}{100}$

45. $\dfrac{2 - 8}{2 - 4}$

46. $\dfrac{27 - 1000}{3 - 10}$

47. $\dfrac{7 - 7}{6 - 7}$

48. $\dfrac{-5 - (-5)}{-5 - 5}$

49. $\dfrac{(-3)^2 - 4(2)(-3)}{-9}$

50. $3\left(-\dfrac{1}{2}\right)\left(-\dfrac{1}{2}\right) - \dfrac{5}{2}\left(-\dfrac{1}{2}\right)$

51. $\dfrac{4}{3}\left(\dfrac{-2}{3}\right) - 6\left(\dfrac{-2}{3}\right)\left(\dfrac{-2}{3}\right)$

52. $\left(\dfrac{-4}{5}\right)\left(\dfrac{-4}{5}\right)\left(\dfrac{-4}{5}\right) + 1$

53. $-2[5 - 3(4 + 2)]$

54. $-6[(7 - 9)(7 - 8) - 1]$

55. $-2(-6)^2$

56. $5(-4)^3$

57. $[-2(-6)]^2$

58. $[5(-4)]^3$

59. $\sqrt{(-4)^2 + (-3)^2}$

60. $\sqrt{(17)^2 - (-15)^2}$

61. $\sqrt[3]{9^3 - 9(19)(3)}$

62. $\sqrt[3]{4^3 - 4(-38)}$

63. $\sqrt{(30)^2 - 4(9)(25)}$

64. $\sqrt{(-5)^2 - 4(4)(-21)}$

65. $\dfrac{512 + 125}{\sqrt[3]{512} + \sqrt[3]{125}}$

66. $\dfrac{64 - 729}{\sqrt[3]{64} - \sqrt[3]{729}}$

67. $1 - \{2 - 3[1 - (1 - 4)^2]\}$

68. $(7 - 3)\{1 - [6 + 2(-3)^2]\}$

69. $(2 - 4)\{2 - 6[2 - 8(3)]\}$

70. $[(-5)^2 - 3(-5) - 2][(-5)^2 + 3(-5) - 2]$

71. $\dfrac{5}{4\left(\dfrac{-5}{2}\right)^2 - 9} - \dfrac{4}{4\left(\dfrac{-5}{2}\right) - 6}$

72. $\dfrac{2(-5 + 7)^2}{3 - 7} + \dfrac{-5 - (-7)}{7 - 3}$

73. $\sqrt{4 - 3(-7)} - \sqrt{25 + 3(-7)}$

74. $[-1 + \sqrt{7(8) - 7}]^2$

(75–80) Evaluate.

75. $3x^2 - 5(x - 4)$ for $x = -5$

76. $2(x^2 + 1)^2 + 3(x^2 + 1)$ for $x = -3$

77. $\dfrac{2x + 1}{4x - 8} - \dfrac{2x + 3}{6x - 12}$ for $x = \dfrac{3}{2}$

78. $1 + \dfrac{4}{3x} - \dfrac{12}{3x + 5}$ for $x = \dfrac{-1}{3}$

79. $\sqrt{5x + 1} - \sqrt{3x - 5}$ for $x = 7$

80. $x - \sqrt{x(x - 5)}$ for $x = 9$

(81–87) Supply the reasons for the statements in each of the following proofs.

81. THEOREM 8

$-a = -1 \cdot a$ and $-1 \cdot a = -a$

Proof

1. $-1a + 1a - (-1 + 1)a$
2. $-1a + 1a = 0a$
3. $-1a + 1a = 0$
4. $-1a + a = 0$
5. $-1a + a = -a + a$
6. $-1a = -a$
7. $-a = -1a$

82. THEOREM 9

$-(a + b) = (-a) + (-b)$ and $(-a) + (-b) = -(a + b)$

Proof

1. $\quad -(a + b) = -1(a + b)$
2. $\qquad\qquad = -1a + (-1b)$
3. $\qquad\qquad = -a + (-b)$
4. $(-a) + (-b) = -(a + b)$

83. THEOREM 10

$-(a - b) = -a + b = b - a$

Proof

1. $-(a - b) = -[a + (-b)]$
2. $\qquad = -a + [-(-b)]$
3. $\qquad = -a + b$
4. $\qquad = b + (-a)$
5. $\qquad = b - a$

84. THEOREM 11

$(-a)b = -ab$

Proof

1. $(-a)b = (-a)b$
2. $\qquad = (-1 \cdot a)b$
3. $\qquad = -1 \cdot ab$
4. $\qquad = -ab$

85. THEOREM 12

$a(-b) = -ab$

Proof

1. $a(-b) = (-b)a$
2. $\qquad = -ba$
3. $\qquad = -ab$

86. THEOREM 13

$(-a)(-b) = ab$

Proof

1. $(-a)(-b) = -(a(-b))$
2. $\qquad = -(-ab)$
3. $\qquad = ab$

87. THEOREM 14   DISTRIBUTIVE THEOREM FOR SUBTRACTION

$a(b - c) = ab - ac$

Proof

1. $a(b - c) = a(b + (-c))$
2. $\qquad = ab + a(-c)$
3. $\qquad = ab + (-ac)$
4. $\qquad = ab - ac$

(88–95)   Prove each of the following theorems.

88. THEOREM 18

$\dfrac{1}{b} \cdot \dfrac{1}{d} = \dfrac{1}{bd}$ for $b \neq 0$ and $d \neq 0$.

89. **THEOREM 19**

$$\frac{a}{c} + \frac{b}{c} = \frac{a+b}{c} \quad \text{for} \quad c \neq 0.$$

90. **THEOREM 20**

$$\frac{a}{b} \cdot \frac{c}{d} = \frac{ac}{bd} \quad \text{for} \quad b \neq 0 \quad \text{and} \quad d \neq 0.$$

91. **THEOREM 21**

If $b \neq 0$, then $\dfrac{-a}{b} = \dfrac{a}{-b} = -\dfrac{a}{b}.$

92. **THEOREM 22**

If $b \neq 0$, then $\dfrac{-a}{-b} = \dfrac{a}{b}.$

93. If $c \neq 0$, then $\dfrac{a}{c} - \dfrac{b}{c} = \dfrac{a-b}{c}.$

94. If $bcd \neq 0$, then $\dfrac{a}{b} \div \dfrac{c}{d} = \dfrac{ad}{bc}.$

95. If $bd \neq 0$, then $\dfrac{1}{b} + \dfrac{1}{d} = \dfrac{b+d}{bd}.$

1.4 ORDER

If two quantities are measured by using real numbers, then the first measurement is equal to the second, greater than the second, or less than the second. These are the only possibilities and exactly one of them can be true.

Up to now we have used the concepts of "greater than," "less than," "positive," and "negative" in an intuitive way. In this section we look at these concepts from a more formal point of view.

We consider "is greater than," expressed in symbols as $>$, to be an undefined relation. Next we define "is less than," expressed in symbols as $<$, and state some axioms about these order relations. Finally, we examine some basic inequalities and their graphs. In this section the letters a, b, and c designate real numbers.

DEFINITION OF $<$

$a < b$ if and only if $b > a$.

ORDER AXIOM 1. TRICHOTOMY AXIOM

Exactly one of the following statements is true:
$$a = b \quad \text{or} \quad a > b \quad \text{or} \quad a < b$$

An important special case of this axiom is the case for $b = 0$; namely,

$a = 0$ or $a > 0$ or $a < 0$

In words, any real number is either equal to zero, is greater than zero, or is less than zero. Now we can define the concepts of "positive" and "negative."

DEFINITION OF A POSITIVE NUMBER

a is a **positive number** if and only if $a > 0$.

DEFINITION OF A NEGATIVE NUMBER

a is a **negative number** if and only if $a < 0$.

The set of positive real numbers is identified by the next two axioms which state that the sum of two positive numbers is a positive number and the product of two positive numbers is a positive number.

ORDER AXIOM 2

Closure Axiom, Addition of Positives. If $a > 0$ and $b > 0$, then $a + b > 0$.

ORDER AXIOM 3

Closure Axiom, Multiplication of Positives. If $a > 0$ and $b > 0$, then $ab > 0$.

The next axiom provides further information regarding the "greater than" relation. It states that a is greater than b if and only if $a - b$ is positive. This provides us with a useful test for determining if one real number is greater than another.

ORDER AXIOM 4

$$a > b \quad \text{if and only if} \quad a - b > 0.$$

In general, an **order relation** is a relation for which the transitive property is valid. Equality is an order relation due to the Transitive Axiom of Equality: If $a = b$ and $b = c$, then $a = c$.

Theorems 1 and 2 that follow show that the transitive property is also valid for the greater than and less than relations and thus are order relations.

THEOREM 1 TRANSITIVE PROPERTY FOR $>$

If $a > b$ and $b > c$, then $a > c$.

THEOREM 2 TRANSITIVE PROPERTY FOR $<$

If $a < b$ and $b < c$, then $a < c$.

Note that the order relations $>$ and $<$ are neither reflexive nor symmetric. For example, $3 > 3$ is false and the statement "if $5 > 3$, then $3 > 5$" is false.

A field for which the axioms and theorems of this section are valid is called an **ordered field.** Both the set of rational numbers and the set of real numbers are ordered fields.

The axiom that distinguishes these two sets is the Completeness Axiom which is valid only for the set of real numbers. Because of this axiom, the set of real numbers is called a **complete ordered field.** In geometric terms, the **Completeness Axiom** states there is a one-to-one correspondence between the set of real numbers and the set of points on a number line. The set of rationals is not complete because there are points on the number line that do not have rational numbers for coordinates.

An **inequality** is a statement that can be expressed in one of the following forms: $x > a$, $x < a$, $x \leq a$, or $x \geq a$.

The statement $x \geq a$ means that x is greater than a or x equals a; $x > a$ or $x = a$. Similarly, $x \leq a$ means that x is less than a or x equals a; $x < a$ or $x = a$.

The statement $a < x < b$ means that a is less than x and x is less than b; that is, $a < x$ and $x < b$. This statement can also be expressed in words by saying "x is between a and b."

The graphs of certain inequalities are called **basic intervals.** These inequalities and their geometric interpretations are summarized below. The graphs are show in Figure 1.3.

Basic Intervals

$x < a$	All points to the left of a, a excluded
$x > a$	All points to the right of a, a excluded
$x \leq a$	All points to the left of a with a included
$x \geq a$	All points to the right of a with a included
$a < x < b$	All points between a and b with a and b excluded
$a \leq x \leq b$	All points between a and b with a and b included
$a < x \leq b$	All points between a and b with b included, a excluded
$a \leq x < b$	All points between a and b with a included, b excluded

EXERCISES 1.4

(1–22) Insert one of the three relation symbols, $=$, $>$, or $<$, between each pair of numbers so that the resulting statement is true.

1.4 ORDER

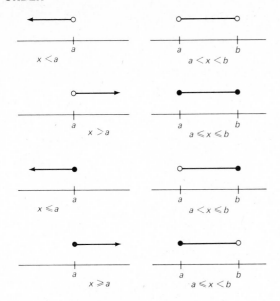

FIG. 1.3. Graphs of basic intervals.

EXAMPLE 1

3 ____ −5

Solution

3 > −5

EXAMPLE 2

−2 ____ −7

Solution

−2 > −7

EXAMPLE 3

−4 ____ 0

Solution

−4 < 0

EXAMPLE 4

−8 ____ −6

Solution

−8 < −6

EXAMPLE 5

9 ____ 0

Solution

9 > 0

EXAMPLE 6

$\frac{1}{2}$ ____ 0.5

Solution

$\frac{1}{2} = 0.5$

1. 10 ____ 8   **2.** 10 ____ 20

3. 14 ____ 17   **4.** 45 ____ 39

5. 26 ____ 0   **6.** 0 ____ 18

7. 0 ____ -8 **8.** -12 ____ 0

9. -15 ____ -20 **10.** -30 ____ -20

11. -25 ____ -10 **12.** -48 ____ -49

13. 13 ____ -9 **14.** -6 ____ 6

15. -7 ____ 4   **16.** 22 ____ -30

17. 0.5 ____ $\frac{1}{4}$ **18.** -0.75 ____ $\frac{-3}{4}$

19. $\frac{1}{8}$ ____ 0.125   **20.** $\frac{-7}{2}$ ____ -3.4

21. -0.5 ____ $-\frac{1}{2}$ **22.** -1.41 ____ -1.42

(23–36) Rewrite each of the following, using relation symbols.

EXAMPLE 7   Rewrite: $x + 5$ is negative.

Solution

$x + 5 < 0$

EXAMPLE 8   Rewrite: $2y$ is positive.

Solution

$2y > 0$

EXAMPLE 9   Rewrite: $r - 2$ is not negative.

Solution

$r - 2 \geq 0$

EXAMPLE 10   Rewrite: $\frac{n}{2}$ is not positive.

Solution

$\frac{n}{2} \leq 0$

EXAMPLE 11   Rewrite: $4x$ is greater than 7.

Solution

$4x > 7$

EXAMPLE 12   Rewrite: $-t$ is less than or equal to 8.

Solution

$-t \leq 8$

EXAMPLE 13   Rewrite: x is between -4 and 2.

Solution

$-4 < x < 2$

23. $2x$ is negative.

24. $4t$ is positive.

25. $n - 3$ is not negative.
26. $y + 6$ is not positive.
27. $x + y$ is less than 9.
28. $3x - 2$ is greater than 5.
29. 10 is greater than or equal to x.
30. 15 is less than or equal to z.
31. $2x$ is between -5 and 5.
32. $3y$ is between -6 and -1.
33. -7 is less than or equal to y and y is less than 2.
34. -8 is less than x and x is less than or equal to -3.
35. π is greater than 3.1415 and less than 3.1416.
36. $\sqrt{3}$ is greater than 1.7320 and less than 1.7321.

(37–40) Measurements are often recorded in the form $x = M \pm e$, where M is the found measurement and e is the possible error in measurement. This is equivalent to $M - e \le x \le M + e$. Express each of the following as a simplified inequality.

EXAMPLE 14
$x = 1.64 \pm 0.02$

Solution
$1.64 - 0.02 \le x \le 1.64 + 0.02$
$1.62 \le x \le 1.66$

37. $x = 15 \pm 4$
38. $x = 10.3 \pm 0.5$
39. $x = 6.50 \pm 0.05$
40. $x = 13.24 \pm 0.01$

(41–50) List the subset of the set of integers for which each of the following is true.

EXAMPLE 15
$x \le -2$

Solution
$\{-2, -3, -4, -5, \ldots\}$

EXAMPLE 16
$x > 6$

Solution
$\{7, 8, 9, 10, \ldots\}$

EXAMPLE 17
$-5 < x \le 2$

Solution
$\{-4, -3, -2, -1, 0, 1, 2\}$

41. $x \le -1$
42. $x \ge -3$
43. $x > -5$
44. $x < -4$
45. $-2 < x < 2$
46. $-3 \le x < 1$
47. $-7 < x \le -2$
48. $3 < x \le 8$
49. $x \ge 0$
50. $x < 0$

(51–64) Graph each of the following on the real number line.

EXAMPLE 18 Graph: $x < 2$

Solution

EXAMPLE 19 Graph: $x \le -3$
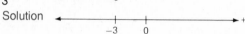
Solution

EXAMPLE 20 Graph: $x > -1$

Solution

EXAMPLE 21 Graph: $x \ge 4$

Solution

EXAMPLE 22 Graph: $2 < x < 6$

Solution

EXAMPLE 23 Graph: $-3 \le x \le 3$

Solution

EXAMPLE 24 Graph: $-5 \le x < -2$

Solution

51. $x > 3$ **52.** $x \ge -2$
53. $x \le -4$ **54.** $x < 6$
55. $x < 0$ **56.** $x \le 0$
57. $x \ge 0$ **58.** $x > 0$
59. $1 < x < 5$ **60.** $0 \le x \le 3$
61. $-4 \le x \le 4$ **62.** $-4 < x < -1$
63. $-7 < x \le -3$ **64.** $-1 \le x < 4$

(65–69) Supply reasons for the statements in each outlined proof.

65. THEOREM

$a < 0$ if and only if $-a > 0$.

1. $a < 0$ if an only if $0 > a$.
2. $0 > a$ if and only if $0 - a > 0$.
3. $0 - a = 0 + (-a)$
4. $0 + (-a) = -a$
5. $0 > a$ if and only if $-a > 0$.

66. THEOREM

For any nonzero real number a, $a^2 > 0$.
(*Note.* $a^2 = a \cdot a$.)

Case 1 $a > 0$

1. If $a > 0$ and $a > 0$, then $a^2 > 0$.

Case 2 $a < 0$

2. $a < 0$ if and only if $-a > 0$.
3. For $-a > 0$, $(-a)(-a) > 0$.
4. $(-a)(-a) = a^2$
5. $a^2 > 0$

67. THEOREM

  $a > b$ if and only if $a + c > b + c$.

1. $a + c > b + c$ if and only if $(a + c) - (b + c) > 0$.
2. $(a + c) - (b + c) = (a + c) + [-b + (-c)]$
3. $\qquad\qquad\quad = a + (-b) + c + (-c)$
4. $\qquad\qquad\quad = a + (-b)$
5. $\qquad\qquad\quad = a - b$
6. $a + c > b + c$ if and only if $a - b > 0$.
7. $a - b > 0$ if and only if $a > b$.
8. $a > b$ if and only if $a + c > b + c$.

68. THEOREM

  If $a > b$ and $c > 0$, then $ac > bc$.

1. $a > b$ if and only if $a - b > 0$.
2. If $a - b > 0$ and $c > 0$, then $(a - b)c > 0$.
3. $(a - b)c = ac - bc$
4. $ac - bc > 0$
5. $ac > bc$

69. THEOREM

  If $a > b$ and $c < 0$, then $ac < bc$.

1. If $c < 0$, then $-c > 0$.
2. $a > b$ if and only if $a - b > 0$.
3. $(a - b)(-c) > 0$
4. $-ac + bc > 0$
5. $bc - ac > 0$
6. $bc > ac$
7. $ac < bc$

(70–74) Prove each of the following. The theorems in Exercises 65–69 may be used as reasons if they apply.

70. $a < b$ if and only if $a + c < b + c$.
71. If $a < b$ and $c > 0$, then $ac < bc$.
72. If $a < b$ and $c < 0$, then $ac > bc$.
73. If $a < 0$ and $b < 0$, then $ab > 0$.
74. If $a < 0$ and $b > 0$, then $ab < 0$.

1.5 SQUARE ROOT RADICALS AND COMPLEX NUMBERS

In Section 1.3 we defined the radical $\sqrt[n]{a}$ for $a > 0$ as the unique positive real number such that $(\sqrt[n]{a})^n = a$. In this section we examine properties of square root radicals and properties of complex numbers.

The following theorems summarize the properties of positive square roots of positive real numbers.

THEOREMS   PROPERTIES OF SQUARE ROOT RADICALS

Let a and b be positive real numbers.

THEOREM

1. $\sqrt{a^2} = a$

EXAMPLE
$\sqrt{5^2} = 5$

THEOREM

2. $\sqrt{(-a)^2} = a$

EXAMPLE
$\sqrt{(-5)^2} = \sqrt{25} = 5$

THEOREM

3. $\sqrt{a}\,\sqrt{b} = \sqrt{ab}$

EXAMPLE
$\sqrt{5}\,\sqrt{2} = \sqrt{10}$

THEOREM

4. $\dfrac{\sqrt{a}}{\sqrt{b}} = \sqrt{\dfrac{a}{b}}$ for $b \neq 0$

EXAMPLE
$\dfrac{\sqrt{5}}{\sqrt{10}} = \sqrt{\dfrac{5}{10}} = \sqrt{\dfrac{1}{2}}$

The proofs of Theorems 1 and 2 follow immediately from the definition. Each side of the equation satisfies the equation $x^2 = a^2$ and each side is a positive real number.

The proofs for Theorems 3 and 4 are outlined in the Exercises.

DEFINITION

An expression involving a square root radical whose radicand is rational is said to be in **simplified form** if:

1. The radicand is a positive integer having no factor that is the square of a prime number.

2. There is no radical in a denominator.

As examples, the simplified form of $\sqrt{75}$ is $5\sqrt{3}$ because $\sqrt{75} = \sqrt{25}\,\sqrt{3} = 5\sqrt{3}$.

The simplified form of $\dfrac{2}{\sqrt{10}}$ is $\dfrac{\sqrt{10}}{5}$ because

$$\frac{2}{\sqrt{10}} = \frac{2\sqrt{10}}{\sqrt{10}\,\sqrt{10}} = \frac{2\sqrt{10}}{10} = \frac{\sqrt{10}}{5}.$$

Since numbers of the form $\sqrt{a}$, $b\sqrt{a}$, $c + b\sqrt{a}$, and $c - b\sqrt{a}$ are real numbers when a is a nonnegative real number, the axioms and the theorems of the real number system apply. As examples,

$$6 + 3\sqrt{5} = 3(2 + \sqrt{5})$$

$$\frac{8 - \sqrt{48}}{2} = \frac{8 - 4\sqrt{3}}{2} = \frac{2(4 - 2\sqrt{3})}{2} = 4 - 2\sqrt{3}$$

$$3\sqrt{2} + 5\sqrt{2} = (3 + 5)\sqrt{2} = 8\sqrt{2}$$

$$3\sqrt{5} - 7\sqrt{5} = (3 - 7)\sqrt{5} = -4\sqrt{5}$$

$$(4 - \sqrt{3})^2 = (4 - \sqrt{3})(4 - \sqrt{3})$$

$$= (4 - \sqrt{3})4 - (4 - \sqrt{3})\sqrt{3}$$

$$= 16 - 4\sqrt{3} - 4\sqrt{3} + 3$$

$$= 19 - 8\sqrt{3}$$

When a, b, and c are rational numbers and $\sqrt{c}$ is an irrational number, the real numbers $a + b\sqrt{c}$ and $a - b\sqrt{c}$ are called **irrational conjugates,** or **conjugate surds.**

For example, $5 - 2\sqrt{3}$ and $5 + 2\sqrt{3}$ are irrational conjugates. Also $3 + \sqrt{6}$ and $3 - \sqrt{6}$ are irrational conjugates. Two irrational conjugates have the property that their product is rational.

Using $(A + B)(A - B) = A^2 - B^2$, then

$$(a + b\sqrt{c})(a - b\sqrt{c}) = a^2 - (b\sqrt{c})^2 = a^2 - b^2c$$

Note that

$$(5 - 2\sqrt{3})(5 + 2\sqrt{3}) = 25 - 4(3) = 13$$

and that

$$(3 + \sqrt{6})(3 - \sqrt{6}) = 9 - 6 = 3$$

Examples in the Exercises show how this property is used to rationalize certain binomial denominators.

COMPLEX NUMBERS

If a real number is negative, then it does not have a square root that is a real number. For example, there is no real number x

for which $x^2 = -1$. The theorem in Exercises 1.4, Problem 66 states that a^2 is positive for every real number except 0. (Note that $0^2 = 0$.) However we can imagine a new number i whose square is -1 and denote i by $\sqrt{-1}$.

DEFINITION

$i = \sqrt{-1}$ and $i^2 = -1$. i is called the **imaginary unit.**

The square roots of the other negative real numbers can be defined in terms of the imaginary unit, i.

DEFINITION

If a is a positive real number, then $\sqrt{-a} = \sqrt{a}\,i = i\sqrt{a}$

As examples, $\sqrt{-9} = \sqrt{9}\sqrt{-1} = 3i$ and $\sqrt{-5} = \sqrt{5}\sqrt{-1} = \sqrt{5}\,i = i\sqrt{5}$. The form $i\sqrt{5}$ is preferred to the form $\sqrt{5}\,i$ to avoid the error of writing $\sqrt{5}\,i$ as $\sqrt{5i}$.

DEFINITION

A number having the form bi where b is any nonzero real number is called a **pure imaginary** number.

DEFINITION

A number having the form $a + bi$ where a and b are any real numbers is called a **complex number.**

If $b = 0$, then $a + bi = a + 0i = a + 0 = a$ and $a + bi$ is a **real number.**

If $b \neq 0$, then $a + bi$ is an **imaginary number.**

The **set of complex numbers** is the union of the set of real numbers with the set of imaginary numbers.

The form $a + bi$ is called the **standard form** of a complex number.

With the following definitions of equality, addition, and multiplication, the set of complex numbers can be shown to be a field. The field axioms can be derived as theorems for the set of complex numbers.

DEFINITION OF EQUALITY

$a + bi = c + di$ if and only if $a = c$ and $b = d$

DEFINITION OF ADDITION

$(a + bi) + (c + di) = (a + c) + (b + d)i$

DEFINITION OF MULTIPLICATION

$$(a + bi)(c + di) = ac - bd + (ad + bc)i$$

Note that by using the properties of a field,
$$
\begin{aligned}
(a + bi)(c + di) &= (a + bi)c + (a + bi)di \\
&= ac + bci + adi + bdi^2 \\
&= ac + (bc + ad)i + bd(-1) \\
&= ac - bd + (ad + bc)i
\end{aligned}
$$
The **additive identity** for complex numbers is $0 + 0i = 0$.
The **additive inverse** of $a + bi$ is $-a - bi$.
Note that
$$(a + bi) + (0 + 0i) = (a + 0) + (b + 0)i = a + bi$$
and that
$$(a + bi) + (-a - bi) = a - a + (b - b)i = 0 + 0i = 0$$
The **multiplicative identity** for complex numbers is $1 + 0i = 1$.

The **multiplicative inverse** of $a + bi$ is $\dfrac{1}{a + bi} = \dfrac{a - bi}{a^2 + b^2}$.

Note that
$$(a + bi)(1 + 0i) = a - 0i + (0 + b)i = a + bi$$
and
$$\frac{1}{a + bi} = \frac{1(a - bi)}{(a + bi)(a - bi)} = \frac{a - bi}{a^2 + b^2} = \frac{a}{a^2 + b^2} + \frac{-b}{a^2 + b^2}i$$
The numbers $a + bi$ and $a - bi$ are called **complex conjugates.** Each is the complex conjugate of the other.

DEFINITION OF SUBTRACTION

$$a + bi - (c + di) = a + bi + (-c - di)$$

DEFINITION OF DIVISION

$$\frac{a + bi}{c + di} = (a + bi)\left(\frac{1}{c + di}\right) = \frac{(a + bi)(c - di)}{c^2 + d^2}$$

While the set of complex numbers is a field, it is not an ordered field. In particular the imaginary unit i cannot be ordered. For an ordered field, every element a of the field must satisfy the following:

1. $a = 0$ or $a < 0$ or $a > 0$.
2. If $a > 0$ and $b > 0$, then $a + b > 0$ and $ab > 0$.

Now, $i \neq 0$ since $0 + 1i \neq 0 + 0i$ and $1 \neq 0$.
If $i > 0$, then $ii > 0$ and $-1 > 0$ since $i^2 = -1$. But -1 is not greater than 0; thus i cannot be greater than 0.
If $i < 0$, then $-i > 0$, $(-i)(-i) > 0$, $i^2 > 0$, and $-1 > 0$. Thus i cannot be less than 0. As a result, i does not satisfy the requirement that $i = 0$ or $i > 0$ or $i < 0$.

Since every complex number cannot be ordered, the set of complex numbers is not an ordered field. However, the set of complex numbers has the very important property that it is closed with respect to the six operations of algebra. This means that every calculation on complex numbers involving one or more of the six operations of algebra (addition, subtraction, multiplication, division, raising to a power, and root extraction) has a solution (answer) that is a complex number. Note that the set of real numbers does not have this property since the answers to some root extraction problems are not real numbers.

EXERCISES 1.5

(1–27) Do the indicated operations and simplify.

EXAMPLE 1
9^2

Solution
$9 \cdot 9 = 81$

EXAMPLE 2
4^3

Solution
$4 \cdot 4 \cdot 4 = 64$

EXAMPLE 3
$\left(\frac{^-5}{4}\right)^2$

Solution
$\left(\frac{^-5}{4}\right)\left(\frac{^-5}{4}\right) = \frac{25}{16}$

EXAMPLE 4
$\left(\frac{^-4}{5}\right)^3$

Solution
$\left(\frac{^-4}{5}\right)\left(\frac{^-4}{5}\right)\left(\frac{^-4}{5}\right) = \frac{^-64}{125}$

EXAMPLE 5
$(-6)^2$

Solution
$(-6)(-6) = 36$

EXAMPLE 6
$(-6)^3$

Solution
$(-6)(-6)(-6) = -216$

EXAMPLE 7
3^5

Solution
$3 \cdot 3 \cdot 3 \cdot 3 \cdot 3 = 243$

EXAMPLE 8

$\left(\frac{2}{3}\right)^4$

> Solution
>
> $\left(\frac{2}{3}\right)\left(\frac{2}{3}\right)\left(\frac{2}{3}\right)\left(\frac{2}{3}\right) = \frac{16}{81}$

EXAMPLE 9

$(-2 \cdot 5)^2$

> Solution
>
> $(-10)^2 = (-10)(-10) = 100$

EXAMPLE 10

$-2 \cdot 5^2$

> Solution
>
> $-2(5)(5) = -2(25) = -50$

1. 8^2	**2.** $(-10)^2$	**3.** 5^3
4. $(-9)^3$	**5.** $\left(\frac{3}{10}\right)^2$	**6.** $\left(\frac{4}{5}\right)^2$
7. $\left(\frac{8}{3}\right)^3$	**8.** $\left(\frac{9}{10}\right)^3$	**9.** 2^4
10. $(-10)^4$	**11.** $(-0.6)^2$	**12.** $(-0.1)^3$
13. $\left(\frac{1}{6}\right)^3$	**14.** 5^3	**15.** 3^5
16. $(-2)^5$	**17.** $(-10)^5$	**18.** 10^6
19. $-3 \cdot 4^2$	**20.** $(-3 \cdot 4)^2$	**21.** $(-5 \cdot 6)^2$
22. $-5(6^2)$	**23.** $-5(-6)^2$	**24.** $-5(-10)^3$
25. $-5 \cdot 10^3$	**26.** $(-5 \cdot 10)^3$	**27.** $(-5)^3 - 10^3$

(28–48) State the positive real number for which each of the following is true. Express rational answers in simplified form. Approximate each irrational answer correct to 3 decimal places, using the "Table of Powers and Roots" in the Endpapers or a hand-held calculator.

EXAMPLE 11

$x^2 = 144$

> Solution
>
> $x = 12$

EXAMPLE 12

$x^2 = \frac{49}{100}$

> Solution
>
> $x = \frac{7}{10}$

EXAMPLE 13

$x^2 = 15$

> Solution
>
> $x = \sqrt{15}, \ \sqrt{15} \approx 3.873$

EXAMPLE 14

$x^3 = 512$

Solution

$x = 8$

EXAMPLE 15

$x^3 = \frac{27}{125}$

Solution

$x = \frac{3}{5}$

EXAMPLE 16

$x^3 = 100$

Solution

$x = \sqrt[3]{100}, \sqrt[3]{100} \approx 4.642$

28. $x^2 = 361$ **29.** $x^2 = 1600$ **30.** $x^2 = 2500$

31. $x^2 = 6084$ **32.** $x^2 = \frac{64}{225}$ **33.** $x^2 = \frac{25}{100}$

34. $x^2 = 7$ **35.** $x^2 = 70$ **36.** $x^2 = 30$

37. $x^2 = 10$ **38.** $x^3 = 729$ **39.** $x^3 = 216$

40. $x^3 = 1000$ **41.** $x^3 = 8000$ **42.** $x^3 = \frac{125}{8}$

43. $x^3 = \frac{343}{512}$ **44.** $x^3 = 18$ **45.** $x^3 = 59$

46. $x^3 = 95$ **47.** $x^3 = 900$ **48.** $x^3 = 25$

(49–72)   Simplify each of the following.

EXAMPLE 17

$\sqrt{(3.20)^2}$

Solution

3.20 by Theorem 1

EXAMPLE 18

$(\sqrt{29})^2$

Solution

29 by definition of $\sqrt{a}$

EXAMPLE 19

$\sqrt{(-12)^2}$

Solution

12 by Theorem 2

EXAMPLE 20

$\sqrt{20}\sqrt{5}$

Solution

$\sqrt{20 \cdot 5} = \sqrt{100} = 10$, using Theorem 3

EXAMPLE 21

$\sqrt{50}$

Solution

$\sqrt{25}\sqrt{2} = 5\sqrt{2}$, using Theorem 3

EXAMPLE 22

$\dfrac{\sqrt{3}}{\sqrt{75}}$

Solution

$\sqrt{\dfrac{3}{75}} = \sqrt{\dfrac{1}{25}} = \dfrac{1}{5}$, using Theorem 4

EXAMPLE 23

$\dfrac{2}{\sqrt{6}}$

Solution

$\dfrac{2\sqrt{6}}{\sqrt{6}\sqrt{6}} = \dfrac{2\sqrt{6}}{6} = \dfrac{\sqrt{6}}{3}$, using Theorem 4

49. $\sqrt{8^2}$ 50. $(\sqrt{7})^2$ 51. $(\sqrt{9.5})^2$

52. $\sqrt{(4.25)^2}$ 53. $\sqrt{(-21)^2}$ 54. $\sqrt{(-105)^2}$

55. $\sqrt{2}\sqrt{18}$ 56. $\sqrt{45}\sqrt{20}$ 57. $\sqrt{40}$

58. $\sqrt{200}$ 59. $\sqrt{98}$ 60. $\sqrt{150}$

61. $\sqrt{180}$ 62. $\sqrt{450}$ 63. $\dfrac{1}{\sqrt{3}}$

64. $\dfrac{1}{\sqrt{5}}$ 65. $\sqrt{\dfrac{1}{10}}$ 66. $\sqrt{\dfrac{2}{5}}$

67. $\dfrac{\sqrt{2}}{\sqrt{72}}$ 68. $\dfrac{\sqrt{5}}{\sqrt{45}}$ 69. $\dfrac{7}{\sqrt{14}}$

70. $\dfrac{4}{\sqrt{8}}$ 71. $\dfrac{\sqrt{6}\sqrt{10}}{\sqrt{5}}$ 72. $\dfrac{\sqrt{28}}{\sqrt{7}\sqrt{3}}$

(73–84) Simplify.

EXAMPLE 24

$(5 - 2\sqrt{3})^2$

Solution

$(5 - 2\sqrt{3})(5 - 2\sqrt{3}) = (5 - 2\sqrt{3})5 - (5 - 2\sqrt{3})2\sqrt{3}$
$= 25 - 10\sqrt{3} - 10\sqrt{3} + 4(3)$
$= 37 - 20\sqrt{3}$

EXAMPLE 25

$(4 - \sqrt{6})(4 + \sqrt{6})$

Solution

$4^2 - (\sqrt{6})^2 = 16 - 6 = 10$

EXAMPLE 26

$\dfrac{6 - \sqrt{20}}{2}$

Solution

$\dfrac{6 - \sqrt{4 \cdot 5}}{2} = \dfrac{6 - 2\sqrt{5}}{2} = \dfrac{2(3 - \sqrt{5})}{2} = 3 - \sqrt{5}$

73. $(4 - \sqrt{5})^2$

74. $(-3 + \sqrt{6})^2$

75. $(7 + 3\sqrt{2})^2$

76. $(8 - 2\sqrt{5})^2$

77. $(6 - \sqrt{5})(6 + \sqrt{5})$

78. $(2 - \sqrt{3})(2 + \sqrt{3})$

79. $(9 + 3\sqrt{2})(9 - 3\sqrt{2})$

80. $(-4 + 5\sqrt{6})(-4 - 5\sqrt{6})$

81. $\dfrac{8 - \sqrt{12}}{2}$

82. $\dfrac{10 + \sqrt{24}}{6}$

83. $\dfrac{-6 + \sqrt{288}}{12}$

84. $\dfrac{-4 - \sqrt{56}}{10}$

(85–100) Rationalize each denominator.

EXAMPLE 27

$\dfrac{1}{5 - 2\sqrt{3}}$

Solution

$\dfrac{1(5 + 2\sqrt{3})}{(5 - 2\sqrt{3})(5 + 2\sqrt{3})} = \dfrac{5 + 2\sqrt{3}}{25 - 4(3)} = \dfrac{5 + 2\sqrt{3}}{13}$

EXAMPLE 28

$\dfrac{12}{3 + \sqrt{6}}$

Solution

$\dfrac{12(3 - \sqrt{6})}{(3 + \sqrt{6})(3 - \sqrt{6})} = \dfrac{12(3 - \sqrt{6})}{9 - 6}$

$= 4(3 - \sqrt{6}) = 12 - 4\sqrt{6}$

85. $\dfrac{1}{4 - \sqrt{10}}$

86. $\dfrac{1}{5 + \sqrt{15}}$

87. $\dfrac{2}{3 + 2\sqrt{2}}$

88. $\dfrac{4}{6 - 4\sqrt{2}}$

89. $\dfrac{4}{3 + \sqrt{5}}$

90. $\dfrac{7}{3 - \sqrt{2}}$

91. $\dfrac{4}{\sqrt{5} - 1}$

92. $\dfrac{\sqrt{3}}{\sqrt{3} + 2}$

93. $\dfrac{5\sqrt{7}}{\sqrt{7} + 2}$

94. $\dfrac{5\sqrt{6}}{\sqrt{6} - 1}$

95. $\dfrac{3 - \sqrt{7}}{3 + \sqrt{7}}$

96. $\dfrac{2 + \sqrt{3}}{2 - \sqrt{3}}$

97. $\dfrac{1}{\sqrt{5} - \sqrt{3}}$

98. $\dfrac{2}{\sqrt{7} + \sqrt{5}}$

99. $\dfrac{3 + 2\sqrt{2}}{3 - 2\sqrt{2}}$

100. $\dfrac{7 - 4\sqrt{3}}{7 + 4\sqrt{3}}$

(101–136) Express each of the following in the *bi* form or the standard *a* + *bi* form.

EXAMPLE 29

$\sqrt{-36}$

Solution

$\sqrt{36}\sqrt{-1} = 6i$

EXAMPLE 30

$\sqrt{-50}$

Solution

$\sqrt{50}\sqrt{-1} = 5\sqrt{2}i = i5\sqrt{2}$

EXAMPLE 31

$(6 - \sqrt{-81}) + (-3 + \sqrt{-25})$

Solution

$(6 - 9i) + (-3 + 5i) = (6 - 3) + (-9 + 5)i = 3 - 4i$

EXAMPLE 32

$(-4 + 7i) - (1 - 2i)$

Solution

$(-4 + 7i) + (-1 + 2i) = -5 + 9i$

EXAMPLE 33

$(4 - 5i)^2$

Solution

$$(4 - 5i)(4 - 5i) = (4 - 5i)4 - (4 - 5i)5i$$
$$= 16 - 20i - 20i + 25i^2$$
$$= 16 - 40i + 25(-1) = -9 - 40i$$

EXAMPLE 34

$\dfrac{10}{3 - 4i}$

Solution

$$\frac{10(3 + 4i)}{(3 - 4i)(3 + 4i)} = \frac{10(3 + 4i)}{9 - 16i^2} = \frac{10(3 + 4i)}{25}$$
$$= \frac{2(3 + 4i)}{5} = \frac{6 + 8i}{5} = \frac{6}{5} + \frac{8}{5}i$$

101. $\sqrt{-4}$ 102. $\sqrt{-9}$

103. $\sqrt{-25}$ 104. $\sqrt{-16}$

105. $\sqrt{-3}$ 106. $\sqrt{-15}$

107. $\sqrt{-24}$ 108. $\sqrt{-72}$

109. $(3 + 5i) + (-2 + 4i)$ 110. $(-6 + 9i) + (-2 - 9i)$

111. $(5 - 3\sqrt{2}i) + (4 + 3\sqrt{2}i)$ 112. $(1 - 2\sqrt{7}i) + (-5 + 3\sqrt{7}i)$

113. $(-8 - \sqrt{-36}) + (-2 + \sqrt{-100})$

114. $(7 + \sqrt{-64}) + (-10 - \sqrt{-49})$

115. $(4 + 8i) - (7 + 4i)$

116. $(-3 + 6i) - (-5 + 3i)$

117. $(10 - \sqrt{-25}) - (-6 - \sqrt{-4})$

118. $(-9 - \sqrt{-9}) - (4 - \sqrt{-1})$

119. $(3 + 5i)^2$ 120. $(7 - 2i)^2$

121. $(-6 - 3i)^2$ 122. $(-1 + i)^2$

123. $(1 - \sqrt{-3})^2$ 124. $(2 + \sqrt{-5})^2$

125. $(4 + 3i)(4 - 3i)$ 126. $(2 - 8i)(2 + 8i)$

127. $(1 - i)(1 + i)$ 128. $(2 + i\sqrt{3})(2 - i\sqrt{3})$

129. $(3 - i5\sqrt{2})(3 + i5\sqrt{2})$ 130. $(9 + i2\sqrt{6})(9 - i2\sqrt{6})$

131. $\dfrac{1}{3 - 4i}$ 132. $\dfrac{2}{2 + 5i}$

133. $\dfrac{10i}{1 + 2i}$ 134. $\dfrac{5i}{4 - 2i}$

135. $\dfrac{2 - 3i}{2 + 3i}$ 136. $\dfrac{5 + i}{5 - i}$

(137–140) Supply the reasons for the given statements in each outlined proof.

137. **THEOREM**

For $a > 0$ and $b > 0$, then $\sqrt{a}\sqrt{b} = \sqrt{ab}$

1. $(\sqrt{a}\sqrt{b})^2 = (\sqrt{a}\sqrt{b})(\sqrt{a}\sqrt{b})$
2. $\qquad\qquad = (\sqrt{a}\sqrt{a})(\sqrt{b}\sqrt{b})$
3. $\qquad\qquad = ab$
4. $(\sqrt{ab})^2 = ab$
5. $\sqrt{a}\sqrt{b} = \sqrt{ab}$, since there is exactly one positive real number x for which $x^2 = ab$.

138. **THEOREM**

For $a > 0$ and $b > 0$, then $\dfrac{\sqrt{a}}{\sqrt{b}} = \sqrt{\dfrac{a}{b}}$.

1. $\left(\dfrac{\sqrt{a}}{\sqrt{b}}\right)^2 = \dfrac{\sqrt{a}\sqrt{a}}{\sqrt{b}\sqrt{b}}$

2. $\qquad\quad = \dfrac{a}{b}$

3. $\left(\sqrt{\dfrac{a}{b}}\right)^2 = \dfrac{a}{b}$

4. $\dfrac{\sqrt{a}}{\sqrt{b}} = \sqrt{\dfrac{a}{b}}$ since there is exactly one positive real number x for which $x^2 = \dfrac{a}{b}$.

139. **THEOREM**   Addition of Complex Numbers is Commutative

$(a + bi) + (c + di) = (c + di) + (a + bi)$

1. $(a + bi) + (c + di) = (a + c) + (b + d)i$
2. $(c + di) + (a + bi) = (c + a) + (d + b)i$
3. $a + c = c + a$ and $b + d = d + b$
4. $(a + bi) + (c + di) = (c + di) + (a + bi)$

140. **THEOREM**   Multiplication of Complex Numbers is Commutative

$(a + bi)(c + di) = (c + di)(a + bi)$

1. $(a + bi)(c + di) = ac - bd + (ad + bc)i$
2. $(c + di)(a + bi) = ca - db + (cb + da)i$
3. $ac = ca$ and $bd = db$
4. $cb + da = da + cb$
5. $cb + da = ad + bc$
6. $(a + bi)(c + di) = (c + di)(a + bi)$

1.6 CHAPTER REVIEW

(1–5) List the elements in each set described and state whether the set is finite or infinite. (1.1)

1. The even natural numbers between 51 and 59.
2. The integers less than 3.
3. The rational numbers between 0 and 1 that have the form $\frac{x}{10}$ where x is a natural number.
4. The irrational numbers that have the form $k\sqrt{2}$ where k is an integer.
5. The composite factors of 17.

(6–8) State whether each number is rational or irrational, and whether or not the number is an integer. (1.1)

6. $8.50324\ldots$
7. $8.50\overline{50}\ldots$
8. 85050

(9–10) Find $A \cup B$ and $A \cap B$. (1.1)

9. $A = \{x \mid x$ is a prime factor of 42$\}$
   $B = \{x \mid x$ is an odd factor of 42$\}$
10. $A = \{x \mid x$ and $\frac{x}{4}$ are integers$\}$

    $B = \{x \mid x$ and $\frac{x}{6}$ are integers$\}$

(11–13) State an equality axiom that justifies each of the following. (1.2)

11. If $7 = 3x - 5$, then $3x - 5 = 7$.
12. $3x - 5 = 3x - 5$
13. If $y = 3x - 5$ and $x = 4$, then $y = 3(4) - 5 = 7$.

(14–22) State a field axiom that justifies each of the following. (1.2)

14. $5(4x) = 20x$
15. $4 + (x - 4) = (x - 4) + 4$
16. $-6x + 6x = 0$
17. $(y + 6)3 = 3(y + 6)$

18. $1(x - 2) = x - 2$

19. $(2k + 7) + (-7) = 2k + [7 + (-7)]$

20. $\frac{1}{5}(5) = 1$

21. $8t + 0 = 8t$

22. $2x(3x + 4) = 6x^2 + 8x$

(23–26) State one or more field axioms that justify each of the following. (1.2)

23. $(2x + 5) + (-5) = 2x$

24. $\frac{1}{9}(9y) = y$

25. $4\left(\dfrac{x - 2}{4}\right) = x - 2$

26. $n + 3(n + 3) = 4n + 9$

(27–36) Do the indicated operations. (1.3)

27. $10 - 2(3 - 7)$

28. $5 - 8 - 2$

29. $(2 - 5)(2 - 4)(2 - 3)$

30. $(75 - 86) - (42 - 53)$

31. $\dfrac{87 - 123}{123 - 87}$

32. $\dfrac{-600}{2(4 - 6)}$

33. $-24\left(\dfrac{3}{8} - \dfrac{5}{6}\right)$

34. $8\left(\dfrac{-3}{2}\right)^2$

35. $\dfrac{8 - 27}{2 - 3}$

36. $\sqrt{(-5)^2 - 4(1)(-6)}$

(37–40) Rewrite each of the following using relation symbols. (1.4)

37. $2x - 8$ is positive.

38. $y + 4$ is negative or equal to zero.

39. $x - 6$ is greater than or equal to y.

40. $5n$ is between -6 and 8.

(41–44) List the subset of the set of integers for which each of the following is true. (1.4)

41. $x < 3$

42. $x > -4$

43. $-4 \le x \le 4$

44. $2 < x \le 8$

(45–47) Graph each of the following on the real number line. (1.4)

45. $x \ge -6$

46. $x < 2$

47. $-1 \le x < 5$

(48–55) Simplify. (1.5)

48. $\sqrt{72}$

49. $\sqrt{5}\sqrt{15}$

50. $\dfrac{2}{\sqrt{6}}$

51. $\sqrt{\dfrac{3}{20}}$

52. $(3 - \sqrt{7})^2$

53. $(5 - 2\sqrt{3})(5 + 2\sqrt{3})$

54. $\dfrac{2 - \sqrt{20}}{6}$

55. $\dfrac{5}{4 + \sqrt{6}}$

(56–62) Express in the *bi* form or the standard $a + bi$ form. (1.5)

56. $\sqrt{-49}$ **57.** $-\sqrt{-6}$

58. $(5 - \sqrt{-16}) - (2 - \sqrt{-9})$ **59.** $(3 + 5i)^2$

60. $(4 - 2i)(4 + 2i)$ **61.** $\dfrac{1}{7 - 4i}$

62. $\dfrac{1 - i}{1 + i}$

ALGEBRAIC EXPRESSIONS

2.1 POLYNOMIALS; OPERATIONS ON POLYNOMIALS

In Section 1.5 the nth power of a real number b was defined as $b^1 = b$ and b^n equals the product of n factors of b for any natural number n, $n > 1$. For a variable x that represents a real number, the power x^n also means n factors of x for $n > 1$ and $x^1 = x$. The powers y^n and z^n have similar meanings for the variables y and z. As examples, $x^2 = xx$, $x^3 = xxx$, and $x^4 = xxxx$, $y^3 = yyy$, and $z^4 = zzzz$.

DEFINITION

A **monomial** is a constant, a variable, an nth power of a variable, or any finite product of these.

As examples, $5, x, y, x^3, -5x^2, 3xy,$ and x^2y^3z are monomials.

DEFINITION

A **polynomial** is a monomial or a sum of monomials.

A **polynomial of degree** n in the variable x has the form
$$a_n x^n + a_{n-1} x^{n-1} + \ldots + a_2 x^2 + a_1 x + a_0$$
where $a_n, a_{n-1}, \ldots, a_2, a_1$, and a_0 are constants, called the **coefficients,** and $a_n \neq 0$. The constant a_n is called the **leading coefficient** and a_0 is called the **constant term.**

As examples, $6x^3 - 4x^2 + 5x + 8$ is a polynomial of degree 3. Its leading coefficient is 6 and its constant term is 8. $x^2 - 5x - 6$ is a polynomial of degree 2, with leading coefficient 1 and constant term -6.

The polynomial $5x + 2$ has degree 1 and the polynomial 4 has degree 0. No degree is assigned to the zero polynomial 0.

DEFINITION

A **bionomial** is a polynomial that has exactly two terms.

For example, $x - 6$, $2x + y$, and $x^2 - 9y^2$ are binomials.

DEFINITION

A **trinomial** is a polynomial that has exactly three terms.

For example, $x^2 - 2x + 1$, $3x + 4y - 2$, and $x^3 + y^2 + z$ are trinomials.

DEFINITION

The **degree of a monomial containing two or more variables** is the sum of the exponents appearing on the variables.

The degree of $4x^2 y^3$ is 5; that is, $2 + 3$. The degree of $7xy^2$ is 3; that is, $1 + 2$ since $x = x^1$.

DEFINITION

The **degree of a polynomial containing two or more variables** is the greatest degree of any of its terms.

For example, the degree of $2x^3 - 4x^2 y^3 + 5y^4$ is 5, the degree of the term $-4x^2 y^3$.

The **notation $P(x)$** is often used to designate a polynomial in the variable x. Then $P(a)$ is the value of $P(x)$ for $x = a$.

EXAMPLE 1

If $P(x) = 4x^2 - 5x + 2$, find $P(3)$.

Solution

$P(3) = 4(3^2) - 5(3) + 2$

$= 36 - 15 + 2 = 23$

Since a polynomial represents a real number when its constants and variable(s) are real numbers, the axioms and theorems for the real number system can be used for operations on polynomials.

Polynomials are added and subtracted by rearranging terms using the commutative and associative axioms and by combining like terms using the distributive axiom. **Like terms** are monomials that have the same literal factors. As examples, $2x$ and $-5x$ are like terms; $3x^2y$ and $4x^2y$ are like terms. However, $5x$ and $5x^2$ are not like terms. Also, $3x^2y$ and $3xy^2$ are not like terms.

EXAMPLE 2 Simplify

$(5x^2 + 3x - 8) + (x^2 - 6x + 2)$

Solution

$= (5x^2 + x^2) + (3x - 6x) + (-8 + 2)$

$= (5 + 1)x^2 + (3 - 6)x + (-6)$

$= 6x^2 - 3x - 6$

EXAMPLE 3 Simplify

$(4x - 7y + 2) - (3x - 4y + 2)$

Solution

$= (4x - 7y + 2) + (-1)(3x - 4y + 2)$

$= (4x - 7y + 2) + (-3x + 4y - 2)$

$= (4x - 3x) + (-7y + 4y) + (2 - 2)$

$= x - 3y$

Polynomials are multiplied by using the distributive axiom one or more times, as shown in Examples 4, 5, and 6.

EXAMPLE 4 Simplify

$5x^2(x^2 - 2xy)$

Solution

$= 5x^2(x^2) - 5x^2(2xy)$

$= 5xx(xx) - 5(2)(xx)(xy)$

$= 5x^4 - 10x^3y$

EXAMPLE 5 Simplify

$(2x^2 - y)(x^2 + 4y)$

Solution

$= (2x^2 - y)x^2 + (2x^2 - y)(4y)$

$= 2x^2(x^2) - yx^2 + 2x^2(4y) - 4y^2$

$= 2x^4 - x^2y + 8x^2y - 4y^2$

$= 2x^4 + 7x^2y - 4y^2$

EXAMPLE 6 Simplify
$(6x - 5)(x^2 - 9)$

Solution

$$= (6x - 5)(x^2) - (6x - 5)9$$
$$= 6x(x^2) - 5x^2 - 6x(9) + 5(9)$$
$$= 6x^3 - 5x^2 - 54x + 45$$

We saw in Chapter 1, for the addition of rational numbers, that

$$\frac{a}{c} + \frac{b}{c} = \frac{a + b}{c} \qquad \text{for } c \neq 0$$

By symmetry,

$$\frac{a + b}{c} = \frac{a}{c} + \frac{b}{c}$$

By extension,

$$\frac{a + b + c}{d} = \frac{a}{d} + \frac{b}{d} + \frac{c}{d}, \qquad \text{for } d \neq 0$$

and, similarly, for any number of terms in the numerator.
This property is used in dividing a polynomial by a monomial.

EXAMPLE 7 Find the indicated quotient.
$$\frac{15x^3 - 20x^2 + 10x}{5x}, \qquad \text{where } x \neq 0$$

Solution

$$= \frac{15x^3}{5x} - \frac{20x^2}{5x} + \frac{10x}{5x}$$
$$= \frac{3 \cdot 5xxx}{5x} - \frac{4 \cdot 5xx}{5x} + \frac{2 \cdot 5x}{5x}$$
$$= \frac{3x^2(5x)}{5x} - \frac{4x(5x)}{5x} + \frac{2(5x)}{5x}$$
$$= 3x^2 - 4x + 2$$

The set of polynomials in the variable x has properties similar to those for the set of integers. Note that $P(x)$ represents an integer if its coefficients are integers and if x is an integer. (See Example 1.) We saw that the quotient of two integers is not always an integer. For example, $\frac{12}{3} = 4$ but $\frac{7}{2} = 3 + \frac{1}{2}$. Similarly, the quotient of two polynomials is not always a polynomial; $\frac{6x^2 + 10x}{2x} = 3x + 5$ since $6x^2 + 10x = 2x(3x + 5)$ but $\frac{6x^2 + 10x + 1}{2x} = 3x + 5 + \frac{1}{2x}$ which is not a polynomial. However there is a formal division process for polynomials that is similar to the long division method of arithmetic. In arithmetic, multiples of the divisor are subtracted from the dividend until the remainder is less than the divisor.

In the division of polynomials, multiples of the divisor are subtracted from the dividend until the **degree** of the remainder is less than the **degree** of the divisor, or until the remainder is zero.

EXAMPLE 8 Divide
$15x^3 - 31x^2 - 9$ by $5x - 2$.

Solution

$$
\begin{array}{r}
3x^2 - 5x - 2 \\
5x - 2\overline{)15x^3 - 31x^2 \quad\quad - 9} \\
\end{array}
$$

$\dfrac{15x^3}{5x} = 3x^2$ and $3x^2(5x - 2) = \quad \dfrac{15x^3 - 6x^2}{}$ (subtract)

$\qquad\qquad\qquad\qquad\qquad\quad -25x^2 \qquad - 9$

$\dfrac{-25x^2}{5x} = -5x$ and $-5x(5x - 2) = \quad \dfrac{-25x^2 + 10x}{}$ (subtract)

$\qquad\qquad\qquad\qquad\qquad\qquad\quad -10x - 9$

$\dfrac{-10x}{5x} = -2$ and $-2(5x - 2) = \quad \dfrac{-10x + 4}{}$ (subtract)

$\qquad\qquad\qquad\qquad\qquad\qquad\qquad\quad -13$

$\dfrac{15x^3 - 31x^2 - 9}{5x - 2} = 3x^2 - 5x - 2 + \dfrac{-13}{5x - 2}$

Note that this last equation has the following form.

$\dfrac{\text{Dividend}}{\text{Divisor}} = (\text{Quotient polynomial}) + \dfrac{\text{Remainder}}{\text{Divisor}}$

or

$\dfrac{P(x)}{D(x)} = Q(x) + \dfrac{R(x)}{D(x)}$

Multiplying each side by $D(x)$,

$$P(x) = Q(x)D(x) + R(x)$$

This equation can be used to check the result of a division.

EXAMPLE 9 Check the result of Example 8 by using
$P(x) = Q(x)D(x) + R(x)$

Solution
$Q(x) = 3x^2 - 5x - 2$
$D(x) = 5x - 2$
$R(x) = -13$

$Q(x)D(x) + R(x) = (3x^2 - 5x - 2)(5x - 2) + (-13)$

$\qquad\qquad\qquad = (3x^2 - 5x - 2)5x + (3x^2 - 5x - 2)(-2) + (-13)$

$\qquad\qquad\qquad = 15x^3 - 25x^2 - 10x - 6x^2 + 10x + 4 - 13$

$\qquad\qquad\qquad = 15x^3 - 31x^2 - 9$

$\qquad\qquad\qquad = P(x)$

EXAMPLE 10 Divide and check.
$\dfrac{x^4 + 2x^3 - 3x^2 - 3x - 2}{x^2 - 4}$

Solution

$$
\begin{array}{r}
x^2 + 2x + 1 \\
x^2 - 4\overline{)x^4 + 2x^3 - 3x^2 - 3x - 2} \\
\underline{x^4 \qquad\quad - 4x^2} \\
2x^3 + x^2 - 3x - 2 \\
\end{array}
$$

$$\begin{array}{r} 2x^3 \qquad\quad - 8x \\ \hline x^2 + 5x - 2 \\ x^2 \qquad\quad - 4 \\ \hline 5x + 2 \end{array}$$

$$\frac{x^4 + 2x^3 - 3x^2 - 3x - 2}{x^2 - 4} = x^2 + 2x + 1 + \frac{5x + 2}{x^2 - 4}$$

Check
$$(x^2 + 2x + 1)(x^2 - 4) + (5x + 2)$$
$$= (x^4 + 2x^3 + x^2) + (-4x^2 - 8x - 4) + 5x + 2$$
$$= x^4 + 2x^3 - 3x^2 - 3x - 2$$

EXERCISES 2.1

For Exercises 1–12, see Example 1.

(1–4)   Find the indicated value if $P(x) = 3x^2 - 2x + 4$.

1. $P(5)$ **2.** $P(-2)$ **3.** $P(0)$ **4.** $P(r)$

(5–8)   Find the indicated value if $Q(x) = x^3 - 4x^2$.

5. $Q(1)$ **6.** $Q(4)$ **7.** $Q(-4)$ **8.** $Q(0)$

(9–12)   Find the indicated value if $P(y) = 4y^2 - 9$.

9. $P(0)$ **10.** $P\left(\frac{3}{2}\right)$ **11.** $P\left(\frac{-3}{2}\right)$ **12.** $P(x + 3)$

(13–26)   Simplify. (Remove parentheses and combine like terms.) See Examples 2 and 3.

13. $(x^2 - 2x) + (3x - 6)$ **14.** $(xy - 4z) - (5y - 4)$
15. $x^3 - 6x^2 - (4x - 24)$ **16.** $x^2 + 2yx - (8xy + 16y^2)$
17. $(x^2 - 4x - 7) + (2x^2 - 5x + 3)$
18. $(y^2 + 6y - 2) + (5y^2 - 9y - 5)$
19. $(t^3 - 3t^2 + 9t) + (3t^2 - 9t + 27)$
20. $(x^3 + 4x^2 + 16x) - (4x^2 + 16x + 64)$
21. $(2y^2 - 8y + 5) - (3y^2 + 8y - 2)$
22. $(4x - 2y - 9) - (4x + 3y + 4)$
23. $(6x + 9y - 12) - (3x + 9y - 15)$
24. $(x^2 + 2xy + y^2) - (x^2 - 2xy + y^2)$
25. $(x^2 - yx) - (xy - y^2) - (x - y)$
26. $(x^3 - 2x^2) - (4x^2 - 8x) + (8x - 16)$

(27–36)   Find the indicated product. See Example 4.

27. $3x(x - 6)$ **28.** $4x(2x + 1)$
29. $2x^2(x + 5)$ **30.** $5x^3(x - 4)$
31. $-6y^2(y - 7)$ **32.** $-9y(2y^2 + 5)$
33. $8x(x^2 - 2x + 1)$ **34.** $-7n(n^2 + 5n - 6)$
35. $-x^2y(x^2 + 6xy + 9y^2)$ **36.** $-2x^2y^2(25x^2 - 10y + y^2)$

(37–56) Find the indicated products. See Examples 5 and 6.

37. $(x + 5)(x - 8)$

38. $(y - 4)(y - 6)$

39. $(2y - 7)(y - 3)$

40. $(3x - 8)(2x + 1)$

41. $(x - 1)(x - 1)$

42. $(4x + 1)(4x + 1)$

43. $(6y + 5)(6y + 5)$

44. $(7y - 8)(7y - 8)$

45. $(9x - 4)(9x + 4)$

46. $(10y - 1)(10y + 1)$

47. $(x^2 - 4)(x^2 - 1)$

48. $(x^2 - 9)(x^2 + 5)$

49. $(4x^2 - y^2)(4x^2 + y^2)$

50. $(4x^2 - y^2)(4x^2 - y^2)$

51. $(x + 5)(x^2 - 16)$

52. $(y - 6)(y^2 - 25)$

53. $(5x + y)(7x - y)$

54. $(4x + 3)(2y - 5)$

55. $(9x^2 - 1)(4y^2 - 1)$

56. $(y^2 - 9)(y^2 - 9)$

(57–62) Find the indicated quotient. See Example 7. (No divisor is zero.)

57. $\dfrac{6x^3 + 12x^2 - 15x}{3x}$

58. $\dfrac{4x^3 - 8x^2 - 24x}{4x}$

59. $\dfrac{8x^3y - 6x^2y^2 - 2xy^3}{2xy}$

60. $\dfrac{-10x^3y + 15x^2y - 5xy}{-5xy}$

61. $\dfrac{24x^4 - 24x^3 + 6x^2}{6x^2}$

62. $\dfrac{-5y^6 + 4y^4 - y^2}{-y^2}$

(63–88) Divide, using formal division. Check your results. See Examples 8 and 9. (No divisor is zero.)

63. $\dfrac{x^2 - 6x - 10}{x - 3}$

64. $\dfrac{6x^2 - x + 1}{2x + 1}$

65. $\dfrac{15x^2 - 4x - 4}{5x + 2}$

66. $\dfrac{12x^2 + 28x - 5}{6x - 1}$

67. $\dfrac{x^3 + 6x^2 + 12x + 8}{x + 2}$

68. $\dfrac{x^3 - 3x^2 + 3x - 1}{x - 1}$

69. $\dfrac{y^4 - 3y^2 - 6}{y^2 - 4}$

70. $\dfrac{2y^4 + y^3 + 2y - 6}{y^2 + 2}$

71. $\dfrac{y^4 + 2y^3 + 10y - 5}{y^2 + 5}$

72. $\dfrac{x^3 - 38x + 12}{x - 6}$

73. $\dfrac{x^4 - 29x^2 + 100}{x^2 - 25}$

74. $\dfrac{y^4 - 18y^2 + 81}{y - 3}$

75. $\dfrac{x^3 + 1}{x + 1}$

76. $\dfrac{x^3 + 8}{x + 2}$

77. $\dfrac{x^3 - 125}{x - 5}$

78. $\dfrac{x^3 - 27}{x - 3}$

79. $\dfrac{x^4 - 1}{x + 1}$

80. $\dfrac{x^4 - 16}{x - 2}$

81. $\dfrac{15x - 10 + 6x^2}{3x - 2}$

82. $\dfrac{24 + 7x - 2x^2}{6 - x}$

83. $\dfrac{y^4 + 12y - 4y^2 - 9}{y^2 + 3 - 2y}$

84. $\dfrac{25 + n^4 - 14n^2}{n^2 - 5 - 2n}$

85. $\dfrac{x^5 + 4x^2 + 2 - 2x^3}{x^2 - 2}$

86. $\dfrac{35 + x^6 - 12x^3}{-5 + x^3}$

87. $\dfrac{8 + 6x^2 - 12x - x^3}{2 - x}$

88. $\dfrac{1 + 3y^2 - 3y - y^3}{y - 1}$

2.2 SYNTHETIC DIVISION

For the division of a polynomial by a **linear divisor,** $x - c$, there is an abbreviated method called **synthetic division.** Consider the long division problem below. Observe the space that has been provided for the missing x^2 term of the dividend.

$$
\begin{array}{r}
2x^3 - x^2 - 3x + 4 \\
x - 3\overline{)2x^4 - 7x^3 \qquad + 13x - 5} \\
\underline{2x^4 - 6x^3} \\
- x^3 \\
\underline{- x^3 + 3x^2} \\
- 3x^2 + 13x \\
\underline{- 3x^2 + 9x} \\
4x - 5 \\
\underline{4x - 12} \\
7
\end{array}
$$

This work may be reduced by not writing a term that repeats the term above.

$$
\begin{array}{r}
2x^3 - x^2 - 3x + 4 \\
x - 3\overline{)2x^4 - 7x^3 \qquad + 13x - 5} \\
\underline{- 6x^3} \\
- x^3 \\
\underline{+ 3x^2} \\
- 3x^2 \\
\underline{+ 9x} \\
4x \\
\underline{- 12} \\
7
\end{array}
$$

Now raising the remaining terms:

$$
\begin{array}{r}
2x^3 - x^2 - 3x + 4 \\
x - 3\overline{)2x^4 - 7x^3 \qquad + 13x - 5} \\
\underline{- 6x^3 + 3x^2 + 9x - 12} \\
- x^3 - 3x^2 + 4x + 7
\end{array}
$$

The work may be reduced even further by omitting the powers of x and writing only the coefficients, providing a 0 for any missing term.

$$
\begin{array}{r}
2 - 1 - 3 \quad 4 \\
1 - 3\overline{)2 - 7 \quad 0 \quad 13 - 5} \\
\underline{- 6 \quad 3 \quad 9 - 12} \\
- 1 - 3 \quad 4 \quad 7
\end{array}
$$

Since this division technique applies only for divisors of the form $x - c$, the coefficient of x in the divisor is always 1 and thus may be omitted. If the first coefficient of the quotient is written in the first position of the last line, then the coefficients of the quotient are repeated on this line and thus may be written on the last line only. If $-c$ is replaced by c and the products obtained by multiplying by c are added instead of subtracting the products obtained by multiplying by $-c$, then the last line of the division remains unchanged. This follows from the fact that $a + (-b) = a - b$. It is convenient to place c on the last line since c multiplies the numbers in this line.

The quotient and remainder may be read from the last line of the synthetic division, with the last number recognized as the remainder and the other numbers excluding c recognized as the coefficients of the quotient polynomial.

$$
\begin{array}{r|rrrrr}
 & 2 & -7 & 0 & 13 & -5 \\
 & & 6 & -3 & -9 & 12 \\
\hline
3)2 & & -1 & -3 & 4 & 7 \\
\end{array}
$$

$\qquad\qquad\downarrow\qquad\qquad\qquad\downarrow$

$\qquad$Quotient$\qquad\qquad$Remainder

$2x^3 - x^2 - 3x + 4$

The format and procedures that follow provide a general description of the synthetic division process for the division of $P(x) = a_n x^n + \cdots + a_1 x + a_0$ by $x - c$.

1. Write in a row the coefficients of $P(x)$ providing a zero for any missing power of x.
2. On the second line below the row of coefficients, bring down the leading coefficient a_n.
3. Multiply a_n by c and add this product to a_{n-1}, writing the sum in the bottom row.
4. Multiply the sum obtained in Step 3 by c and add this product to the next coefficient of $P(x)$, writing the sum in the bottom row.
5. Repeat Steps 3 and 4 until the remainder is obtained.

EXAMPLE 1 Use synthetic division to perform the indicated division:

$$\frac{3x^4 - 2x^2 + 4}{x - 2}$$

Solution

$$
\begin{array}{r|rrrrr}
 & 3 & 0 & -2 & 0 & 4 \\
 & & 6 & 12 & 20 & 40 \\
\hline
2)3 & & 6 & 10 & 20 & 44 \\
\end{array}
$$

$3x^4 - 2x^2 + 4 = (x - 2)(3x^3 + 6x^2 + 10x + 20) + 44$

Therefore

$$\frac{3x^4 - 2x^2 + 4}{x - 2} = 3x^3 + 6x^2 + 10x + 20 + \frac{44}{x - 2}$$

EXAMPLE 2   Divide $7x^3 + 12x^2 - 5x + 4$ by $x + 2$.

Solution

$$x - c = x + 2 = x - (-2) \quad \text{and} \quad c = -2$$

$$
\begin{array}{r}
7 \quad\quad 12 \quad\quad -5 \quad\quad 4 \\
\underline{ -14 \quad\quad 4 \quad\quad 2} \\
-2\overline{)\,7 \quad\quad -2 \quad\quad -1 \quad\quad 6}
\end{array}
$$

$$7x^3 + 12x^2 - 5x + 4 = (x + 2)(7x^2 - 2x - 1) + 6$$

Therefore

$$\frac{7x^3 + 12x^2 - 5x + 4}{x + 2} = 7x^2 - 2x - 1 + \frac{6}{x + 2}$$

EXERCISES 2.2

Use synthetic division to express each of the following as a polynomial or as the sum of a polynomial and a fraction having the degree of the numerator less than the degree of the denominator.

1. $\dfrac{x^3 - 6x^2 + 10x - 3}{x - 4}$

2. $\dfrac{x^3 - 3x^2 - 20x + 7}{x - 6}$

3. $\dfrac{y^3 - 2y^2 - 16y + 2}{y + 3}$

4. $\dfrac{y^3 + 7y^2 + 9y - 4}{y + 5}$

5. $\dfrac{x^3 - 4x^2 - 10}{x - 5}$

6. $\dfrac{x^3 - 15x + 4}{x + 4}$

7. $\dfrac{t^4 - t^2 - 12}{t + 2}$

8. $\dfrac{n^4 - 13n^2 + 36}{n - 3}$

9. $\dfrac{x^2 + 1}{x + 1}$

10. $\dfrac{x^3 - 8}{x - 2}$

11. $\dfrac{x^3 + 125}{x + 5}$

12. $\dfrac{x^4 - 2x^2 + 1}{x - 1}$

13. $\dfrac{2x^3 - x^2 + 6x - 3}{x - \frac{1}{2}}$

14. $\dfrac{3x^3 + 2x^2 + 3x + 2}{x + \frac{2}{3}}$

15. $\dfrac{5x^3 - 23x^2 + 4}{x + \frac{2}{5}}$

16. $\dfrac{2x^3 - x^2 - 25}{x - \frac{5}{2}}$

2.3 FACTORING

In Section 1.1 we discussed factors and prime factors of positive integers. The *Fundamental Theorem of Arithmetic* states that there is exactly one way to express a composite positive integer (natural number) as a product of prime factors, disregarding the order in which the factors occur. For example,

$$1400 = 2 \cdot 700$$
$$= 2(2 \cdot 350)$$
$$= 2 \cdot 2 \cdot 2(175)$$
$$= 2^3 \cdot 5(35)$$
$$= 2^3 \cdot 5(5 \cdot 7)$$
$$= 2^3 \cdot 5^2 \cdot 7$$

The last product, $2^3 \cdot 5^2 \cdot 7$, is the prime factorization of 1400. Note that there are other ways to write 1400 in a factored form, but only one way to write it as a product where each factor is prime, if we disregard the order in which the prime factors are written.

There is a factorization theorem, similar to the Fundamental Theorem of Arithmetic, for polynomials whose coefficients are integers. First we need to explain what is meant by a polynomial factor.

DEFINITION OF POLYNOMIAL FACTOR

Let A, B, and P be polynomials with integers for coefficients. Then A is a factor of P if and only if $P = AB$.

Note that $P = 1 \cdot P = -1(-P)$ for any polynomial P. The factors 1, -1, P, and $-P$ are called **trivial factors** of P.

DEFINITION OF PRIME POLYNOMIAL

A **polynomial with integers for coefficients is prime** if and only if its degree is greater than zero and it has no polynomial factor with integers for coefficients other than the trivial factors.

As examples, $2x + 5$, $x^2 + 4$, and $x^2 + x + 2$ are prime polynomials.

On the other hand, $3x + 15$ is not prime because $3x + 15 = 3(x + 5)$ and $x^2 - 25$ is not prime because $x^2 - 25 = (x - 5)(x + 5)$ and 7 is not a prime *polynomial* because 7 is a constant (an integer). Note that the degree of 7 is 0.

PRIME FACTORIZATION THEOREM

Every polynomial of positive degree, with integers for coefficients, is either prime or can be expressed in exactly one way, disregarding the order in which the factors are written, as the product of a constant (integer) and one or more prime polynomial factors or as the product of prime polynomial factors.

This unique factorization is called the **completely factored form** of the polynomial.

The following table of factoring forms is useful for finding the completely factored form of a polynomial.

The common monomial form is justified by the distributive axiom and its related theorems. The other forms can be verified by performing the indicated multiplications.

Table of Factoring Forms

1. Common Monomial Factor	$AB + AC = A(B + C)$ $AB + AC + AD = A(B + C + D)$
2. Difference of Squares	$A^2 - B^2 = (A - B)(A + B)$
3. Sum of Cubes	$A^3 + B^3 = (A + B)(A^2 - AB + B^2)$
4. Difference of Cubes	$A^3 - B^3 = (A - B)(A^2 + AB + B^2)$
5. Perfect Square Trinomial	$A^2 + 2AB + B^2 = (A + B)^2$
6. Simple Trinomial	$X^2 + (A + B)X + AB = (X + A)(X + B)$
7. General Trinomial	$AX^2 + BX + C = (aX + b)(cX + d)$

where $ac = A$, $bd = C$, and $ad + bc = B$.

Examples 1 through 8 illustrate the process of factoring when only one of the basic seven forms needs to be used.

EXAMPLE 1
$6x^4 + 54x^2$

Solution

$6x^2(x^2) + 6x^2(9) = 6x^2(x^2 + 9)$ 　　　　　Form 1

EXAMPLE 2
$5y^3 + 15y^2 + 5y$

Solution

$5y(y^2) + 5y(3y) + 5y(1) = 5y(y^2 + 3y + 1)$ 　　　　　Form 1

EXAMPLE 3
$4x^2 - 81y^2$

Solution

$(2x)^2 - (9y)^2 = (2x - 9y)(2x + 9y)$ 　　　　　Form 2

EXAMPLE 4
$x^3 + 125$

Solution

$x^3 + 5^3 = (x + 5)(x^2 - 5x + 25)$ 　　　　　Form 3

EXAMPLE 5
$216n^3 - 1$

Solution
$(6n)^3 - 1^3 = (6n - 1)(36n^2 + 6n + 1)$ Form 4

EXAMPLE 6
$16x^2 - 40x + 25$

Solution
$(4x)^2 + 2(4x)(-5) + (-5)^2 = (4x - 5)^2$ Form 5

EXAMPLE 7
$t^4 + 2t^2 - 63$

Solution
$(t^2)^2 + (9 - 7)t^2 + 9(-7) = (t^2 + 9)(t^2 - 7)$ Form 6

EXAMPLE 8
$3x^2 - 13x - 10$

Solution
$(3 \cdot 1)x^2 + (-5 \cdot 3 + 2 \cdot 1)x + 2(-5) = (3x + 2)(x - 5)$ Form 7

When Form 7 is used, as in Example 8, the terms of the factors are usually found by a "trial and error" process. The possible factored forms are listed so that the product of the coefficients of the first terms is the leading coefficient and the product of the last terms is the constant term. For Example 8, the possibilities are

$(3x + 1)(x - 10)$ $(3x + 2)(x - 5)$
$(3x - 1)(x + 10)$ $(3x - 2)(x + 5)$
$(3x + 10)(x - 1)$ $(3x + 5)(x - 2)$
$(3x - 10)(x + 1)$ $(3x - 5)(x + 2)$

Each of these possibilities is tested by comparing the product with the given polynomial. Only one of these possibilities, with the exception of the order in which the factors are written, can be the factored form because the prime factorization is unique. If none of the possibilities produces the given polynomial, then the given polynomial is either prime or another form may apply.

Factorization is usually easiest when the forms are tried in the order they are listed in the table. First find the greatest monomial factor, if there is one. Note, in Example 1, that $6x^4 + 54x^2 = 3x(2x^3 + 18x)$ but $2x^3 + 18x$ is not prime and the terms $2x^3$ and $18x$ still have a common monomial factor. When using Form 1, it is important to check the terms of the nonmonomial factor for additional common factors.

When the polynomial to be factored is a trinomial, first check for common monomial factors and then determine whether or not the polynomial is a perfect square. When it is a perfect square, the lengthy "trial and error" process used for the general trinomial can be avoided.

2.3 FACTORING

Sometimes it is necessary to use two or more factoring forms to find the completely factored form of a polynomial. This is illustrated in the next two examples.

EXAMPLE 9 Completely factor $5x^3 - 30x^2 + 45x$.

Solution

$$5x^3 - 30x^2 + 45x = 5x(x^2 - 6x + 9) \qquad \text{Common monomial factor}$$
$$= 5x(x - 3)^2 \qquad \text{Perfect square trinomial}$$

EXAMPLE 10 Completely factor $x^4 - 45x^2 - 196$.

Solution   There is no common monomial factor and the trinomial is not a perfect square so Form 6 is applied.

$$x^4 - 45x^2 - 196 = (x^2 - 49)(x^2 + 4) \qquad \text{Simple trinomial}$$
$$= (x - 7)(x + 7)(x^2 + 4) \qquad \text{Difference of squares}$$

Sometimes it is necessary to group certain terms of a polynomial in order to recognize one of the factoring forms. First, we look for three terms that form a perfect square trinomial. Two terms must be perfect squares and the third term must be twice the product of their square roots. This type of factorization is shown in Examples 11 and 12.

EXAMPLE 11 Completely factor $x^4 + 12x^2 + 36 - 64y^2$.

Solution   First note that $x^4 = (x^2)^2$, $36 = 6^2$, and $12x^2 = 2(x^2)(6)$. We enclose these three terms in parentheses and factor this group.

$$x^4 + 12x^2 + 36 - 64y^2 = (x^4 + 12x^2 + 36) - 64y^2$$
$$= (x^2 + 6)^2 - 64y^2$$

Now a difference in squares is observed, and

$$x^4 + 12x^2 + 36 - 64y^2 = (x^2 + 6 - 8y)(x^2 + 6 + 8y)$$

EXAMPLE 12 Completely factor $x^4 - x^2 + 10x - 25$.

Solution   Here it is necessary to observe that

$$-(a^2 + 2ab + b^2) = -a^2 - 2ab - b^2$$

and

$$-(a^2 - 2ab + b^2) = -a^2 + 2ab - b^2$$

Then

$$x^4 - x^2 + 10x - 25 = x^4 - (x^2 - 10x + 25) \qquad \text{Observing the}$$
perfect square

$$= x^4 - (x - 5)^2 \qquad \text{Factoring the group}$$

$$= [x^2 - (x - 5)][x^2 + (x - 5)] \qquad \text{Observing the form } A^2 - B^2$$

$$= (x^2 - x + 5)(x^2 + x - 5) \qquad \text{Simplifying}$$

If it is not possible to factor a polynomial by grouping three terms to form a perfect square trinomial, it might be possible to group terms in pairs and recognize a common factor of each group.

ALGEBRAIC EXPRESSIONS

EXAMPLE 13 Completely factor $x^3 - 4x^2 + 6x - 24$.

Solution No three terms form a perfect square so grouping in pairs is tried.

$$\begin{aligned} x^3 - 4x^2 + 6x - 24 &= (x^3 - 4x^2) + (6x - 24) \\ &= x^2(x - 4) + 6(x - 4) \\ &= (x^2 + 6)(x - 4) \\ \\ &= (x - 4)(x^2 + 6) \end{aligned}$$

Grouping in pairs
Factoring each group
Recognizing
$BA + CA = (B + C)(A)$
with $A = x - 4$

EXAMPLE 14 Factor $x^4 + 3x^3 - 8x - 24$.

Solution

$$\begin{aligned} x^4 + 3x^3 - 8x - 24 &= (x^4 + 3x^3) - (8x + 24) \\ &= x^3(x + 3) - 8(x + 3) \\ &= (x^3 - 8)(x + 3) \\ \\ &= (x - 2)(x^2 + 2x + 4)(x + 3) \\ \\ &= (x - 2)(x + 3)(x^2 + 2x + 4) \end{aligned}$$

Grouping in pairs
Factoring each group
Using
$BA - CA = (B - C)(A)$
with $A = x + 3$
Recognizing a
difference of cubes

In the completely factored form, it is customary to write monomial factors first, then binomials, and then trinomials. Also polynomials are arranged by increasing degree from left to right.

The grouping technique can also be used to find the factors of a general trinomial.

For the product

$$(ax + b)(cx + d) = acx^2 + (ad + bc)x + bd$$

note that the product of the leading coefficient and the constant term is $(ac)(bd) = abcd = (ad)(bc)$ and that $ad + bc$ is the coefficient of x. Therefore given a general trinomial, we multiply the leading coefficient and the constant term and then find two numbers whose product is the result of this multiplication and whose sum is the coefficient of the middle term. This procedure is illustrated in Example 15.

EXAMPLE 15 Factor $12x^2 + 8x - 15$.

Solution Forming the product, $12(-15) = -180$. Factoring 180 as a product of two integers, the possibilities are

$1 \cdot 180, \quad 2 \cdot 90, \quad 3 \cdot 60,$
$4 \cdot 45, \quad 5 \cdot 36, \quad 6 \cdot 30,$
$9 \cdot 20, \quad 10 \cdot 18, \quad 12 \cdot 15.$

From these pairs of factors, we select a pair, and we take one factor positive and the other negative to produce the negative product of -180 and the sum of $+8$; namely -10 and $+18$. Now we rewrite $8x$ as $-10x + 18x$, and

$$12x^2 + 8x - 15 = 12x^2 - 10x + 18x - 15$$
$$= 2x(6x - 5) + 3(6x - 5)$$
$$= (2x + 3)(6x - 5)$$

EXERCISES 2.3

(1–16) Completely factor by using one of the basic forms. See Examples 1–8.

1. $x^2 - 100$
    2. $4x^2 - 100x$

3. $x^2 - 7x + 10$
    4. $9x^2 - 12x + 4$

5. $8x^3 - 40x^2 + 16x$
    6. $64x^4 - 1$

7. $16y^2 + 56y + 49$
    8. $y^4 + 9y^2 - 36$

9. $27x^3 - y^3$
    10. $15x^4 + 15x^3 - 15x^2$

11. $4t^4 - t^2 - 5$
    12. $1000n^3 + k^3$

13. $9x^2y - 36xy^2$
    14. $8x^3 - 125y^3$

15. $1 + 64y^3$
    16. $12x^4 - 7x^2y + y^2$

(17–28) Completely factor by using one or more of the basic forms. See Examples 9 and 10.

17. $4x^4 - 40x^3 + 36x^2$
    18. $x^4 - 15x^2 - 16$

19. $x^4 - 50x^2 + 625$
    20. $x^4 - 8x^3 + 16x^2$

21. $60x^3 - 24x^2 - 36x$
    22. $a^3x^3 - 216a^3$

23. $64x^3y^2 + 8y^5$
    24. $3x^4 - 24x^2 + 48$

25. $x^6 - 64$
    26. $160x^4 + 720x^2 + 810$

27. $7x^5 - 27x^3 - 4x$
    28. $abx^6y + 1000aby$

(29–38) Completely factor. See Examples 11 and 12.

29. $x^2 + 14x + 49 - y^2$
    30. $y^2 - x^2 - 16x - 64$

31. $x^4 - x^2 + 2x - 1$
    32. $x^4 - 4x^2 + 4 - x^2$

33. $x^2 - 10xy + 25y^2 - 16x^2y^2$
    34. $36 - x^2 - 6xy - 9y^2$

35. $x^4 + 4$; $(x^4 + 4 = x^4 + 4x^2 + 4 - 4x^2)$

36. $y^4 + y^2 + 1$; $(y^4 + y^2 + 1 = y^4 + 2y^2 + 1 - y^2)$

37. $y^4 - 19y^2 + 9$
    38. $x^4 + 8x^2y^2 + 36y^4$

(39–54) Completely factor. See Examples 13, 14, and 15.

39. $x^3 + 2x^2 + 4x + 8$
    40. $x^3 - 3x^2 + 9x - 27$

41. $y^3 + 6y^2 - y - 6$
    42. $y^3 + 2y^2 - 25y - 50$

43. $x^3 - xy^2 - x^2y + y^3$
    44. $x^4 - 4x^2y^2 - 9x^2 + 36y^2$

45. $x^2y^2 + 3x^2y - 2xy^2 - 6xy$
    46. $6ab - 30a - 24b + 120$

47. $10rs - 35 + 10r - 35s$
    48. $u^3v^3 - uv + u^3v - uv^3$

49. $6x^2 + 23x + 20$
    50. $8x^2 - 47x + 35$

51. $10x^2 - 37x - 36$
    52. $12x^2 + x - 20$

53. $16y^2 + 194y - 75$
    54. $25y^2 - 999y - 40$

(55–70) Completely factor.

55. $20x^4 + 34x^3 - 12x^2$ **56.** $4x^4 - 72x^2 + 324$
57. $x^5 - 29x^3 + 100x$ **58.** $8x^3 - 1000y^3$
59. $x^4 + 2x^3 + x + 2$ **60.** $x^4 - 2x^3 + 8x - 16$
61. $x^2y^2 - 2xy + 1 - x^2$ **62.** $y^4 - y^2 + 4y - 4$
63. $r^4 + a^2r^2 + a^4$ **64.** $r^4 + r^2 + 25$
65. $u - u^2 - u^3 + u^4$ **66.** $1 - 2n + n^2 - n^4$
67. $x^4 + (2a - b^2)x^2 + a^2$ **68.** $x^6 - a^6$
69. $ax^2y + bxy^2 + bx^2y + axy^2$ **70.** $x^5 - b^2x^3 - a^3x^2 + a^3b^2$

2.4 RATIONAL EXPRESSIONS, SUMS AND DIFFERENCES

DEFINITION

An **algebraic expression** is a constant, a variable, or an indicated result obtained by performing one or more of the six algebraic operations. Examples of algebraic expressions are listed below.

$$\left.\begin{array}{l} x + 2y \\ 3x^2 - 5x - 2 \\ 4x^2 + y^2 - 16 \end{array}\right\} \text{ Polynomials}$$

$$\left.\begin{array}{l} \dfrac{1}{x} \\[2mm] \dfrac{5}{x - 2} + \dfrac{x}{x + 3} \\[2mm] \dfrac{x^2 + 4xy + 6y^2}{x - 3y} \end{array}\right\} \text{ Rational Expressions}$$

$$\left.\begin{array}{l} x + \sqrt{3x - 4} \\ \sqrt[3]{x^2 + y^2} \end{array}\right\} \text{ Radical Expressions}$$

$$\left.\begin{array}{l} 2^x \\ 10^{x+y} \end{array}\right\} \text{ Exponential Expressions}$$

Polynomials, discussed in the previous two sections, are the simplest types of algebraic expressions. This section and the following one deal with rational expressions, those that can be expressed as quotients of polynomials. Other types of algebraic expressions will be treated in subsequent chapters.

If A and B are polynomials whose coefficients are integers or rational numbers, then their quotient $\frac{A}{B}$ designates a rational number as long as $B \neq 0$. For this reason an expression that can be written in the form $\frac{A}{B}$ is called a **rational expression** when A and B are polynomials with any real numbers for coefficients.

The **set of quotients of polynomials** is a field and it has properties similar to those for rational numbers. For the following, A, B, C, and D are polynomials with $B \neq 0$ and $D \neq 0$.

DEFINITION OF EQUAL QUOTIENTS

$\frac{A}{B} = \frac{C}{D}$ if and only if $AD = BC$

FUNDAMENTAL THEOREM OF RATIONAL EXPRESSIONS

$\frac{A}{B} = \frac{AD}{BD}$ and $\frac{AD}{BD} = \frac{A}{B}$

DEFINITION OF ADDITION

$\frac{A}{D} + \frac{B}{D} = \frac{A + B}{D}$

SUBTRACTION THEOREM

$\frac{A}{D} - \frac{B}{D} = \frac{A - B}{D}$

A **quotient** of **polynomials** is also called an **algebraic fraction,** or more simply, a **fraction.** It should be noted that an algebraic fraction is not defined for any value of a variable that causes a denominator to become zero. For example, $\frac{x - 6}{x - 5}$ is not defined for $x = 5$, and $\frac{4}{x + 3}$ is not defined for $x = -3$.

DEFINITION

If A and B are polynomials with integers for coefficients, then $\frac{A}{B}$ is in **simplified form** if and only if A and B have no common prime polynomial factor and no common prime number factor. The simplified form of $\frac{A}{1}$ is A.

EXAMPLE 1 Simplify $\dfrac{4x^2 - 36}{2x^2 + 4x - 30}$,   where $x \neq 3$ and $x \neq -5$.

Solution The polynomials are first factored to determine common factors of numerator and denominator.

$$\frac{4x^2 - 36}{2x^2 + 4x - 30} = \frac{4(x^2 - 9)}{2(x^2 + 2x - 15)}$$

$$= \frac{2 \cdot 2(x - 3)(x + 3)}{2(x - 3)(x + 5)}$$

$$= \frac{2(x + 3)\,2(x - 3)}{(x + 5)\,2(x - 3)}$$

$$= \frac{2(x + 3)}{x + 5} \qquad \text{Simplified form}$$

$$= \frac{2x + 6}{x + 5} \qquad \begin{array}{l}\text{Alternate}\\ \text{simplified form}\end{array}$$

Note that the final result is obtained by using the Fundamental Theorem to remove the greatest common factor $2(x - 3)$.

The Fundamental Theorem is also used to build up fractions for the operations of addition and subtraction.

EXAMPLE 2 Add and simplify.

$$\frac{x - 4}{x - 5} + \frac{x + 2}{x + 5} + \frac{-10}{x^2 - 25}, \qquad \text{where } x \neq 5 \;\; \text{and} \;\; x \neq -5$$

Solution

$$\frac{x - 4}{x - 5} + \frac{x + 2}{x + 5} + \frac{-10}{(x - 5)(x + 5)} \qquad \text{Factoring}$$

$$= \frac{(x - 4)(x + 5)}{(x - 5)(x + 5)} + \frac{(x + 2)(x - 5)}{(x + 5)(x - 5)} + \frac{-10}{(x - 5)(x + 5)} \qquad \begin{array}{l}\text{Building-up,}\\ \text{using the}\\ \text{Fundamental}\\ \text{Theorem}\end{array}$$

$$= \frac{(x^2 + x - 20) + (x^2 - 3x - 10) + (-10)}{(x + 5)(x - 5)} \qquad \begin{array}{l}\text{Using the}\\ \text{definition of}\\ \text{addition}\end{array}$$

$$= \frac{2x^2 - 2x - 40}{(x + 5)(x - 5)} \qquad \begin{array}{l}\text{Simplifying the}\\ \text{numerator}\end{array}$$

$$= \frac{2(x + 4)(x - 5)}{(x + 5)(x - 5)} \qquad \begin{array}{l}\text{Factoring the}\\ \text{numerator}\end{array}$$

$$= \frac{2(x + 4)}{x + 5} \qquad \begin{array}{l}\text{Using the}\\ \text{Fundamental}\\ \text{Theorem to}\\ \text{simplify}\end{array}$$

The simplified form can be written either as $\dfrac{2(x + 4)}{x + 5}$ or as $\dfrac{2x + 8}{x + 5}$.

EXAMPLE 3 Subtract and simplify.

$$\frac{x + 1}{x^2 - 9} - \frac{x - 5}{x^2 - 6x + 9} \qquad \text{where } x \neq 3 \quad \text{and} \quad x \neq -3$$

Solution

$$\frac{x + 1}{x^2 - 9} - \frac{x - 5}{x^2 - 6x + 9} = \frac{x + 1}{(x - 3)(x + 3)} + \frac{(-1)(x - 5)}{(x - 3)(x - 3)}$$

$$= \frac{(x + 1)(x - 3) - (x - 5)(x + 3)}{(x + 3)(x - 3)(x - 3)}$$

$$= \frac{(x^2 - 2x - 3) - (x^2 - 2x - 15)}{(x + 3)(x - 3)(x - 3)}$$

$$= \frac{x^2 - 2x - 3 - x^2 + 2x + 15}{(x + 3)(x - 3)(x - 3)}$$

$$= \frac{12}{(x + 3)(x - 3)^2}$$

For some problems, it is important to recognize that a binomial of the form $b - a$ can also be written as $(-1)(a - b)$;

$$b - a = -1(a - b)$$

This feature occurs in Example 4.

EXAMPLE 4 Simplify.

$$\frac{x + 2}{2x - 8} - \frac{x + 1}{3x - 12} + \frac{2}{4 - x}, \qquad \text{where } x \neq 4$$

Solution

$$\frac{x + 2}{2x - 8} - \frac{x + 1}{3x - 12} + \frac{2}{4 - x} = \frac{x + 2}{2(x - 4)} + \frac{-1(x + 1)}{3(x - 4)} + \frac{2}{-1(x - 4)}$$

$$= \frac{(x + 2)3}{2(x - 4)3} + \frac{-1(x + 1)2}{3(x - 4)2} + \frac{2(-6)}{-1(x - 4)(-6)}$$

$$= \frac{(3x + 6) + (-2x - 2) + (-12)}{6(x - 4)}$$

$$= \frac{x - 8}{6(x - 4)}$$

While $\dfrac{x - 8}{6(x - 4)}$ can also be written as $\dfrac{x - 8}{6x - 24}$, it is usually best to leave the denominator in factored form in order to make any further calculation easier to do.

EXERCISES 2.4

(1–20) Simplify. No variable may equal a value for which a denominator becomes zero. See Example 1.

1. $\dfrac{3x - 15}{2x - 10}$

2. $\dfrac{x^2 + 2x}{2x + 4}$

3. $\dfrac{x^2 - 9x}{x^2 + 9x}$

4. $\dfrac{5x^2 + 5}{5x + 5}$

5. $\dfrac{x - 6}{6 - x}$

6. $\dfrac{4 - x^2}{x^2 - 4}$

7. $\dfrac{3y^3}{3y^3 + 9y^2}$

8. $\dfrac{4y^2 - 1}{2y - 4y^2}$

9. $\dfrac{16 - y^2}{y^2 - 8y + 16}$

10. $\dfrac{y^3 - 125}{y - 5}$

11. $\dfrac{y + 6}{y^3 + 216}$

12. $\dfrac{y^3 - 2y^2 - 8y}{y^3 - 6y^2 + 8y}$

13. $\dfrac{4 - 100x^2}{5x - 1}$

14. $\dfrac{7x - 49}{49 - x^2}$

15. $\dfrac{5x^2 - 5x - 100}{5x^2 - 15x - 50}$

16. $\dfrac{3n^2 + 5n - 2}{2n^2 - 5n - 3}$

17. $\dfrac{n^3 - n^2 - 4n + 4}{n^3 - 3n^2 + 2n}$

18. $\dfrac{xy - 4x - 3y + 12}{xy + 2x - 3y - 6}$

19. $\dfrac{x^4 - 2x^3 + 8x - 16}{4x^4 - 8x^3 + 16x^2}$

20. $\dfrac{t^4 - t^2 - 12t - 36}{6t^2 - 6t - 36}$

(21–48) Simplify. (Add or subtract as indicated and reduce final fraction to lowest terms.) No variable may equal a value for which a denominator becomes zero. See Examples 2, 3, and 4.

21. $\dfrac{1}{x} + \dfrac{1}{x + 3}$

22. $\dfrac{1}{6x} + \dfrac{1}{9x}$

23. $\dfrac{1}{4x} - \dfrac{1}{5x^2}$

24. $\dfrac{1}{x + 4} - \dfrac{1}{x - 4}$

25. $\dfrac{1}{y - 2} - \dfrac{1}{y + 2}$

26. $\dfrac{1}{y - 3} + \dfrac{1}{3 - y}$

27. $\dfrac{1}{5 - x} + \dfrac{1}{x - 5}$

28. $\dfrac{1}{1 - x} - \dfrac{1}{x - 1}$

29. $\dfrac{x}{3x - 6} + \dfrac{x}{6x - 12}$

30. $\dfrac{1}{4x + 20} - \dfrac{1}{6x + 30}$

31. $\dfrac{1}{y} + \dfrac{2}{y^2} - \dfrac{1}{y^3}$

32. $\dfrac{3}{xy} - \dfrac{4}{yz} + \dfrac{5}{xz}$

33. $\dfrac{x + 1}{x - 6} - \dfrac{x}{x^2 - 12x + 36}$

34. $\dfrac{y}{y^2 + 8y + 16} - \dfrac{y}{y^2 - 16}$

35. $\dfrac{x + y}{x - y} - \dfrac{x - y}{x + y}$

36. $\dfrac{1 + t}{(1 - t)^2} + \dfrac{1 - t}{t^2 - 1}$

37. $t + 9 - \dfrac{t^2 - 90}{t - 9}$

38. $1 - \dfrac{(1 - n)^2}{(1 + n)^2}$

39. $\dfrac{1}{x^2 + x - 2} + \dfrac{1}{x^2 - 4x + 3} - \dfrac{1}{x^2 - x - 6}$

40. $\dfrac{3}{x - 6} + \dfrac{3}{x + 6} - \dfrac{36}{36 - x^2}$

41. $\dfrac{3}{y + 3} - \dfrac{2}{3y + 5} - \dfrac{7y + 9}{3y^2 + 14y + 15}$

42. $\dfrac{2}{y^2 + 2} - \dfrac{y^2}{(y^2 + 2)^2} - \dfrac{4}{(y^2 + 2)^2}$

43. $\dfrac{1}{4t^2 - 4t} - \dfrac{1}{5t^2 - 5t} + \dfrac{1}{20 - 20t}$

44. $\dfrac{1}{3t^3 - 6t^2} + \dfrac{1}{3t^3 + 6t^2} + \dfrac{1}{3t^3 - 12t}$

45. $\dfrac{x + 3}{x - 5} - \dfrac{x + 5}{x - 3} - \dfrac{16}{x^2 - 8x + 15}$

46. $\dfrac{1}{8x^2 + 4x} + \dfrac{1}{8x^2 - 4x} - \dfrac{4x^2}{4x^2 - 1}$

47. $\dfrac{1}{x^2 + xy} + \dfrac{1}{xy + y^2} + \dfrac{1}{x^2 - xy}$

48. $\dfrac{1}{x^2y + xy^2 + xyz} + \dfrac{1}{x^2z + xyz + xz^2} + \dfrac{1}{xyz + y^2z + yz^2}$

2.5 RATIONAL EXPRESSIONS, PRODUCTS AND QUOTIENTS

The operations of multiplication and division for rational expressions are similar to those for rational numbers. If A, B, C, and D are polynomials, then we have the following definition and theorem.

DEFINITION

Product of Rational Expressions

$$\frac{A}{B} \cdot \frac{C}{D} = \frac{AC}{BD} \qquad \text{for } B \neq 0 \quad \text{and} \quad D \neq 0$$

THEOREM

Quotient of Rational Expressions

$$\frac{\dfrac{A}{B}}{\dfrac{C}{D}} = \frac{A}{B} \cdot \frac{D}{C} = \frac{AD}{BC} \qquad \text{for } B \neq 0, \quad C \neq 0, \quad \text{and} \quad D \neq 0$$

As a general rule, when polynomials are multiplied or divided, each numerator and each denominator should be factored in order to simplify the final fraction.

EXAMPLE 1 Simplify

$$\frac{x^2 - 9}{12x^3 + 24x^2} \cdot \frac{4x^4 + 8x^3}{x^2 - 6x + 9}$$

where $x \neq 0$, $x \neq -2$, $x \neq 3$

Solution

$$= \frac{(x - 3)(x + 3)}{12x^2(x + 2)} \cdot \frac{4x^3(x + 2)}{(x - 3)(x - 3)}$$

$$= \frac{x(x + 3)4x^2(x - 3)(x + 2)}{3(x - 3)4x^2(x - 3)(x + 2)}$$

$$= \frac{x(x + 3)}{3(x - 3)}$$

EXAMPLE 2 Simplify

$$\frac{9x^2 - 6x}{3x^2 + 4x - 4} \cdot \frac{x^4 - x^2 - 12}{x^3 - 2x^2 + 3x - 6}, \quad \text{where } x \neq -2,\ x \neq 2,\ x \neq \frac{2}{3}$$

Solution

$$= \frac{3x(3x - 2)}{(x + 2)(3x - 2)} \cdot \frac{(x^2 - 4)(x^2 + 3)}{(x - 2)(x^2 + 3)}$$

$$= \frac{3x}{x + 2} \cdot \frac{(x + 2)(x - 2)}{1(x - 2)}$$

$$= \frac{3x(x + 2)}{1(x + 2)}$$

$$= 3x$$

EXAMPLE 3 Simplify

$$\frac{x^3 + 125}{4x^2 + 18x - 10} \div \frac{x^3 - 5x^2 + 25x}{2x^3 + 11x^2 - 6x}, \quad \text{where } x \neq -5 \text{ and } x \neq \frac{1}{2}$$

Solution

$$= \frac{x^3 + 125}{4x^2 + 18x - 10} \cdot \frac{2x^3 + 11x^2 - 6x}{x^3 - 5x^2 + 25x}$$

$$= \frac{(x + 5)(x^2 - 5x + 25)}{2(2x^2 + 9x - 5)} \cdot \frac{x(2x^2 + 11x - 6)}{x(x^2 - 5x + 25)}$$

$$= \frac{(x + 5)(x^2 - 5x + 25)}{2(2x - 1)(x + 5)} \cdot \frac{(2x - 1)(x + 6)}{(x^2 - 5x + 25)}$$

$$= \frac{(x + 6)(x + 5)(2x - 1)(x^2 - 5x + 25)}{2(x + 5)(2x - 1)(x^2 - 5x + 25)}$$

$$= \frac{x + 6}{2}$$

A **simple fraction** is a fraction that has no indicated divisions in its numerator or denominator. By contrast, a **complex fraction** is one that has one or more indicated divisions in its numerator or denominator, or both. As examples, $\frac{3}{4}$ and $\frac{x^2 - 4}{x^2 + 6x + 9}$ are simple fractions. The fractions $\dfrac{\frac{1}{2} + \frac{1}{3}}{6}$ and $\dfrac{1}{\frac{1}{x} - \frac{1}{y}}$ are complex fractions. A complex fraction can be reduced to a simple fraction, or to a polynomial, by performing the indicated operations in the main

numerator and denominator, when possible, and then doing the
final division.

EXAMPLE 4 Simplify

$$\frac{\dfrac{1}{2}+\dfrac{1}{3}}{6}$$

Solution

$$\frac{1}{2}+\frac{1}{3}=\frac{3}{6}+\frac{2}{6}=\frac{5}{6}$$

$$\frac{\dfrac{1}{2}+\dfrac{1}{3}}{6}=\frac{\dfrac{5}{6}}{6}=\frac{5}{6}\div 6=\frac{5}{6}\cdot\frac{1}{6}=\frac{5}{36}$$

EXAMPLE 5 Simplify

$$\frac{x-y}{\dfrac{1}{x}-\dfrac{1}{y}},\qquad\text{where }x\neq 0,\quad y\neq 0$$

Solution

$$\frac{1}{x}-\frac{1}{y}=\frac{y}{xy}-\frac{x}{xy}=\frac{y-x}{xy}$$

$$\frac{x-y}{\dfrac{1}{x}-\dfrac{1}{y}}=(x-y)\div\left(\frac{1}{x}-\frac{1}{y}\right)$$

$$=(x-y)\div\frac{y-x}{xy}$$

$$=\frac{x-y}{1}\cdot\frac{xy}{y-x}$$

$$=\frac{-1(y-x)xy}{(y-x)}$$

$$=-xy$$

EXAMPLE 6 Simplify

$$\frac{\dfrac{1}{x}-\dfrac{1}{6}}{\dfrac{1}{x^2}-\dfrac{1}{36}},\qquad\text{where }x\neq 0\quad\text{and}\quad x\neq 6\quad\text{and}\quad x\neq -6$$

Solution

Main numerator

$$\frac{1}{x}-\frac{1}{6}=\frac{6-x}{6x}$$

Main denominator

$$\frac{1}{x^2}-\frac{1}{36}=\frac{36-x^2}{36x^2}$$

Given quotient

$$\frac{6-x}{6x}\div\frac{36-x^2}{36x^2}=\frac{6-x}{6x}\cdot\frac{36x^2}{(6-x)(6+x)}$$

$$=\frac{6x(6x)(6-x)}{(6+x)(6x)(6-x)}$$

$$=\frac{6x}{6+x}=\frac{6x}{x+6}$$

EXAMPLE 7 Simplify

$$1 + \cfrac{1}{1 - \cfrac{2}{1 + x}}, \qquad \text{where } x \neq -1 \quad \text{and} \quad x \neq 1$$

Solution The complex fraction is

$$\cfrac{1}{1 - \cfrac{2}{1 + x}}$$

The main denominator is

$$1 - \frac{2}{1 + x} = \frac{1 + x}{1 + x} - \frac{2}{1 + x} = \frac{x - 1}{1 + x} = \frac{x - 1}{x + 1}$$

Now the given rational expression can be written as

$$1 + \cfrac{1}{\cfrac{x - 1}{x + 1}} = 1 + \left(1 \div \frac{x - 1}{x + 1}\right)$$

$$= 1 + \left(1 \cdot \frac{x + 1}{x - 1}\right)$$

$$= 1 + \frac{x + 1}{x - 1}$$

$$= \frac{x - 1}{x - 1} + \frac{x + 1}{x - 1}$$

$$= \frac{2x}{x - 1}$$

A simplification of a rational expression can be checked by replacing each variable by a constant, by calculating the value of the original expression and the value of the final expression, and then by determining if these values are equal. Any value may be selected for a variable as long as the value does not cause a denominator to become zero.

EXAMPLE 8 Check Example 1.

Solution Any value can be selected for x except 0, -2, and 3. For $x = 2$,

$$\frac{x^2 - 9}{12x^3 + 24x^2} \cdot \frac{4x^4 + 8x^3}{x^2 - 6x + 9} = \frac{-5}{192} \cdot \frac{128}{1} = \frac{-10}{3}$$

$$\frac{x(x + 3)}{3(x - 3)} = \frac{2(2 + 3)}{3(2 - 3)} = \frac{10}{-3} = \frac{-10}{3}$$

EXERCISES 2.5

(1–20) Simplify. (Do the indicated calculations, expressing the result in simplified form.) No variable has a value that makes a denominator zero. See Examples 1, 2, and 3.

1. $\dfrac{21x^2}{20y^3} \cdot \dfrac{10y^3}{42x^3}$

2. $\dfrac{63x^2y}{24z^2} \cdot \dfrac{40z^2}{18xy^2}$

3. $\dfrac{60x^2}{99y^2z} \div \dfrac{80yz}{55x}$

4. $\dfrac{25a^2b}{36rs} \div \dfrac{35ab^2}{42rs}$

5. $\dfrac{x-3}{x-4} \cdot \dfrac{x^2-16}{x^2-9}$

6. $\dfrac{5x-5}{3x-6} \cdot \dfrac{6x-18}{6x-6}$

7. $\dfrac{x+2}{x-2} \cdot \dfrac{2-x}{4-x^2}$

8. $\dfrac{49-y^2}{9-y^2} \cdot \dfrac{y^2-9}{y^2-7y}$

9. $\dfrac{4x+20}{x^2+8x+16} \cdot \dfrac{x^2+4x}{x^2+7x+10}$

10. $\dfrac{6x^2+9x}{2x^2+3x-9} \cdot \dfrac{4x^2-12x+9}{12x^2-27}$

11. $\dfrac{y^2-3y+2}{5y^2-26y+5} \div \dfrac{y^2-y-2}{5y^2+24y-5}$

12. $\dfrac{4y^2+4y+1}{2y^2-3y-2} \div \dfrac{2y^2+5y+2}{2y^2+3y-2}$

13. $\dfrac{x^4+3x^2-4}{45x^4+90x^3} \cdot \dfrac{54x^4+108x^3}{x^3-x^2+4x-4}$

14. $\dfrac{x^3+8}{x^4+8x^2+16-4x^2} \cdot \dfrac{x^3-8}{x^3+2x^2-2x-4}$

15. $\dfrac{xy-3x+7y-21}{xy+7x+5y+35} \div \dfrac{xy-3x-5y+15}{xy+7x-7y-49}$

16. $\dfrac{x^4-x^2-4x-4}{4x^4+4x^3} \div \dfrac{x^3-2x^2-x+2}{8x^4}$

17. $\dfrac{x-1}{x-2} \cdot \dfrac{x-2}{x-3} \cdot \dfrac{3-x}{1-x}$

18. $\dfrac{a-b}{a-c} \cdot \dfrac{b-c}{b-a} \cdot \dfrac{c-a}{c-b}$

19. $\left(\dfrac{ax-bx}{ax-ay} \cdot \dfrac{ax+ay}{ab-b^2}\right) \div \dfrac{xy+y^2}{bx-by}$

20. $\left(\dfrac{ac}{2a-b} \cdot \dfrac{a^2-b^2}{2ac+2bc}\right) \div \dfrac{a^2-ab}{4ac-2bc}$

(21–42) Simplify. (Do the indicated operations, expressing the result in simplified form.) See Examples 4–7. No variable has a value that makes a denominator zero.

21. $\dfrac{\dfrac{3}{5}}{x}$

22. $\dfrac{\dfrac{3}{5}}{x}$

23. $\dfrac{\dfrac{1}{3}+\dfrac{1}{4}}{1-\dfrac{1}{12}}$

24. $\dfrac{4+1}{1+\dfrac{1}{4}}$

25. $\dfrac{x-\dfrac{1}{x}}{1-x}$

26. $\dfrac{x-2}{1-\dfrac{x^2}{4}}$

27. $\dfrac{2x}{\dfrac{x}{30} - \dfrac{x}{40}}$

28. $\dfrac{x + 1 + \dfrac{1}{x - 1}}{x}$

29. $\dfrac{\dfrac{1}{x} - \dfrac{1}{y}}{\dfrac{1}{x} + \dfrac{1}{y}}$

30. $\dfrac{\dfrac{1}{x} + \dfrac{1}{y}}{\dfrac{1}{xy}}$

31. $\dfrac{\dfrac{1}{b} - a}{\dfrac{1}{a} - b}$

32. $\dfrac{1 - \dfrac{n}{2}}{\dfrac{n}{2} - 1}$

33. $\dfrac{\dfrac{n + 2}{n - 1} - \dfrac{n - 1}{n + 2}}{\dfrac{1}{n + 2} + \dfrac{1}{n - 1}}$

34. $\dfrac{\dfrac{a}{a - b} - \dfrac{b}{a + b}}{\dfrac{a}{a - b} + \dfrac{b}{a + b}}$

35. $\dfrac{x - 5 - \dfrac{8}{x + 2}}{x - 2 - \dfrac{5}{x + 2}}$

36. $\dfrac{x - 4 - \dfrac{9}{x - 4}}{x + 2 + \dfrac{9}{x - 4}}$

37. $1 - \dfrac{1}{1 + \dfrac{1}{x}}$

38. $x + \dfrac{1}{1 - \dfrac{1}{x}}$

39. $\dfrac{1}{1 - \dfrac{1}{1 + \dfrac{1}{x}}}$

40. $\dfrac{1}{x + \dfrac{1}{1 - \dfrac{1}{x}}}$

41. $\dfrac{1}{1 - \dfrac{1}{1 - \dfrac{1}{1 - y}}}$

42. $y - \dfrac{1}{1 - \dfrac{1}{1 + \dfrac{1}{y - 1}}}$

(43–50) Simplify each of the following by multiplying the main numerator and main denominator by the LCM (least common multiple) of the denominators occurring in each. See Examples 9 and 10.

EXAMPLE 9 Simplify

$\dfrac{x - y}{\dfrac{1}{x} - \dfrac{1}{y}}$

Solution Multiplying the main numerator and main denominator by xy,

$$\frac{xy(x - y)}{xy\left(\dfrac{1}{x} - \dfrac{1}{y}\right)} = \frac{xy(x - y)}{y - x} = \frac{-xy(y - x)}{y - x} = -xy$$

EXAMPLE 10 Simplify

$$\frac{\dfrac{1}{4} - \dfrac{1}{x}}{\dfrac{x}{16} - \dfrac{1}{x}}$$

Solution Multiplying the main numerator and the main denominator by $16x$,

$$\frac{16x\left(\dfrac{1}{4} - \dfrac{1}{x}\right)}{16x\left(\dfrac{x}{16} - \dfrac{1}{x}\right)} = \frac{4x - 16}{x^2 - 16} = \frac{4(x - 4)}{(x + 4)(x - 4)} = \frac{4}{x + 4}$$

43. $\dfrac{\dfrac{1}{y} - x}{\dfrac{1}{x} - y}$

44. $\dfrac{\dfrac{1}{x} - \dfrac{1}{5}}{x - 5}$

45. $\dfrac{2 - x}{\dfrac{1}{2} - \dfrac{1}{x}}$

46. $\dfrac{\dfrac{1}{a} + \dfrac{1}{b}}{\dfrac{1}{ab}}$

47. $\dfrac{2d}{\dfrac{d}{30} + \dfrac{d}{60}}$

48. $\dfrac{n - \dfrac{1}{n}}{1 + \dfrac{1}{n}}$

49. $\dfrac{x + 3 - \dfrac{7}{x - 3}}{x - 3 - \dfrac{1}{x - 3}}$

50. $\dfrac{x + 11 + \dfrac{9}{x + 5}}{x + 5 - \dfrac{9}{x + 5}}$

2.6 CHAPTER REVIEW

1. Find the indicated value for $P(x) = 2x^2 - 5x - 3$. (2.1)
  (a) $P(3)$ (b) $P(-1)$ (c) $P(y - 2)$

(2–3) Simplify. (2.1)

2. $(3x^2 + 4x - 7) + (x^2 - 5x - 2)$
3. $(5x - 3y + 4) - (2x - 3y + 8)$

(4–6) Find the indicated product. (2.1)

4. $-3xy(x^2 - 2xy - y^2)$
5. $(5x^2 - 2)(x^2 + 6)$

6. $(x^2 + 4)(y^2 - 5)$

(7–8) Divide, using formal division. (2.1)

7. $\dfrac{6x^3 + x^2 + 3x - 15}{3x - 4}$ **8.** $\dfrac{y^3 + y + 10}{y + 2}$

(9–10) Divide, using synthetic division. (2.2)

9. $\dfrac{x^3 - 50x - 7}{x + 7}$ **10.** $\dfrac{t^4 - 6t^3 + t^2 - 8t + 12}{t - 6}$

(11–18) Completely factor. (2.3)

11. $36x^4 - 96x^3 + 64x^2$

12. $x^4 - 16y^4$

13. $x^4 - 26x^2 + 25$

14. $27x^6 - x^3$

15. $x^4 - 4x^2 + 4 - 9y^2$

16. $36x^4 - x^2 - 10x - 25$

17. $x^4 - 7x^3 + x - 7$

18. $2x^3 - x^2 - 162x + 81$

(19–26) Simplify. No variable equals a value for which a denominator becomes zero. (2.4)

19. $\dfrac{x^2 - 5x}{x^2 - 2x}$ **20.** $\dfrac{2y^2 + 5y - 3}{2y^2 - 7y + 3}$

21. $\dfrac{2y^2 + 12y + 72}{y^3 - 216}$ **22.** $\dfrac{x^2 - 7x}{49 - 7x}$

23. $\dfrac{1}{2x - 10} + \dfrac{1}{3x - 15} - \dfrac{1}{4x - 20}$

24. $y - 4 - \dfrac{y^2 - 20}{y + 4}$

25. $\dfrac{2}{x^2 + 3x} + \dfrac{1}{x^2 - 3x} - \dfrac{2}{x^2 - 9}$

26. $\dfrac{x + 2}{x^2 - 1} + \dfrac{x - 2}{(x + 1)^2}$

(27–34) Simplify. No variable equals a value for which a denominator becomes zero. (2.5)

27. $\dfrac{x^2 - 9}{9x^2 - 1} \cdot \dfrac{1 - 3x}{3 - x}$ **28.** $\dfrac{x^3 + 2x^2}{4x - 8} \cdot \dfrac{x^2 - 4x + 4}{x^2 - 2x - 8}$

29. $\dfrac{5x^2 - 24x - 5}{5x^2 + 24x - 5} \div \dfrac{x^2 - 5x}{5x^2 - x}$ **30.** $\dfrac{x^2 - y^2}{xy} \div \dfrac{x - y}{x^2 y^2}$

31. $\dfrac{y - \dfrac{16}{y}}{4 - \dfrac{16}{y}}$ **32.** $\dfrac{x^2 - \dfrac{x + 4}{3}}{\dfrac{3x - 4}{6}}$

33. $x - \dfrac{1}{1 - \dfrac{x}{x + 1}}$ **34.** $\dfrac{y}{1 - \dfrac{1}{1 + \dfrac{y}{5}}}$

(35–36) Evaluate for the indicated value of the variable. (2.5)

35. $\dfrac{2x}{x+2} + \dfrac{5x^2}{x^2 - x - 6}$ For $x = \dfrac{1}{2}$

36. $\dfrac{1}{t} - \dfrac{2}{5t^2} - 10t$ For $t = \dfrac{3}{5}$

3

EQUATIONS AND INEQUALITIES

In Chapters 1 and 2 we learned about algebraic numbers and algebraic expressions and how operations were performed on these numbers and expressions.

In this chapter we compare algebraic expressions and learn how to find the set of numbers for which one algebraic expression becomes equal to, greater than, or less than another algebraic expression.

Unless it is stated otherwise the set of real numbers is the universal set of numbers; that is, every constant is a real number and every variable designates a real number.

3.1   LINEAR EQUATIONS AND INEQUALITIES

DEFINITION

An **open statement,** equation or inequality, is a statement that contains one or more variables.

DEFINITION

A **solution** of an open statement in one variable is a number from the universal set of numbers for which the open statement becomes a true statement when the variable is replaced by this number.

As examples, 5 is a solution of $x + 4 = 9$, and 6 is a solution of $x \geq 2$.

DEFINITION

The **solution set** of an open statement in one variable is the set of all solutions of the statement.

DEFINITION

An **identity** is a statement whose solution set contains every member in the universal set for which each expression is defined.

For example, $3(x - 5) = 3x - 15$ and $\dfrac{x + 4}{(x - 1)(x - 3)} = \dfrac{x + 4}{(x - 1)(x - 3)}$ are identities. The first is true for all values of x and the second is true for all values of x except 1 and 3.

If a statement is false or undefined for all values of the variable(s), then its solution set is the empty set, $\varnothing$. Such a statement is called a **contradiction**. As an example, $x + 5 = x$ is a contradiction.

DEFINITION

A statement that is true for one or more values of the variable(s) and false for an infinite number of values is called a **conditional statement.**

For example, the following statements are conditional:

$3x - 5 = 7$	linear equation
$x^2 - 3x = 10$	quadratic equation
$2x + 6 \leq 4$	linear inequality
$x^2 + x > 6$	quadratic inequality

This chapter is concerned with finding the solution sets of conditional statements such as these, and other types. In this section we shall learn about conditional linear equations and linear inequalities in one variable. A major concern in solving a conditional statement (that is, finding its solution set) is the process of finding a simpler equivalent statement.

DEFINITION

Equivalent statements are statements that have the same solution set.

The next theorem provides us with some methods for finding equivalent equations. It tells us that the solution set of an equation is not changed if we add or subtract the same algebraic expression to each side of an equation or if we multiply or divide each side by the same nonzero expression.

THEOREM ON EQUIVALENT EQUATIONS

Let A, B, and C be algebraic expressions.

1. $A = B$ is equivalent to $A + C = B + C$.
2. $A = B$ is equivalent to $A - C = B - C$.
3. If $C \neq 0$, then $A = B$ is equivalent to $AC = BC$.
4. If $C \neq 0$, then $A = B$ is equivalent to $\dfrac{A}{C} = \dfrac{B}{C}$.

The proof of this theorem parallels the proofs that were given for Theorems 3 and 4 in Section 1.3.

DEFINITION

A **linear equation in one variable** x is an equation equivalent to $x = a$.

In general, the solution of a linear equation involves the use of the distributive axiom to remove parentheses and collect like terms, and the application of one or more parts of the theorem on equivalent equations.

EXAMPLE 1 Solve $5x - 3(x - 2) = 2(15 - x)$.

Solution

$5x - 3x + 6 = 30 - 2x$	Removing parentheses
$2x + 6 = 30 - 2x$	Collecting like terms
$4x = 24$	Adding $2x - 6$ to each side
$x = 6$	Multiplying each side by $\frac{1}{4}$

Check Replacing x by 6 in the original equation and doing the operations in the indicated order,
$$5(6) - 3(6 - 2) = 2(15 - 6)$$
$$30 - 12 = 2(9)$$
$$18 = 18 \quad \textit{true}$$
Therefore the solution set is $\{6\}$.

An equation that contains one or more letters, variables or constants, is called a **literal equation.** Such an equation can be solved for one of the letters by treating this letter as a variable and considering the others to be constants.

EXAMPLE 2   Solve $A = \dfrac{h(a + b)}{2}$ for b.

Solution

$$\frac{h(a + b)}{2} = A$$

$h(a + b) = 2A$ Multiplying each side by 2

$ha + hb = 2A$ Removing parentheses

$hb = 2A - ha$ Subtracting ha from each side

$b = \dfrac{2A - ha}{h}$ for $h \neq 0$ Dividing each side by h

EXAMPLE 3   Solve $2x - 3y = 6$ for y.

Solution

$-3y = -2x + 6$ Subtracting $2x$ from each side

$y = \dfrac{-2x + 6}{-3}$ Dividing each side by -3

$y = \dfrac{2x - 6}{3}$ Multiplying numerator and denominator by -1

or $y = \dfrac{2}{3}x - 2$

EXAMPLE 4   Solve $xy - 6 = 2x + 3y$ for y.

Solution   Subtracting $3y$ from each side and adding 6 to each side,

$xy - 3y = 2x + 6$

Factoring the left side in order to isolate y,

$y(x - 3) = 2x + 6$

Dividing each side by $x - 3$,

$y = \dfrac{2x + 6}{x - 3}$ for $x \neq 3$

DEFINITION

A **linear inequality in one variable** x is a statement that is equivalent to $x < a$, $x > a$, $x \leq a$, or $x \geq a$.

We examined these forms and their geometric meanings in Section 1.4.

Solving a linear inequality is similar to solving a linear equation with two notable exceptions.

1. If the sides of an inequality are exchanged, the order is reversed (from $<$ to $>$, or from $>$ to $<$).
2. If each side of an equality is multiplied or divided by a negative number, then the order is reversed.

As examples:

If $3 < x$, then $x > 3$.

If $-x > 2$, then $x < -2$.

Since $-5 < 2$, then $(-3)(-5) > (-3)2$ and $15 > -6$.

The following theorem provides us with methods that can be used to solve inequalities.

THEOREM ON EQUIVALENT INEQUALITIES

Let A, B, and C be algebraic expressions.

I. **1.** $A > B$ is equivalent to $A + C > B + C$.
   2. $A < B$ is equivalent to $A + C < B + C$.

II. If $C > 0$, then

      3. $A > B$ is equivalent to $AC > BC$.
      4. $A < B$ is equivalent to $AC < BC$.

III. If $C < 0$, then

      5. $A > B$ is equivalent to $AC < BC$.
      6. $A < B$ is equivalent to $AC > BC$.

EXAMPLE 5 Solve $10 - 4(x - 3) > 2x - 2$.

Solution Removing parentheses,
$$10 - 4x + 12 > 2x - 2$$
Combining like terms,
$$-4x + 22 > 2x - 2$$
Adding $-2x - 22$ to each side,
$$-6x > -24$$
Multiplying each side by $\frac{-1}{6}$ and reversing the order,
$$x < 4$$
The solution set is $\{x \mid x < 4\}$.
The graph of the solution set is shown at the left.

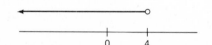

Note that the solution set of Example 5 is an infinite set of real numbers. A partial check can be made by selecting two or more numbers from the solution set found and substituting in the original inequality.

Check For $x = 3$,
$$10 - 4(3 - 3) > 2(3) - 2$$
$$10 - 0 > 6 - 2$$
$$10 > 4 \quad true$$
For $x = -2$,
$$10 - 4(-2 - 3) > 2(-2) - 2$$
$$10 + 20 > -4 - 2$$
$$30 > -6 \quad true$$
It is also a good idea to check the value for the end point of the interval even when this value does not belong to the solution set. For Example 5 we need to show that the original inequality becomes false for $x = 4$.

$$10 - 4(4 - 3) > 2(4) - 2$$
$$10 - 4(1) > 8 - 2$$
$$6 > 6 \quad \textit{false} \quad \text{since } 6 = 6$$

EXAMPLE 6   Solve $3x - (5x + 4) \le 5(x + 2)$.

Solution

$$-2x - 4 \le \quad 5x + 10$$
$$\underline{-5x + 4 \quad -5x + \ \ 4} \qquad \text{Adding } -5x + 4 \text{ to each side}$$
$$-7x \quad\quad \le 14$$
$$x \ge -2 \qquad\qquad \text{Multiplying each side by } \tfrac{-1}{7}$$
$$\text{and reversing the order}$$

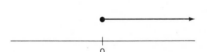

The solution set is $\{x \mid x \ge -2\}$.
The graph of the solution set is shown at the left.
Check   Recall that $x \ge -2$ means $x > -2$ or $x = -2$.
For $x = -2$,
$$3(-2) - (5(-2) + 4) \le 5(-2 + 2)$$
$$-6 - (-6) \le 5(0)$$
$$0 \le 0 \quad \textit{true} \quad \text{since } 0 = 0 \text{ is true}$$
For $x = -1$,
$$3(-1) - (5(-1) + 4) \le 5(-1 + 2)$$
$$-3 - (-1) \le 5(1)$$
$$-2 \le 5 \quad \textit{true} \quad \text{since } -2 < 5 \text{ is true}$$

EXAMPLE 7   Solve $(x - 3)^2 \le (x + 3)^2$.

Solution   This inequality appears to be nonlinear. However
the square terms cancel out leaving a linear inequality.
$$x^2 - 6x + 9 \le x^2 + 6x + 9 \qquad \text{Squaring binomials}$$
$$-6x + 9 \le 6x + 9 \qquad \text{Subtracting } x^2 \text{ from each side}$$
$$\underline{6x - 9 \quad 6x - 9} \qquad \text{Adding } 6x - 9 \text{ to each side}$$
$$0 \le 12x$$
$$0 \le x \qquad\qquad \text{Multiplying each side by } \tfrac{1}{12}$$
$$x \ge 0 \qquad\qquad \text{Exchanging sides of inequality}$$

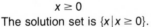

The solution set is $\{x \mid x \ge 0\}$.
The graph of the solution set is shown at the left.

EXERCISES 3.1

(1–10)   Solve for x. See Example 1.

1. $5x + 7 = 8x - 5$ 2. $6 - x = 4x + 21$
3. $9 - 2(x - 3) = x$ 4. $8x - (x + 5) = 9x + 7$
5. $3(x - 1) - 5(x + 3) = 0$ 6. $(x + 5)(x - 2) = (x + 4)(x - 4)$
7. $(x + 1)^2 - (x - 1)^2 = 2$ 8. $4x(2x + 5) = (2x - 5)(4x - 5)$
9. $(2x - 1)(5x + 1) - (2x + 1)(5x - 3) = 8$
10. $3(x - 2)^2 - (3x + 1)(x - 1) = 10$

(11–30) Solve for the variable indicated. See Examples 2, 3, and 4.

11. $2x + y = 10$ for y
12. $x - 3y = 9$ for x
13. $5x - y - 5 = 0$ for y
14. $3x - 4y = 12$ for y
15. $8y - 3x = 4$ for x
16. $2x + y - z = 6$ for y
17. $5x - 2y - z = 10$ for z
18. $15 - 2x + 3y - 2z = 0$ for x
19. $4x^2 - 4x = 8 - y$ for y
20. $xy - 2y^2 = 8$ for x
21. $P = 2a + 2b$ for b
22. $V = \frac{1}{2}a^2b$ for b
23. $A = \dfrac{hb^2}{3}$ for h
24. $L = a + (n - 1)d$ for n
25. $r = s(1 + nt)$ for n
26. $A = P + Prt$ for P
27. $K = \frac{1}{2}mv^2 + mg$ for m
28. $x - 5b = 5 - bx$ for x
29. $rx + sx = s^2 - r^2$ for x
30. $ax + by + c = 0$ for y

(31–50) State the solution set and graph the solution set. See Examples 5, 6, and 7.

31. $5x - 12 > 8$
32. $3x + 4 \geq 1$
33. $4x + 9 \leq 1$
34. $2x - 7 < 3$
35. $-2x \geq 12$
36. $-5x \leq -30$
37. $8 - x < 3$
38. $23 - 4x > 27$
39. $3(x - 1) \leq 17 - (8 - x)$
40. $6x - 2(x + 3) \geq x + 9$
41. $2x - 5(x - 2) > 2(x - 5)$
42. $9 - 4(x + 1) < 3(x + 11)$
43. $(x + 3)(x - 5) \leq x^2 + 1$
44. $(x - 4)(x + 7) \geq (x + 4)(x - 7)$
45. $(x + 2)^2 \geq (x - 2)^2$
46. $(x + 1)^2 - (x - 1)^2 > 5x$
47. $x(x - 5) + 5 < x(x - 3) + 3$
48. $(x - 2)^2 + 2 \leq (x - 1)^2 + 1$
49. $(x + 2)(3x - 2) > (x + 4)(3x - 4)$
50. $(x^2 + 1)(x - 1) < x(x^2 - x + 2)$

3.2 FACTORABLE POLYNOMIAL EQUATIONS, FRACTIONAL EQUATIONS

DEFINITION

A **polynomial equation** is an equation that can be expressed in the form $P = 0$ where P is a polynomial.

For example, $x^2 = 7x - 10$ is a polynomial equation because it is equivalent to $x^2 - 7x + 10 = 0$. Since $x^2 - 7x + 10 = (x - 2)(x - 5)$, it follows that $x^2 - 7x + 10 = 0$ if and only if $(x - 2)(x - 5) = 0$. We will see in this section that this last equation is equivalent to the statement $x - 2 = 0$ or $x - 5 = 0$ and thus the solution set of the original equation is $\{2, 5\}$. This follows from the fact that a product of real numbers is zero only when one of its factors is zero.

ZERO PRODUCT THEOREM

For real numbers a and b:
If $ab = 0$, then $a = 0$ or $b = 0$.

Proof

If $a = 0$, then the theorem is obviously true.

If $a \neq 0$, then $\dfrac{1}{a}$ is a real number and $\dfrac{1}{a}(ab) = \dfrac{1}{a}(0)$ and $b = 0$.
Thus $a = 0$ or $b = 0$.

If A and B are polynomials in one variable with real numbers for coefficients, then it follows that the set of real numbers for which $AB = 0$ is the same as that for which $A = 0$ or $B = 0$. In other words, $AB = 0$ is equivalent to $A = 0$ or $B = 0$.

EXAMPLE 1 Solve $2x^2 - 6x = 36$.

Solution Writing an equivalent equation whose right side is zero,
$2x^2 - 6x - 36 = 0$
Factoring,
$2(x + 3)(x - 6) = 0$
Using the Zero Product Theorem,
$2(x + 3) = 0$ or $x - 6 = 0$
Solving,
$x = -3$ or $x = 6$
The solution set is $\{-3, 6\}$.
Check For $x = -3$,
$2(-3)^2 - 6(-3) = 36$
$\qquad\quad 18 + 18 = 36 \qquad true$
For $x = 6$,
$\quad\ 2(6^2) - 6(6) = 36$
$\qquad\quad 72 - 36 = 36$
$\qquad\qquad\quad 36 = 36 \qquad true$

EXAMPLE 2 Solve $x^3 + 2x^2 - 4x - 8 = 0$.

Solution Factoring,

$$x^2(x + 2) - 4(x + 2) = 0$$
$$(x^2 - 4)(x + 2) = 0$$

Using the Zero Product Theorem,

$$x^2 - 4 = 0 \quad \text{or} \quad x + 2 = 0$$
$$(x - 2)(x + 2) = 0 \quad \text{or} \quad x + 2 = 0$$
$$x - 2 = 0 \quad \text{or} \quad x + 2 = 0 \quad \text{or} \quad x + 2 = 0$$
$$x = 2 \quad \text{or} \quad x = -2 \quad \text{or} \quad x = -2$$

The solution set is $\{2, -2\}$.

Note that the solution -2 was obtained twice; thus it is called a **double root** while 2 is called a simple root. If a solution is obtained three times, it is called a **triple root.** In general, a root that is not simple is called a **multiple root.**

For a polynomial equation, the number of its solutions is equal to the degree of the polynomial if we agree to count roots with their multiplicity. If the degree of the polynomial is two, then the equation has two solutions if we agree to count a double root twice. Similarly, if the degree of the polynomial is three, then the equation has three solutions if we agree to count a double root twice and a triple root three times.

If an equation involves fractions, an equivalent equation without fractions can be obtained by the following procedure.

Procedure for Fractional Equations

1. Express each side as a single simplified fraction.
2. Build up each of the two resulting fractions so that they have the same denominator and so that the least number of build-up factors are used. The resulting denominator is called the LCD, the least common denominator.
3. Multiply each side by the LCD, simplify, and solve.

An equation that has a variable in a denominator is undefined for any value of the variable that causes the denominator to become zero. Such a value is called a **restricted value** and *cannot* be a solution of the equation. Multiplying each side of an equation by the LCD produces an equation that has the same solution set as the original equation only for those values of x for which the LCD does *not* become zero.

EXAMPLE 3 Solve $\dfrac{2}{x + 6} - \dfrac{1}{2x - 8} = \dfrac{4}{x^2 + 2x - 24}$.

Solution

$$\frac{2 \cdot 2(x-4)}{(x+6)2(x-4)} - \frac{1(x+6)}{2(x-4)(x+6)} = \frac{4}{(x-4)(x+6)}$$

$$\frac{3x-22}{2(x-4)(x+6)} = \frac{2 \cdot 4}{2(x-4)(x+6)}$$

Multiplying each side by the LCD, $2(x-4)(x+6)$,

$$3x - 22 = 8$$
$$3x = 30 \quad \text{and} \quad x = 10$$

Check

$$\frac{2}{10+6} - \frac{1}{20-8} = \frac{4}{100+20-24}$$

$$\frac{1}{8} - \frac{1}{12} = \frac{4}{96}$$

$$\frac{3}{8 \cdot 3} - \frac{2}{12 \cdot 2} = \frac{4}{96}$$

$$\frac{1}{24} = \frac{1}{24} \qquad \textit{true}$$

The solution set is $\{10\}$.

EXAMPLE 4 Solve $\dfrac{x}{x+2} + \dfrac{5}{x^2-x-6} = \dfrac{x-2}{x-3}$.

Solution

$$\frac{x(x-3)}{(x+2)(x-3)} + \frac{5}{(x+2)(x-3)} = \frac{x-2}{x-3}$$

$$\frac{x^2-3x+5}{(x+2)(x-3)} = \frac{(x-2)(x+2)}{(x-3)(x+2)}$$

Multiplying each side by the LCD, $(x-3)(x+2)$,

$$x^2 - 3x + 5 = x^2 - 4$$
$$-3x = -9$$
$$x = 3$$

Check

$$\frac{3}{3+2} + \frac{5}{9-3-6} = \frac{3-2}{3-3}$$

$$\frac{3}{5} + \frac{5}{0} = \frac{1}{0} \qquad \textit{undefined}$$

The original equation is not defined for $x = 3$.
Thus the solution set is the empty set, $\varnothing$.

Example 4 illustrates that multiplication by the LCD does not produce an equation having the same solution set as the original equation whenever a solution of the resulting equation causes the LCD to be zero.

Values for which the LCD is zero are called **restricted values.** To find these values, we solve LCD = 0. For Example 4, LCD = $(x+2)(x-3)$ and if $(x+2)(x-3) = 0$, then

$$x + 2 = 0 \quad \text{or} \quad x - 3 = 0$$
$$x = -2 \quad \text{or} \quad x = 3$$

The restrictions for Example 4 are $x \neq -2$ and $x \neq 3$. Since division by zero is undefined, the original equation is undefined for $x = -2$ and for $x = 3$.

EXAMPLE 5 Solve $\dfrac{3y + 20}{4y + 16} - \dfrac{y + 2}{4y - 16} = \dfrac{y - 16}{y^2 - 16}$.

Solution

$$\frac{3y + 20}{4(y + 4)} - \frac{y + 2}{4(y - 4)} = \frac{y - 16}{(y + 4)(y - 4)}$$

$$\frac{(3y + 20)(y - 4) - (y + 2)(y + 4)}{4(y + 4)(y - 4)} = \frac{4(y - 16)}{4(y + 4)(y - 4)}$$

LCD $= 4(y + 4)(y - 4)$.

Solving LCD $= 0$,

$4(y + 4)(y - 4) = 0$

$y = -4 \quad \text{or} \quad y = 4$

The restrictions are $y \neq -4$ and $y \neq 4$.

Multiplying each side by the LCD,

$3y^2 + 8y - 80 - (y^2 + 6y + 8) = 4y - 64 \quad \text{and} \quad y \neq -4 \quad \text{and} \quad y \neq 4$

$$2y^2 - 2y - 24 = 0$$

$$2(y - 4)(y + 3) = 0$$

$(y = 4 \quad \text{or} \quad y = -3) \quad \text{and} \quad (y \neq -4 \quad \text{and} \quad y \neq 4)$

The solution set is $\{-3\}$.

Check

$$\frac{-9 + 20}{-12 + 16} - \frac{-3 + 2}{-12 - 16} = \frac{-3 - 16}{9 - 16}$$

$$\frac{11}{4} - \frac{1}{28} = \frac{19}{7}$$

$$\frac{77 - 1}{28} = \frac{19 \cdot 4}{7 \cdot 4}$$

$$\frac{76}{28} = \frac{76}{28} \quad \textit{true}$$

EXERCISES 3.2

(1–20) Solve for the variable. See Examples 1 and 2.

1. $x^2 + 21 = 10x$
2. $x^2 - 9 = 6x - 9$
3. $(x - 2)^2 = 9$
4. $x^2 = 3x + 40$
5. $(y + 1)(y + 4) = 4$
6. $y(y - 8) = 8(2 - y)$
7. $2y(y + 2) = y + 20$
8. $3(y^2 - 1) = 5(y - 1)$
9. $n^2 = 25$
10. $(t + 1)^2 = 49$
11. $x^3 + 2x^2 = 9x + 18$
12. $x^3 + 25 = x^2 + 25x$
13. $x^3 - 10x^2 + 24x = 0$
14. $x^3 + 3x^2 - 9x = 27$
15. $x^4 - 10x^2 + 9 = 0$
16. $x^4 - 20x^2 + 64 = 0$
17. $144x^4 - 48x^3 + 4x^2 = 0$
18. $5x^4 + 19x^3 - 4x^2 = 0$
19. $(x - 2)(x - 3)(x - 4) = 0$
20. $(x + 1)(x - 2)(x - 3)(x + 4) = 0$

(21–50) Solve and check. See Examples 3, 4, and 5.

21. $\dfrac{4}{x} = 12$

22. $\dfrac{-x}{20} = 5$

23. $\dfrac{1}{x} + \dfrac{1}{10} = \dfrac{1}{6}$

24. $\dfrac{1}{6x} - \dfrac{1}{9x} = \dfrac{1}{6}$

25. $\dfrac{3}{x-2} - \dfrac{2}{x+3} = \dfrac{20}{x^2+x-6}$

26. $\dfrac{x}{x+2} - \dfrac{x}{x-4} = \dfrac{60}{x^2-2x-8}$

27. $\dfrac{4}{5y} - \dfrac{1}{2y} = 1$

28. $\dfrac{1}{2y} - \dfrac{1}{3y} = 1$

29. $\dfrac{8}{y-1} = \dfrac{12}{y+1}$

30. $\dfrac{x-3}{x+7} = \dfrac{x+7}{x-3}$

31. $\dfrac{x-2}{x+1} = \dfrac{x-4}{x+3}$

32. $\dfrac{t}{3t-24} - \dfrac{t}{t-8} = 2$

33. $\dfrac{5}{t-3} - \dfrac{9}{t^2-9} = \dfrac{8}{t+3}$

34. $\dfrac{2}{x} - \dfrac{x+12}{x^2-x} = \dfrac{6}{1-x}$

35. $\dfrac{2x}{x-5} + 2 = \dfrac{10}{x-5}$

36. $\dfrac{x}{x+1} + \dfrac{3}{x^2-x-2} = \dfrac{x-1}{x-2}$

37. $\dfrac{t}{t-4} + \dfrac{1}{4-t} = 2$

38. $1 - \dfrac{1}{t-3} = \dfrac{3}{3-t}$

39. $\dfrac{1}{x} + \dfrac{1}{x+2} = \dfrac{1}{2x}$

40. $\dfrac{1}{x} + \dfrac{1}{4} = x + 4$

41. $\dfrac{1}{x^2-x} + \dfrac{1}{x^2+x} = \dfrac{x}{x^2-1}$

42. $\dfrac{2x}{x^2-1} - \dfrac{1}{x^2+x} = \dfrac{4}{x^2-x}$

43. $\dfrac{x+4}{2x-4} + \dfrac{x-2}{x-4} = \dfrac{9}{2}$

44. $\dfrac{2}{x} + \dfrac{1}{6} = \dfrac{2}{x-6}$

45. $\dfrac{2x}{x+1} + \dfrac{12}{x^2-2x-3} = \dfrac{x}{x-3}$

46. $\dfrac{x}{x-6} + \dfrac{x}{x+6} = \dfrac{22x-60}{x^2-36}$

47. $\dfrac{1}{y-1} - \dfrac{1}{y+1} = \dfrac{1}{40}$

48. $\dfrac{y+2}{y-2} = \dfrac{y-10}{2y-10}$

49. $\dfrac{1}{x} + \dfrac{1}{2} + \dfrac{1}{5} = \dfrac{1}{x+7}$

50. $\dfrac{1}{t-5} + \dfrac{10}{t+5} = \dfrac{1}{9}$

(51–64) Solve for x. State any restrictions necessary so no denominator is zero.

51. $x^2 - 7ax + 10a^2 = 0$

52. $k^2x^2 + 3kx = 18$

53. $x^2 - (a+b)x + ab = 0$

54. $(x-a)^2 + a = (x-b)^2 + b$

55. $\dfrac{1}{x} = \dfrac{1}{a} + \dfrac{1}{b}$

56. $\dfrac{1}{x} + \dfrac{1}{c} = x + c$

57. $\dfrac{1}{x} - \dfrac{1}{c} = \dfrac{1}{d} - \dfrac{1}{x}$

58. $\dfrac{x-a}{x-b} - \dfrac{b}{a} = 0$

59. $\dfrac{a}{x-b} = \dfrac{b}{x-a}$

60. $\dfrac{b}{ax} - \dfrac{a}{bx} = \dfrac{1}{a} + \dfrac{1}{b}$

61. $\dfrac{25x^2}{25x^2 - k^2} = \dfrac{5x - k}{5x + k}$

62. $\dfrac{a-b}{x-1} = \dfrac{a+b}{x+1}$

63. $\dfrac{x+a}{x-b} - \dfrac{x-a}{x+b} = \dfrac{a^2 - b^2}{x^2 - b^2}$

64. $\dfrac{1}{x} + \dfrac{1}{c} + \dfrac{1}{d} = \dfrac{1}{x+c+d}$

3.3 ABSOLUTE VALUE EQUATIONS AND INEQUALITIES

Since a positive real number r can be expressed as $+r$ and its opposite as $-r$, every nonzero real number a can be considered as having two parts, a sign and a positive real number, called the **absolute value of a,** written in symbols as $|a|$.

DEFINITION OF ABSOLUTE VALUE OF a, $|a|$

If $a > 0$, $|a| = a$.
If $a < 0$, $|a| = -a$.
If $a = 0$, $|a| = 0$.

As examples,

$$|5| = 5 \quad \text{and} \quad |-5| = -(-5) = 5$$

Note that the absolute value of a number is always positive or zero. Geometrically, the absolute value of a number can be interpreted as the distance its corresponding point is from the origin. The points corresponding to 5 and -5 are each 5 units from the origin, and 5 and -5 each have an absolute value of 5.

The statement $|x| \geq 0$ has the set of real numbers for its solution set.

In contrast, the solution set for $|x| < 0$ is the empty set, $\varnothing$.

In this section we are concerned with conditional equations and inequalities involving absolute value.

The following theorems follow directly from the definition of absolute value. They are useful in solving absolute value equations and inequalities.

THEOREMS

For all real numbers a and b,

1. $|ab| = |a| \cdot |b|$
2. $\left|\dfrac{a}{b}\right| = \dfrac{|a|}{|b|}$ for $b \neq 0$
3. $|a - b| = |b - a|$

THEOREM 4

If P is a polynomial with real numbers for coefficients and if a is a positive real number, then $|P| = a$ has the same solution set as $(P = a$ or $P = -a)$.

EXAMPLE 1   Solve $|x| = 8$.

Solution   Using the definition:
If $x \geq 0$, then $|x| = x$ and $x = 8$.
If $x < 0$, then $|x| = -x$ and $-x = 8$ and $x = -8$.
The solution set is $\{8, -8\}$.
Alternate Solution   Using Theorem 4, the solution set of $|x| = 8$ is the solution set for $(x = 8$ or $x = -8)$; that is, $\{8, -8\}$.

EXAMPLE 2   Solve $|x| = -4$.

Solution $\varnothing$, the empty set.
Note that Theorem 4 requires the a in $|P| = a$ to be positive. Since no absolute value can be negative, the equation $|x| = -4$ has no solution.

EXAMPLE 3   Solve $|2x - 3| = 7$.

Solution
$2x - 3 = 7$ or $2x - 3 = -7$
$2x = 10$ or $2x = -4$
$x = 5$ or $x = -2$
The solution set is $\{5, -2\}$.

Check
For $x = 5$, $|2 \cdot 5 - 3| = |10 - 3| = |7| = 7$.
For $x = -2$, $|2(-2) - 3| = |-4 - 3| = |-7| = 7$.

EXAMPLE 4   Solve $|x| > 8$.

Solution   Using the definition:
Case 1   For $x \geq 0$, $|x| = x$ and $x > 8$.
Case 2   For $x < 0$, $|x| = -x$ and $-x > 8$ and $x < -8$.
Since both cases are possible, the solution set is the union of the solution sets for each case.
The solution set is $\{x \mid x > 8$ or $x < -8\}$.
The graph of the solution set is

EXAMPLE 5   Solve $|x| \leq 8$.

Solution   Using the definition, there are two cases.

Case 1 $x \geq 0$ and $x \leq 8$. The graph for these values of x is

The "and" statement can be written $0 \leq x \leq 8$.

Case 2 $x < 0$ and $-x \leq 8$ and $x \geq -8$. The graph for $x < 0$ and $x \geq -8$ is

This "and" statement can be written $-8 \leq x < 0$.

Forming the union of the solution sets for Case 1 and Case 2, we obtain

$\{x \mid -8 \leq x < 0 \text{ or } 0 \leq x \leq 8\}$

The graph of this set is

As a result, we can state the solution set as follows:

$\{x \mid -8 \leq x \leq 8\}$

Note that the solution set of $|x| \leq 8$ is the "complement" of the solution set of $|x| > 8$. This means that the set of values for which $|x| \leq 8$ is true is the set of values for which $|x| > 8$ is false.

Theorem 5 provides an easier method for solving absolute value inequalities. It is proved by paralleling the discussion in Examples 4 and 5, with 8 replaced by a, a positive real number.

THEOREM 5

If P is a polynomial with real numbers for coefficients and if a is any positive real number, then

$|P| < a$ has the same solution set as $-a < P < a$

and

$|P| > a$ has the same solution set as $P < -a$ or $P > a$.

Similarly, the solution set for $|P| \leq a$ is $-a \leq P \leq a$ and the solution set for $|P| \geq a$ is $P \leq -a$ or $P \geq a$.

EXAMPLE 6 Solve $|3x - 5| < 7$.

Solution

$$-7 < 3x - 5 < 7$$
$$\underline{+5 \qquad +5 \quad +5}$$
$$-2 < \qquad 3x < 12$$
$$\frac{-2}{3} < \qquad \frac{3x}{3} < \frac{12}{3}$$
$$\frac{-2}{3} < \qquad x < 4$$

The solution set is $\left\{ x \middle| \dfrac{-2}{3} < x < 4 \right\}$.

The graph of the solution set is

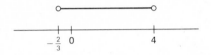

The statement $\quad -7 < 3x - 5 < 7$ has the meaning
$$-7 < 3x - 5 \quad and \quad 3x - 5 < 7$$

Solving,
$$3x - 5 > -7 \qquad and \quad 3x < 12$$
$$x > -\frac{2}{3} \qquad and \quad x < 4$$

which is equivalent to $\dfrac{-2}{3} < x < 4$. This work can be condensed

as shown in the solution of Example 6.

EXAMPLE 7 Solve $|9 - 5x| \geq 6$.

Solution   Since $|9 - 5x| = |5x - 9|$ by Theorem 3,
it is convenient to rewrite the given statement
as $|5x - 9| \geq 6$.
$$5x - 9 \leq -6 \quad or \quad 5x - 9 \geq 6 \qquad \text{by Theorem 5}$$
$$5x \leq 3 \qquad or \qquad 5x \geq 15$$
$$x \leq \frac{3}{5} \qquad or \qquad x \geq 3$$

The solution set is $\left\{ x \middle| x \leq \dfrac{3}{5} \text{ or } x \geq 3 \right\}$.

No condensation is possible for this "or" statement.

The graph of the solution set is

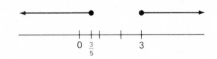

EXERCISES 3.3

(1–20) Solve for the variable. See Examples 1, 2, and 3.

1. $|x| = 6$
2. $|x| = 10$
3. $|x| = -7$
4. $|x| = -9$
5. $|x - 4| = 6$
6. $|x + 7| = 3$
7. $|10 - x| = 2$
8. $|8 - x| = 5$
9. $|2x - 3| = 9$
10. $|2x - 7| = 1$
11. $|12 - 5x| = 8$
12. $|15 - 4x| = 5$
13. $|-6 - 3t| = 12$
14. $|-10t - 5| = 35$
15. $5 + |y - 3| = 8$
16. $7 - |4 - y| = 2$
17. $8 - 2|1 - y| = 4$
18. $6 = 9 - 3|2y + 1|$
19. $10 = 6 - |y + 3|$
20. $12 - 5|2 - 3y| = 22$

(21–40) Solve and graph the solution set. See Examples 4, 5, 6, and 7.

21. $|x| < 7$
22. $|x| > 4$
23. $|x| \geq 9$
24. $|x| \leq 5$
25. $|x - 4| \leq 6$
26. $|x + 6| \geq 2$
27. $|8 - x| > 10$
28. $|12 - x| < 3$
29. $|2x + 3| \geq 5$
30. $|2x - 1| \leq 9$
31. $|9 - 2x| < 15$
32. $|18 - 3x| > 12$
33. $|3x - 8| \leq 7$
34. $|5x + 6| \geq 14$
35. $|-4x - 8| \geq |-12|$
36. $|7 - 2x| \leq 10 - |-5|$
37. $6 - 2|x - 1| > 2$
38. $12 - 5|2 - x| < 2$
39. $1 > 9 - 4|-x - 3|$
40. $2 < 20 - 3|-x + 5|$

(41–50) State the solution set for each of the following.

41. $|x| \geq 0$
42. $|x| < 0$
43. $|x| = -x$
44. $|x| = x$
45. $|x| = |-x|$
46. $|-x| = -x$
47. $|2x - 5| < -3$
48. $|3x - 1| > -5$
49. $|-x - 6| \geq 0$
50. $|12 - x| < 0$

3.4 QUADRATIC EQUATIONS

DEFINITION

A **quadratic equation** in the variable x is an equation that is equivalent to $ax^2 + bx + c = 0$ where a, b, and c are complex numbers (real or imaginary), and $a \neq 0$.

3.4 QUADRATIC EQUATIONS

We solved some quadratic equations in Section 3.2 by the factoring method. Although the factoring method is usually the easiest, not all quadratic equations can be solved by this method, especially when the coefficients of the factors are restricted to be integers. Such an equation is $x^2 - 6x + 4 = 0$ which we will solve in this section by the method known as "completing the square." This method is second in simplicity to factoring when the coefficient of x^2 is 1 and the coefficient of x is an even integer.

Noting that
$$(ax + b)^2 - c = (ax + b)^2 - (\sqrt{c})^2$$
$$= (ax + b - \sqrt{c})(ax + b + \sqrt{c})$$
it follows that the solutions of $(ax + b)^2 - c = 0$ are those for which
$$ax + b - \sqrt{c} = 0 \quad \text{or} \quad ax + b + \sqrt{c} = 0$$
and
$$ax + b = \sqrt{c} \quad \text{or} \quad ax + b = -\sqrt{c}$$
These results are stated as the following theorem.

THEOREM

If $(ax + b)^2 = c$, where a, b, and c are any complex numbers, then
$$ax + b = \sqrt{c} \quad \text{or} \quad ax + b = -\sqrt{c}$$

EXAMPLE 1 Solve $x^2 - 6x + 4 = 0$ by completing the square.

Solution Isolating the variable terms on the left side,
$$x^2 - 6x = -4$$
Completing the square by adding the square of one-half the coefficient of x to each side,
$$x^2 - 6x + (-3)^2 = -4 + 9$$
$$(x - 3)^2 = 5$$
This is equivalent to
$$x - 3 = \sqrt{5} \quad \text{or} \quad x - 3 = -\sqrt{5}$$
and
$$x = 3 + \sqrt{5} \quad \text{or} \quad x = 3 - \sqrt{5}$$
The solution set is $\{3 + \sqrt{5}, 3 - \sqrt{5}\}$.

Check For $x = 3 + \sqrt{5}$,
$$x^2 - 6x + 4 = (3 + \sqrt{5})^2 - 6(3 + \sqrt{5}) + 4$$
$$= 14 + 6\sqrt{5} - 18 - 6\sqrt{5} + 4 = 0$$
For $x = 3 - \sqrt{5}$,
$$x^2 - 6x + 4 = (3 - \sqrt{5})^2 - 6(3 - \sqrt{5}) + 4$$
$$= 14 - 6\sqrt{5} - 18 + 6\sqrt{5} + 4 = 0$$

The solutions of Example 1 may also be written in a condensed form as $3 \pm \sqrt{5}$. In general, $a \pm b$ means $a + b$ or $a - b$ for all complex numbers a and b.

A formula for the **roots** (solutions) of any quadratic equation can be obtained by applying the method of completing the square to the general quadratic equation.

Multiplying each side of $ax^2 + bx + c = 0$ by $\dfrac{1}{a}$ since $a \neq 0$, we obtain

$$x^2 + \frac{b}{a}x + \frac{c}{a} = 0$$

Isolating the variable terms by adding $\dfrac{-c}{a}$ to each side,

$$x^2 + \frac{b}{a}x = \frac{-c}{a}$$

Completing the square by adding $\left(\dfrac{1}{2} \cdot \dfrac{b}{a}\right)^2$ to each side,

$$x^2 + \frac{b}{a}x + \frac{b^2}{4a^2} = \frac{-c}{a} + \frac{b^2}{4a^2}$$

Forming the square on the left side and simplifying the right side,

$$\left(x + \frac{b}{2a}\right)^2 = \frac{b^2 - 4ac}{4a^2}$$

Forming the square roots,

$$x + \frac{b}{2a} = \frac{\sqrt{b^2 - 4ac}}{2a} \quad \text{or} \quad x + \frac{b}{2a} = \frac{-\sqrt{b^2 - 4ac}}{2a}$$

Solving for x,

$$x = \frac{-b + \sqrt{b^2 - 4ac}}{2a} \quad \text{or} \quad x = \frac{-b - \sqrt{b^2 - 4ac}}{2a}$$

A condensed form of this final result is known as the **quadratic formula.**

THEOREM   QUADRATIC FORMULA

If $ax^2 + bx + c = 0$ where $a \neq 0$, then
$$x = \frac{-b \pm \sqrt{b^2 - 4ac}}{2a}$$

EXAMPLE 2   Solve $3x^2 = 2x + 4$ by using the quadratic formula.

Solution   Rewriting as $ax^2 + bx + c = 0$,
$3x^2 - 2x - 4 = 0$
Identifying a, b, and c,
$a = 3$, $b = -2$, $c = -4$
Substituting in the quadratic formula and simplifying,

$$x = \frac{-(-2) \pm \sqrt{(-2)^2 - 4(3)(-4)}}{2(3)}$$

$$= \frac{2 \pm \sqrt{4 + 48}}{6}$$

$$= \frac{2 \pm \sqrt{52}}{6}$$

$$= \frac{2 \pm \sqrt{4 \cdot 13}}{6}$$

$$= \frac{2 \pm 2\sqrt{13}}{6}$$

$$= \frac{2(1 \pm \sqrt{13})}{2(3)}$$

$$= \frac{1 \pm \sqrt{13}}{3}$$

The solution set is $\left\{ \dfrac{1 + \sqrt{13}}{3}, \dfrac{1 - \sqrt{13}}{3} \right\}$.

Check For $x = \dfrac{1 + \sqrt{13}}{3}$,

$$3x^2 = 3\left(\frac{1 + \sqrt{13}}{3}\right)^2 = \frac{3(14 + 2\sqrt{13})}{9} = \frac{14 + 2\sqrt{13}}{3}$$

$$2x + 4 = 2\left(\frac{1 + \sqrt{13}}{3}\right) + 4 = \frac{2 + 2\sqrt{13}}{3} + \frac{12}{3} = \frac{14 + 2\sqrt{13}}{3}$$

If a quadratic equation has integers for coefficients, then only one irrational root needs to be checked since the **irrational roots are** always **conjugates,** having the form $a + b\sqrt{c}$ and $a - b\sqrt{c}$. Since $\dfrac{1 + \sqrt{13}}{3}$ is a solution, its conjugate $\dfrac{1 - \sqrt{13}}{3}$ must also be a solution.

Imaginary roots also **occur in conjugate pairs** for quadratic equations with integers for coefficients. If $a + bi$ is a solution, then so is $a - bi$, and only one of these solutions needs to be checked.

EXAMPLE 3 Use the quadratic formula to solve $2x^2 + 2x + 5 = 0$.

Solution $a = 2$, $b = 2$, and $c = 5$

$$x = \frac{-2 \pm \sqrt{4 - 4(10)}}{4}$$

$$= \frac{-2 \pm \sqrt{-36}}{4}$$

$$= \frac{-2 \pm 6i}{4} = \frac{2(-1 \pm 3i)}{2(2)}$$

$$= \frac{-1 \pm 3i}{2}$$

The solution set is $\left\{ \dfrac{-1 + 3i}{2}, \dfrac{-1 - 3i}{2} \right\}$.

Check For $x = \dfrac{-1 + 3i}{2}$,

$$2\left(\frac{-1 + 3i}{2}\right)^2 + 2\left(\frac{-1 + 3i}{2}\right) + 5 = \frac{-8 - 6i}{2} + (-1 + 3i) + 5$$

$$= -4 - 3i - 1 + 3i + 5 = 0$$

EXAMPLE 4 Solve $x^2 - (r + s)x + rs = 0$.

Solution Factoring,

$(x - r)(x - s) = 0$

and

$x - r = 0$ or $x - s = 0$

and

$x = r$ or $x = s$

Note in Example 4 that the roots are r and s, the product of the roots, rs, is the constant term of the equation, and the sum of the roots, $r + s$, is the opposite of the coefficient of x.

For $ax^2 + bx + c = 0$ with $a \neq 0$,

$$x^2 + \frac{b}{a}x + \frac{c}{a} = 0$$

Using the quadratic formula,

The sum of the roots $= \dfrac{-b + \sqrt{b^2 - 4ac}}{2a} + \dfrac{-b - \sqrt{b^2 - 4ac}}{2a} = -\dfrac{b}{a}$

The product of the roots $= \left(\dfrac{-b + \sqrt{b^2 - 4ac}}{2a} \right) \left(\dfrac{-b - \sqrt{b^2 - 4ac}}{2a} \right)$

$$= \frac{(-b)^2 - (b^2 - 4ac)}{4a^2} = \frac{c}{a}$$

Thus the sum of the roots is the opposite of the coefficient of x and the product of the roots is the constant term when the coefficient of x^2 is 1. The following theorem summarizes these results.

THEOREM SUM AND PRODUCT OF ROOTS

For $ax^2 + bx + c = 0$ with $a \neq 0$ the sum of the roots is $\dfrac{-b}{a}$ and the product of the roots is $\dfrac{c}{a}$.

The preceding theorem provides a simple method for checking the solutions of a quadratic equation.

EXAMPLE 5 By using the sum and product of roots theorem, check that $\dfrac{5 - \sqrt{6}}{2}$ and $\dfrac{5 + \sqrt{6}}{2}$ are the solutions of $4x^2 - 20x + 19 = 0$.

Solution Dividing by the leading coefficient 4, $x^2 - 5x + \dfrac{19}{4} = 0$. Using the theorem, the sum of the roots is $-(-5) = 5$. Adding the solutions, $\dfrac{5 - \sqrt{6}}{2} + \dfrac{5 + \sqrt{6}}{2} = \dfrac{5 + 5}{2} = 5$. The sum checks. Using the theorem, the product of the roots is $\dfrac{+19}{4}$. Multiplying the solutions, $\left(\dfrac{5 - \sqrt{6}}{2} \right) \left(\dfrac{5 + \sqrt{6}}{2} \right) = \dfrac{25 - 6}{4} = \dfrac{19}{4}$. The product checks.

The radicand, $b^2 - 4ac$, in the quadratic formula is called the **discriminant** because it can be used to determine what sort of roots the quadratic equation has. In particular we have the following theorem.

THEOREM NATURE OF ROOTS OF QUADRATIC EQUATION

Let $\dfrac{-b + \sqrt{D}}{2a}$ and $\dfrac{-b - \sqrt{D}}{2a}$ be the roots of the quadratic equation

$ax^2 + bx + c = 0$ where $D = b^2 - 4ac$ and a, b, and c are integers with $a \neq 0$.

1. If $D = 0$, there is one rational root and it is a double root.
2. If $D = n^2$ where n is a nonzero integer, then the roots are unequal and rational.
3. If $D > 0$ and D is not the square of an integer, then the roots are unequal, irrational, and conjugate.
4. If $D < 0$, then the roots are unequal, imaginary, and conjugate.

EXAMPLE 6 Without solving the equation, determine the nature of the roots of each of the following equations:
(a) $x^2 - 4x + 2 = 0$
(b) $x^2 - 4x + 3 = 0$
(c) $x^2 - 4x + 4 = 0$
(d) $x^2 - 4x + 5 = 0$

Solution
(a) $D = b^2 - 4ac = (-4)^2 - 4(1)(2) = 16 - 8 = 8$
Since $D > 0$ and $D \neq n^2$, the roots are unequal, irrational, and conjugate.
(b) $D = b^2 - 4ac = (-4)^2 - 4(1)(3) = 16 - 12 = 4 = 2^2$
Since $D = n^2$, the roots are unequal and rational.
(c) $D = b^2 - 4ac = (-4)^2 - 4(1)(4) = 16 - 16 = 0$
Since $D = 0$, there is one rational double root.
(d) $D = b^2 - 4ac = (-4)^2 - 4(1)(5) = 16 - 20 = -4$
Since $D < 0$, the roots are unequal, imaginary, and conjugate.

An equation that is not quadratic may become quadratic after a substitution. Such equations are said to be **quadratic in form.** For example, $x^4 + 4x^2 - 45 = 0$ is not quadratic. However, if we let $y = x^2$, then the equation becomes $y^2 + 4y - 45 = 0$ which is quadratic in y. This can then be solved by the methods used for quadratic equations.

EXAMPLE 7 Solve $x^4 + 4x^2 - 45 = 0$.

Solution Let $y = x^2$; then
$y^2 + 4y - 45 = 0$
Factoring,
$(y - 5)(y + 9) = 0$
$y - 5 = 0$ or $y + 9 = 0$
$y = 5$ or $y = -9$
Now replacing y by x^2,
$x^2 = 5$ or $x^2 = -9$
$x = \pm\sqrt{5}$ or $x = \pm 3i$
The solution set is $\{\sqrt{5}, -\sqrt{5}, 3i, -3i\}$.
Note that the solution set contains four solutions, matching the degree of the given equation.

EXAMPLE 8   Solve $x^6 + 7x^3 - 8 = 0$.

Solution   Let $y = x^3$. Then

$y^2 = x^3 \cdot x^3 = x^6$

and

$y^2 + 7y - 8 = 0$

Factoring,

$(y + 8)(y - 1) = 0$

$y + 8 = 0$ or $y - 1 = 0$

Replacing y by x^3,

$x^3 + 8 = 0$ or $x^3 - 1 = 0$

Factoring,

$\quad\quad (x + 2)(x^2 - 2x + 4) = 0$ or $(x - 1)(x^2 + x + 1) = 0$

$x + 2 = 0$ or $x^2 - 2x + 4 = 0$ or $x - 1 = 0$ or $x^2 + x + 1 = 0$

$x + 2 = 0$	$x^2 - 2x + 4 = 0$	$x - 1 = 0$	$x^2 + x + 1 = 0$
$x = -2$	Completing the square (since $a = 1$ and b is even),	$x = 1$	Using the quadratic formula,

$$x^2 - 2x = -4$$
$$x = \frac{-1 \pm \sqrt{1 - 4}}{2}$$

$$x^2 - 2x + 1 = -3$$
$$x = \frac{-1 \pm i\sqrt{3}}{2}$$

$$(x - 1)^2 = -3$$
$$x - 1 = \pm\sqrt{-3}$$
$$x = 1 \pm i\sqrt{3}$$

The solution set is

$$\left\{ 1, -2, 1 + i\sqrt{3}, 1 - i\sqrt{3}, \frac{-1 + i\sqrt{3}}{2}, \frac{-1 - i\sqrt{3}}{2} \right\}.$$

   Note that there are six solutions and the degree of the given equation is 6.

EXERCISES 3.4

(1–10)   Add a constant term to each of the following so the resulting polynomial is a perfect square trinomial. Factor the resulting polynomial.

1. $x^2 + 8x$ **2.** $x^2 - 10x$

3. $x^2 - 12x$ **4.** $x^2 + 2x$

5. $y^2 + 3y$ **6.** $y^2 + y$

7. $t^2 - t$ **8.** $y^2 - 5y$

9. $x^2 + 2dx$ **10.** $x^2 - dx$

(11–22)   Solve by completing the square. Check. See Example 1.

11. $x^2 + 4x - 2 = 0$ **12.** $x^2 + 8x + 9 = 0$

13. $x^2 - 10x + 12 = 0$ **14.** $x^2 - 2x - 1 = 0$

15. $y^2 - 6y - 9 = 0$ **16.** $y^2 - 12y + 12 = 0$

17. $z^2 + 14z + 29 = 0$ **18.** $z^2 + 16z - 36 = 0$

19. $t^2 - 18t + 49 = 0$ **20.** $t^2 - 20t - 4800 = 0$

21. $x^2 + 10x - 2000 = 0$ **22.** $x^2 + 40x + 400 = 0$

(23–40) Solve by using the quadratic formula. Check by using the sum and product of roots theorem. See Examples 2, 3, and 5.

23. $2x^2 - 10x + 11 = 0$	**24.** $3x^2 - 6x + 2 = 0$
25. $5x^2 + 4x + 1 = 0$	**26.** $x^2 + 3x + 4 = 0$
27. $x^2 + x - 1 = 0$	**28.** $x^2 - x + 1 = 0$
29. $3x^2 = 2x - 5$	**30.** $4x^2 = 3x + 2$
31. $4y(y - 1) = 7$	**32.** $2y^2 + 8 = 7y$
33. $10t^2 + 11t = 6$	**34.** $t(t - 3) = 2$
35. $6(1 - 2z^2) = z$	**36.** $200 - z = 2z(z + 4)$
37. $x^2 + 4 = 3x$	**38.** $x^2 - 3 = 4x$
39. $2x(5 - x) = 9$	**40.** $y^2 = 5(y - 5)$

(41–54) Using the discriminant, determine if (a) there is a double root or two unequal roots, and (b) the roots are rational, irrational, or imaginary. Do not solve the equation. See Example 6.

41. $x^2 - 6x + 3 = 0$	**42.** $x^2 - 6x + 5 = 0$
43. $x^2 - 6x + 8 = 0$	**44.** $x^2 - 6x + 9 = 0$
45. $x^2 - 6x + 10 = 0$	**46.** $x^2 - 6x - 9 = 0$
47. $x^2 - 5x - 4 = 0$	**48.** $x^2 + 5x - 6 = 0$
49. $2x^2 + 5x - 3 = 0$	**50.** $2x^2 - 5x + 4 = 0$
51. $3x^2 - 2x + 5 = 0$	**52.** $3x^2 - 2x + 1 = 0$
53. $4x^2 - 20x + 25 = 0$	**54.** $2x^2 - 20x + 25 = 0$

(55–72) Solve and check. See Examples 7 and 8.

55. $x^4 - 13x^2 + 36 = 0$	**56.** $x^4 - 15x^2 - 16 = 0$
57. $x^4 - 50x^2 + 625 = 0$	**58.** $x^4 + 3x^2 - 54 = 0$
59. $x^4 = x^2 + 20$	**60.** $x^4 = 9(2x^2 - 9)$
61. $x^6 - 28x^3 + 27 = 0$	**62.** $x^6 + 117x^3 = 1000$
63. $x^6 = 63x^3 + 64$	**64.** $x^6 = 64$
65. $x^6 = 1$	**66.** $x^4 = 1$
67. $x^4 = 81$	**68.** $\dfrac{1}{x^4} - \dfrac{17}{x^2} + 16 = 0$
69. $\dfrac{1}{x^4} + \dfrac{5}{x^2} = 36$	**70.** $2\left(x + \dfrac{1}{x}\right)^2 - 9\left(x + \dfrac{1}{x}\right) + 10 = 0$
71. $6\left(x - \dfrac{1}{x}\right)^2 - 7\left(x - \dfrac{1}{x}\right) = 24$	**72.** $x^8 - 41x^4 = 400$

(73–80) Solve for x, treating the other letters as constants.

73. $x^2 + 2bx + c = 0$	**74.** $x^2 - 2rx + r^2 - s^2 = 0$
75. $x^2 + 2ax - 3a^2 = 0$, where $a > 0$	**76.** $x^2 + 2bx + b^2 - c = 0$, where $c > 0$
77. $x^2 - 2ax - a^2 = 0$, where $a > 0$	**78.** $x^2 - (a + b)xy + aby^2 = 0$
79. $x^2 + 2ry - r^2y^2 = 1$	**80.** $x^2 - 2xy + y^2 - k^2 = 0$

(81–92) Determine the value of k for each of the following. See Examples 9 and 10.

EXAMPLE 9 One root of $2x^2 + kx - 63 = 0$ is 7.

Solution Dividing each side of the equation by 2, the leading coefficient,

$$x^2 + \frac{k}{2}x - \frac{63}{2} = 0$$

The sum of the roots is $\frac{-k}{2}$, the opposite of the coefficient of x. The product of the roots is $\frac{-63}{2}$, the constant term. Letting r = the unknown root,

$$7r = \frac{-63}{2} \quad \text{and} \quad r + 7 = \frac{-k}{2}$$

$$r = \frac{-9}{2} \quad \text{and} \quad \frac{-9}{2} + 7 = \frac{-k}{2}$$

$$\frac{5}{2} = \frac{-k}{2} \quad \text{and} \quad k = -5$$

EXAMPLE 10 $5x^2 - 20x + k = 0$ has a double root.

Solution The sum of the roots is $\frac{20}{5} = 4$ and the product is $\frac{k}{5}$. Since the roots r and s are equal,

$r = s$, then
$r + s = r + r = 2r$ and $rs = r^2$
Then
$$2r = 4 \quad \text{and} \quad r = 2$$
$$r^2 = \frac{k}{5} = 2^2 \quad \text{and} \quad k = 20$$

Alternate Solution When a quadratic equation has a double root, its discriminant is zero. Therefore
$$b^2 - 4ac = (-20)^2 - 4(5k) = 0$$
$$20k = 400 \quad \text{and} \quad k = 20$$

81. One root of $x^2 + kx + 20 = 0$ is 5.
82. One root of $x^2 - 3x + k = 0$ is 7.
83. One root of $x^2 - 8x + k = 0$ is $4 - \sqrt{3}$.
84. One root of $x^2 + kx + 13 = 0$ is $3 + 2i$.
85. $3x^2 - 6x + k = 0$ has a double root.
86. One root of $x^2 + kx + 4 = 0$ is $3 + \sqrt{5}$.
87. One root of $x^2 + kx + 5 = 0$ is $-2 - i$.
88. $2x^2 + 3x + k = 0$ has a double root.
89. The roots of $2x^2 + bx + k = 0$ are $4 \pm \sqrt{6}$.
90. The roots of $2x^2 + kx + c = 0$ are $5 \pm 3i$.
91. The roots of $kx^2 + 8x + k = 0$ are equal.
92. The roots of $kx^2 + 1 = kx$ are equal.

(93–94) State for what values of k, the given equation has (a) a double real root, (b) unequal real roots, and (c) unequal imaginary roots.

93. $x^2 + 2x + k = 0$ **94.** $x^2 - 4x + k = 0$

(95–100) Write an equation of the form $ax^2 + bx + c = 0$, where a, b, and c are integers, having the given numbers as solutions.

95. $7, -4$ **96.** $5, -8$ **97.** $6 \pm \sqrt{30}$

98. $-4 \pm \sqrt{14}$ **99.** $-5 \pm 3i$ **100.** $2 \pm 6i$

3.5 QUADRATIC INEQUALITIES

A **quadratic inequality** is one that is equivalent to $P > 0$, $P < 0$, $P \geq 0$, or $P \leq 0$ where P is a polynomial of degree 2, with real numbers for coefficients.

Examples of quadratic inequalities are $x^2 - 3x < 10$, $x^2 \geq 4$, and $2x^2 \leq 5x$. Note that $x^2 - 3x < 10$ is equivalent to $x^2 - 3x - 10 < 0$. Since $x^2 - 3x - 10 = (x + 2)(x - 5)$, then $x^2 - 3x - 10 < 0$ is equivalent to $(x + 2)(x - 5) < 0$. To solve this inequality, we need to find the set of real numbers x for which the indicated product is negative. Since a product of real numbers may be positive, negative, or zero, we first determine the values of x for which the indicated product is zero.

Solving $(x + 2)(x - 5) = 0$, $x = -2$ or $x = 5$. These solutions determine two points on a real number line, and the two points determine the three intervals:

$$x < -2, \quad -2 < x < 5, \quad \text{and} \quad x > 5$$

See Figure 3.1.

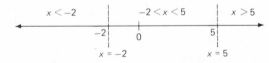

FIG. 3.1. Intervals determined by two points.

Note, for $x < -2$, $x + 2$ is negative and for $x > -2$, $x + 2$ is positive. Also, for $x < 5$, $x - 5$ is negative and for $x > 5$, $x - 5$ is positive. Consequently,

for $x < -2$, $(x + 2)(x - 5)$ is positive   (negative times negative)

for $-2 < x < 5$, $(x + 2)(x - 5)$ is   negative   (positive times negative)

for $x > 5$, $(x + 2)(x - 5)$ is positive   (positive times positive)

Thus the solution set for $(x + 2)(x - 5) < 0$ is $\{x \,|\, -2 < x < 5\}$. Also the solution set for $(x + 2)(x - 5) > 0$ is $\{x \,|\, x < -2 \text{ or } x > 5\}$.

In general we have the following theorem which is useful for solving quadratic inequalities.

THEOREM QUADRATIC INEQUALITIES

Let a and b be real numbers with $a < b$. Then

1. $(x - a)(x - b) < 0$ is equivalent to $a < x < b$.
2. $(x - a)(x - b) > 0$ is equivalent to $x < a$ or $x > b$.

Noting that a and b are the roots of $(x - a)(x - b) = 0$, we can also express the theorem as follows:

For a negative product of the form $(x - a)(x - b)$, x is between the roots.

For a positive product, x is less than the smaller root or x is greater than the larger root.

The proof of the theorem parallels the special case preceding the theorem.

EXAMPLE 1 Solve $x^2 + 24 \geq 10x$.

Solution Forming an equivalent inequality whose right side is zero,
$x^2 - 10x + 24 \geq 0$
Factoring,
$(x - 4)(x - 6) \geq 0$
Using the theorem with $a = 4$ and $b = 6$ since $4 < 6$,
$x \leq 4$ or $x \geq 6$
The solution set is $\{x \mid x \leq 4 \text{ or } x \geq 6\}$.

EXAMPLE 2 Solve $2x^2 < x + 15$.

Solution   Alternate Method.

1. Solve the corresponding equation.
$$2x^2 = x + 15$$
$$2x^2 - x - 15 = 0$$
$$(2x + 5)(x - 3) = 0$$
$$x = \frac{-5}{2} \quad \text{or} \quad x = 3$$

2. Find the intervals determined by the roots of the equation.

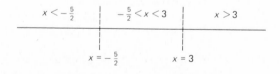

$$x < -\frac{5}{2} \qquad -\frac{5}{2} < x < 3 \qquad x > 3$$

$$x = -\frac{5}{2} \qquad x = 3$$

3. Test a value in each interval in the original inequality.

For $x < \frac{-5}{2}$, select $x = -3$ and
$$2(-3)^2 < -3 + 15$$
$$18 < 12 \qquad \textit{false}$$

For $\frac{-5}{2} < x < 3$, select $x = 0$ and
$$2(0) < 0 + 15$$
$$0 < 15 \qquad \textit{true}$$

For $x > 3$, select $x = 10$ and
$$2(10)^2 < 10 + 15$$
$$200 < 25 \qquad \textit{false}$$

4. Select each interval for which a true statement was obtained. Also test the endpoints of each interval.

The solution set is $\left\{ x \mid \frac{-5}{2} < x < 3 \right\}$.

EXAMPLE 3   Solve $\dfrac{x}{x-1} \geq 2$.

Solution   We first note that $x - 1 = 0$ for $x = 1$ and thus $x \neq 1$.

Secondly, $x - 1$ is negative for $x < 1$ and positive for $x > 1$. Since multiplication by a negative number reverses the order and multiplication by a positive number does not, we can avoid separating the problem into two cases by multiplying each side of the inequality by the square of the denominator which is always positive for $x \neq 1$.

Multiplying by $(x - 1)^2$,

$$(x - 1)^2 \frac{x}{x-1} \geq 2(x-1)^2$$
$$x(x - 1) \geq 2(x^2 - 2x + 1)$$
$$x^2 - x \geq 2x^2 - 4x + 2$$
$$-x^2 + 3x - 2 \geq 0$$

Multiplying each side by -1 and reversing the order,
$$x^2 - 3x + 2 \leq 0$$

Factoring,
$$(x - 1)(x - 2) \leq 0$$

Using the theorem with $a = 1$ and $b = 2$ and noting the restriction $x \neq 1$, $1 < x \leq 2$.

The solution set is $\{x \mid 1 < x \leq 2\}$.

EXAMPLE 4 Solve $2x^2 + 11 \geq 10x$.

Solution Solving the corresponding equation by using the quadratic formula,

$$2x^2 + 11 = 10x$$
$$2x^2 - 10x + 11 = 0$$
$$x = \frac{10 \pm \sqrt{100 - 4(22)}}{4}$$
$$x = \frac{5 \pm \sqrt{3}}{2}$$

Using the theorem with $a = \dfrac{5 - \sqrt{3}}{2}$ and $b = \dfrac{5 + \sqrt{3}}{2}$, since $\dfrac{5 - \sqrt{3}}{2} < \dfrac{5 + \sqrt{3}}{2}$, the solution is

$$x \leq \frac{5 - \sqrt{3}}{2} \quad \text{or} \quad x \geq \frac{5 + \sqrt{3}}{2}$$

The solution may be checked, or the alternate method shown in Example 2 may be used to solve the given inequality, by testing values in the intervals determined by the roots of the equation. For this inequality, the intervals are

$$x < \frac{5 - \sqrt{3}}{2}, \quad \frac{5 - \sqrt{3}}{2} < x < \frac{5 + \sqrt{3}}{2}, \quad x > \frac{5 + \sqrt{3}}{2}$$

EXAMPLE 5 Solve (a) $9x^2 - 12x + 4 \geq 0$ and (b) $9x^2 - 12x + 4 < 0$.

Solution Note that $9x^2 - 12x + 4 = (3x - 2)^2$.

(a) The inequality becomes $(3x - 2)^2 \geq 0$. Since the square of a real number is always positive or zero, the solution set is the set of real numbers, R.

(b) The inequality becomes $(3x - 2)^2 < 0$. Since the square of a real number is never negative, the solution set is the empty set, $\varnothing$.

EXAMPLE 6 Solve (a) $x^2 - 8x + 25 \geq 0$ and (b) $x^2 - 8x + 25 < 0$.

Solution Solving $x^2 - 8x + 25 = 0$ by the quadratic formula,

$$x = \frac{8 \pm \sqrt{64 - 4(25)}}{2} = \frac{8 \pm \sqrt{-36}}{2} = 4 \pm 3i$$

Since the roots are imaginary, there are no real numbers for which the quadratic polynomial is zero. However, by completing the square, we can express the inequalities as follows:

(a) $x^2 - 8x + 25 \geq 0$
$$x^2 - 8x + 16 \geq -25 + 16$$
$$(x - 4)^2 \geq -9$$

3.5 QUADRATIC INEQUALITIES

Since a positive number or zero is always greater than a negative number, the inequality is true for all real numbers. The solution set is the set of real numbers, R.

(b) $(x - 4)^2 < -9$

Since a positive number or zero is never less than a negative number, the inequality is false for all real numbers. The solution set is the empty set, $\emptyset$.

EXERCISES 3.5

(1–18) Solve and check. See Examples 1 and 2.

1. $(x - 3)(x - 6) < 0$
2. $(x + 4)(x - 7) > 0$
3. $x^2 - 5x \geq 14$
4. $x^2 + 4x \leq 5$
5. $2x^2 + 3 > 7x$
6. $8 \leq 5x^2 + 18x$
7. $6y \geq y^2 + 8$
8. $20y^2 < 9y - 1$
9. $x^2 < 9$
10. $x^2 \geq 25$
11. $y^2 \geq 9y$
12. $y^2 \leq 5y$
13. $x^2 + 1 > 2x$
14. $x^2 + 9 < 6x$
15. $n^2 \leq 4(n - 1)$
16. $n^2 \geq 12(n - 3)$
17. $x^2(x^2 - 1) > 0$
18. $x^2(x^2 + 1) \leq 0$

(19–30) Solve and check. See Example 3.

19. $\frac{1}{x} > 5$
20. $\frac{2}{x - 1} \leq 1$
21. $\frac{x - 1}{x - 3} \leq 0$
22. $\frac{x}{x - 2} > 0$
23. $\frac{x - 1}{x} \geq 1$
24. $\frac{x - 3}{x + 3} < 1$
25. $\frac{x}{x + 4} < 2$
26. $\frac{x}{2} \leq \frac{2}{x}$
27. $\frac{1}{10} \leq \frac{x}{x - 3}$
28. $\frac{x + 1}{x + 4} < 3$
29. $\frac{5}{x} \leq \frac{x}{5}$
30. $\frac{x + 5}{x - 5} \geq 2$

(31–42) Solve and check. See Examples 4, 5, and 6.

31. $x^2 + 34 \geq 12x$
32. $x^2 + 6x \leq 1$
33. $3x^2 + 5x < 4$
34. $2x^2 > x + 5$
35. $36x > 4x^2 + 81$
36. $42x \leq 9x^2 + 49$
37. $16x^2 + 9 \geq 24x$
38. $36x^2 + 5 < 12x$
39. $9x^2 + 2 \geq 6x$
40. $4x + 3 \leq 2x^2$
41. $x + 1 \geq x^2$
42. $5x^2 - 1 \geq 2x$

(43–54) Solve each inequality. See Example 7.

EXAMPLE 7 Solve $x^3 + 63 \geq 7x^2 + 9x$.

Solution Solving the corresponding equation,
$$x^3 + 63 = 7x^2 + 9x$$
$$x^3 - 7x^2 - 9x + 63 = 0$$
$$x^2(x - 7) - 9(x - 7) = 0$$
$$(x^2 - 9)(x - 7) = 0$$
$$(x + 3)(x - 3)(x - 7) = 0$$
$$x = -3 \quad \text{or} \quad x = 3 \quad \text{or} \quad x = 7$$

Find the intervals determined by the roots of the equation,

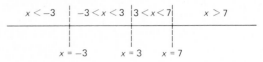

Select a value in each interval and check in the given inequality.

For $x < -3$, select $x = -4$ and
$$(-4)^3 + 63 \geq 7(-4)^2 + 9(-4)$$
$$-1 \geq 76 \quad \text{false}$$

For $-3 < x < 3$, select $x = 0$ and
$$0 + 63 \geq 0 + 0$$
$$63 \geq 0 \quad \text{true}$$

For $3 < x < 7$, select $x = 5$ and
$$5^3 + 63 \geq 7(5^2) + 9(5)$$
$$188 \geq 220 \quad \text{false}$$

For $x > 7$, select $x = 10$ and
$$10^3 + 63 \geq 7(10^2) + 9(10)$$
$$1063 \geq 790 \quad \text{true}$$

Since $x = -3$, $x = 3$, and $x = 7$ satisfy the equality, the solution set is $\{x \mid -3 \leq x \leq 3 \text{ or } x \geq 7\}$.

43. $(x + 4)(x - 2)(x - 5) > 0$

44. $(x + 6)(x + 3)(x - 4) < 0$

45. $x^3 - 5x^2 - 4x + 20 \leq 0$

46. $x^3 + 6x^2 \geq 9x + 54$

47. $x^4 \geq 13x^2 - 36$

48. $x^4 + 16 \leq 17x^2$

49. $x^3 + 7x^2 < 5x + 35$

50. $x^4 > 13x^2 - 30$

51. $6x^2 + 7 \leq x^4$

52. $x^3 + 4x \leq 6x^2 + 24$

53. $x^4 + 8x^2 + 16 \geq 0$

54. $12x^2 > x^4 + 36$

3.6 APPLICATIONS

Problems stated in words and requiring numbers for solutions occur in almost every walk of life. We shall examine some of these problems in this section. Many problems fall into categories for which there are standard methods that yield one or more equations that provide a solution. The procedure described below provides a means for solving stated problems.

Procedure for Word Problems

1. Read the problem slowly at least two times.
2. Make a sketch for the problem, if possible.
3. Let a letter represent an unknown number.
4. Express other unknown numbers in terms of this letter.
5. Summarize the numerical information using a table or sketch.
6. Recall any formulas that apply to the problem.
7. Form an equation by using information that states that two numbers are the same.
8. Solve the equation and determine the number required.
9. Check in the original word problem.

GEOMETRIC PROBLEMS

Many applied problems involve one or more geometrical shapes such as a square, a rectangle, or a triangle. The following formulas are useful for problems concerning an area A or a perimeter P.

Square	Rectangle	Triangle	Right Triangle

$P = 4s$ $P = 2a + 2b$ $P = a + b + c$ $P = a + b + c$

$A = s^2$ $A = ab$ $A = \dfrac{bh}{2}$ $A = \dfrac{ab}{2}$

$c^2 = a^2 + b^2$

EXAMPLE 1 A rancher wants to enclose a rectangular area of land with exactly 82 lineal feet of fencing. The length of the area is to be 9 feet longer than the width. Find the length and the width of the rectangular area.

Solution   Let

$w =$ the width in feet

Then

$w + 9 =$ the length in feet

Using $2a + 2b = P$, then

$2w + 2(w + 9) = 82$

$4w + 18 = 82$

$w = 16$ feet,   width

$w + 9 = 25$ feet,   length

MIXTURE PROBLEMS

There are many applications in which two or more quantities are combined to form a mixture. Basic ideas involved are the following:

(Unit value) · (Amount) = Value

Value of mixture = Sum of values of components

A table or chart, such as those shown in Examples 2 and 3, is useful for summarizing the data.

EXAMPLE 2   A person wants to put part of $6000 in a passbook savings account that earns 6% interest and the rest of the $6000 in an account that earns 8% interest. How much should be put in each account so that the total income is $410?

Solution   Let x = number of dollars at 6%.
Then $6000 - x$ = number of dollars at 8%.

ITEM	UNIT VALUE ·	AMOUNT =	INCOME
6%	0.06	x	$0.06x$
8%	0.08	$6000 - x$	$0.08(6000 - x)$
Mixture		6000	410

$$0.06x + 0.08(6000 - x) = 410$$
$$6x + 8(6000 - x) = 41{,}000$$
$$x = 3500 \text{ dollars} \quad \text{and} \quad 6000 - x = 2500 \text{ dollars}$$

Therefore $3500 should be put into the 6% account and $2500 into the 8% account.

EXAMPLE 3   How many pounds of walnuts costing 70 cents a pound should be combined with 40 pounds of almonds costing 80 cents a pound and 30 pounds of pecans costing 90 cents a pound to produce a mixture that will cost 75 cents a pound?

Solution   Let x = number of pounds of walnuts.

Item	Unit value ·	Amount =	Value
Walnuts	70	x	$70x$
Almonds	80	40	80(40)
Pecans	90	30	90(30)
Mixture	75	$x + 70$	$75(x + 70)$

$$70x + 80(40) + 90(30) = 75(x + 70)$$
$$x = 130 \text{ pounds of walnuts}$$

UNIFORM MOTION PROBLEMS

Uniform motion problems involve objects moving at a constant rate of speed. The formula
$$d = rt$$
Distance = (Rate) · (Time)
is involved in these problems. It is useful to make a sketch of the motions, find expressions for the rate, time, and distance of each object, and then use the sketch to form an equation relating the distances. This technique is shown in Examples 4 and 5.

EXAMPLE 4 At 7 a.m. a truck leaves a warehouse and travels due east to a city 420 miles away. At 9 a.m. a car leaves the city and travels due west on the same highway in order to intercept the truck. If the rate of the truck is 45 mph and the rate of the car is 65 mph, at what clock time does the car intercept the truck?

Solution Let $x =$ the travel time elapsed for the truck.

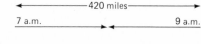

Truck: $r = 45$ Car: $r = 65$
        $t = x$ $t = x - 2$
        $d = 45x$ $d = 65(x - 2)$

The distances are obtained by using $d = rt$. The equation is obtained by using the sketch and noting that the sum of the distances traveled by the truck and the car is equal to the distance between the warehouse and the city.

$$45x + 65(x - 2) = 420$$
$$x = 5 \text{ hours}$$

The clocktime is $7 + 5 = 12$ noon.

EXAMPLE 5 At 6 p.m. a helicopter leaves an airfield and travels due north. One hour later an airplane leaves an airport 60 miles north of the airfield and also travels due north. The helicopter and the airplane both arrive at the same airport at 9 p.m. If the rate of the airplane is 40 mph faster than that of the helicopter, find the rate of each.

Solution Let
$$x = \text{rate of helicopter in mph}$$
$$x + 40 = \text{rate of airplane in mph}$$

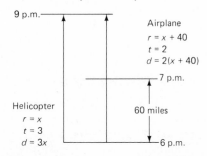

The distance traveled by the helicopter is the sum of the distance traveled by the airplane and the take-off differences.

$$3x = 2(x + 40) + 60$$
$$x = 140 \text{ mph}$$
$$x + 40 = 180 \text{ mph}$$

The helicopter's speed is 140 mph and the airplane's is 180 mph.

CURRENT-WIND PROBLEMS

When an object is traveling in water or in air, there is usually a water current or wind that contributes to the rate of speed of the object.

If

a = the rate of the object when there is no current or wind

and

b = the rate of the current or wind

then

$a - b$ = the object's rate traveling against the current or wind

$a + b$ = the object's rate traveling with the current or wind

This type of problem is illustrated in Example 6.

EXAMPLE 6 A boat whose rate in still water is 22 mph travels downstream for 3 hours. The return trip upstream is made in 4 hours and 20 minutes. Find the rate of the current.

Solution Let

x = rate of current

Then

$22 + x$ = rate downstream

and

$22 - x$ = rate upstream

Since the distance downstream equals the distance upstream,

$$3(22 + x) = \left(4 + \tfrac{20}{60}\right)(22 - x)$$

$$3(22 + x) = \tfrac{13}{3}(22 - x)$$

$$9(22 + x) = 13(22 - x)$$

$$x = 4 \text{ mph}$$

WORK PROBLEMS

In problems involving work, it is assumed that the total amount of work done by two or more persons or machines is the sum of the amount of work done by the persons or machines. Basic ideas are the following:

$$\text{Rate of work} = \frac{1}{\text{Time to do whole job}}$$

$$\text{Amount of work} = (\text{Rate}) \times (\text{Time})$$

Examples 7 and 8 illustrate typical work problems.

EXAMPLE 7   A painter working alone can do a certain job in 10 hours. His son, working alone, takes 15 hours to do the same job. How long would it take them to do the same job if they worked together?

Solution   Let x = time in hours they work together.

$$\text{Rate of painter} = \frac{1}{10} \qquad \text{Rate of son} = \frac{1}{15}$$

$$\text{Work done by painter} = \frac{x}{10} \qquad \text{Work done by son} = \frac{x}{15}$$

Since the total work done = 1, the whole job,

$$\frac{x}{10} + \frac{x}{15} = 1$$
$$3x + 2x = 30$$
$$5x = 30$$
$$x = 6 \text{ hours}$$

EXAMPLE 8   Of three machines, the time of the slowest to process a certain amount of data is three times that of the fastest. The time of the other machine to process the same amount of data is twice that of the fastest. When the three machines work together, they can process this same amount of data in 12 minutes. Find the time it takes each machine, working alone, to process the data.

Solution   Let

$$x = \text{time of fastest and rate} = \frac{1}{x}$$

$$3x = \text{time of slowest and rate} = \frac{1}{3x}$$

$$2x = \text{time of other and rate} = \frac{1}{2x}$$

Adding the work done in 12 minutes by each of the machines,

$$\frac{12}{x} + \frac{12}{3x} + \frac{12}{2x} = 1$$
$$x = 22 \text{ minutes, fastest}$$
$$3x = 66 \text{ minutes, slowest}$$
$$2x = 44 \text{ minutes, other}$$

EXERCISES 3.6

A. LINEAR EQUATIONS

(1–6)   Geometric Problems

1. Exactly 9 yards of fringe are to be sewn around the four sides of a

rectangular-shaped tablecloth. If the length of the tablecloth is 18 inches longer than it is wide, find the width and length of the table-cloth. Note that 1 yard = 36 inches.

2. A farmer has 550 lineal meters of fencing he wants to use to enclose three sides of a rectangular field (one length and two widths), the fourth side being along the bank of a river. The length of the field is to be 25 meters longer than the width. Find the length and width of the field.

3. An 80 × 100 centimeter rectangular picture is to have a frame of uniform width. If the perimeter of the framed picture is 400 centimeters, find the width of the frame.

4. A piece of steel wire, 36 inches long, is to be bent into the shape of a triangle. Two sides are equal in length and the third side is 6 inches longer than each of the equal sides. Find the length of each side.

5. A glass window has the shape of a square surmounted by an isosceles triangle (a triangle having two equal sides), as shown in the figure. Each of the two equal sides of the triangle is 6 inches longer than a side of the square. The perimeter of the window is 182 inches.

(a) Find the length of each side of the triangle.
(b) If the height of the triangle is 20 inches, find the area of the window.

6. An office area is to be rectangular in shape with two aisles of equal width, as shown in the figure. Find the width of each aisle so that the area between the aisles is 750 square feet.

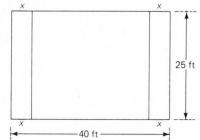

(7–11) Mixture Problems

7. A man wants to invest $40,000, some of this in bonds earning 8% interest and the rest in mortgages earning 10% interest. How much should he invest in each to obtain a total income of $3500?

8. How many cubic centimeters of water should a chemist add to 60 cubic centimeters of a 12% ammonia solution to dilute it to a 5% ammonia solution?

9. A landscaper wants to buy four times as many shrubs costing $5 each as trees costing $9 each. How many of each can he buy if he spends exactly $580?

10. How many kilograms of select coffee worth $12 a kilogram should be mixed with 60 kilograms of an ordinary coffee worth $8 a kilogram to obtain a mixture worth $9 a kilogram?

11. How many liters of pure alcohol and how many liters of a 20% alcohol solution should be mixed to obtain 80 liters of a 25% alcohol solution?

(12–16) Uniform Motion

12. The time for a jet plane, averaging 650 mph, to travel from San Francisco to Chicago is 4 hours less than the time of an old propeller-driven plane that averaged 250 mph. Find the time of the trip for each plane and find the distance between the two airports.

13. Two airplanes left the same airport at the same time, one traveling due north and the other due south. At the end of 3 hours they were 1500 miles apart. If the southbound airplane traveled 40 mph faster than the northbound airplane, find the rate of each.

14. A local train left a station and traveled due west at a rate of 24 mph. Two hours later, an express train left the same station and traveled due west at a rate of 56 mph. Find the distance from the station when the express train passes the local train.

15. A Coast Guard boat leaves a dock to overtake a ship that is 8 miles from the dock. The rate of the boat is 34 mph and the rate of the ship is 18 mph. How far from the dock does the boat overtake the ship?

16. Peter leaves his home at 10 a.m. and cycles toward Sarah's home. At 11 a.m. Sarah leaves her home by car to deliver an urgent message to Peter. Sarah intercepts Peter at 11:30 a.m. If the distance between their homes is 34 miles, and if Sarah drives 20 mph faster than Peter cycles, find the rate at which Peter cycles.

(17–20) Current-Wind Problems

17. The rate of an airplane when no wind is blowing is 600 mph. Traveling against a headwind, the airplane takes 4 hours to make a certain trip. The return trip, traveling with the same wind, now a tailwind, takes $3\frac{1}{2}$ hours. Find the speed of the wind.

18. A boat travels 2 hours downstream. The return trip upstream takes 3 hours. If the rate of the current is 6 mph, find the distance the boat travels downstream.

19. A person who can row 8 mph in still water takes 20 minutes to row a certain distance upstream. The return trip downstream takes 12 minutes. Find the rate of the current.

20. An airplane traveling against a wind of 25 mph takes $3\frac{1}{2}$ hours to make a certain trip. Traveling with a wind of 30 mph, the airplane takes 3 hours to travel the same distance. Find the rate of the plane when no wind is blowing.

(21–25) Work Problems

21. One machine takes 36 minutes to sort a certain number of cards. Another machine takes 45 minutes to sort the same number of

cards. How long would it take both machines, working together, to sort this number of cards?

22. It takes a cabinet maker and his assistant, working together, 8 days to do a certain job. The assistant, working alone, takes twice as much time as the cabinet maker, working alone, to do the same job. Find the time each takes, working alone, to do the job.

23. Three machinists take 4 hours, 6 hours, and 12 hours, respectively, to do a certain job, when working alone. How long would it take them to do this job if they worked together?

24. An inlet pipe takes 10 hours to fill a tank, while an outlet pipe takes 14 hours to empty the tank. If both pipes were left open, how long would it take to fill the tank?

25. An old machine takes twice as long to do a certain job as a new machine. An even older machine takes 4 times as long as the new machine to do the same job. Working together, the three machines can do the job in 8 minutes. Find the time it takes the new machine, working alone, to do the job.

B. QUADRATIC EQUATIONS

1. Find the length of the diagonal of a square whose perimeter is 24 feet.
2. Find the length and width of a rectangle whose perimeter is 54 meters and whose area is 180 square meters.
3. Equal squares are to be cut from the four corners of a rectangular piece of sheet metal that is 60 centimeters long and 40 centimeters wide. The ends are then to be bent so that an open box is formed. If the area of the bottom of the box is to be 1500 square centimeters, find the length of each side of the squares.
4. A lawn in a park has the shape of a rectangle, 50 feet long and 32 feet wide. A path of uniform width is to be made around the lawn so that the area of the path is 720 square feet. Find the width of the path.
5. A piece of sheet metal has the shape of a right triangle whose hypotenuse is 17 decimeters. The longer leg of the triangle is 1 decimeter less than twice the shorter leg. Find the lengths of the legs of the triangle.
6. A bus and a car make the same 300-mile trip. The car travels 10 mph faster than the bus and makes the trip in 1 hour less time. Find the average speed of the bus.
7. An eastbound airplane and a westbound airplane each traveled 500 miles to the same airport. Both airplanes started at the same time but the eastbound plane arrived at the airport $\frac{1}{2}$ hour later than the westbound airplane since it traveled 50 mph slower. Find the average rate of each airplane.
8. A person made a 200-mile business trip and then returned on a different route that was 300 miles long. The average rate on the return trip was 10 mph faster than the average rate going but the return trip took 1 hour longer. Find the time for each trip.

9. An airplane traveled 360 miles with a headwind and then immediately returned along the same route. Its total flight time was 5 hours. If the rate of the plane in still air was 150 mph, find the speed of the wind, assuming no wind change for the entire trip.

10. A motorboat traveled downstream for 60 kilometers and then returned the same distance upstream. The return trip took 1 hour longer than the trip downstream. The average rate of the water current was 5 kph. Find the rate of the boat in still water.

11. One machine takes 5 hours longer to do a certain job than another machine. When both machines work together, they do the job in 6 hours. Find the time for each machine to do the job, working alone.

12. A larger pipe takes 18 minutes less time to fill a tank than does a smaller pipe. Together the two pipes can fill the tank in 40 minutes. Find the time required for the larger pipe alone to fill the tank.

13. A new worker takes 5 days longer to do a certain job than an experienced worker. Together they can do this job in 4 days less than the time it takes the experienced worker to do the job alone. Find the time it takes the experienced worker to do the job, working alone.

14. One construction crew takes 15 hours to do a certain job. A second crew, working alone, takes 8 hours longer than a third crew, working alone, to do the same job. When the three crews work together, they can do the job in 5 hours. Find the time it takes the slowest crew, working alone, to do the job.

15. A grocer bought some boxes of rice for $27. Two weeks later the price increased 5 cents a box and the number of boxes he could buy for $27 was 6 less than before. Find the original number of boxes he bought.

    Let x = original number of boxes. Then $\dfrac{2700}{x}$ = original cost per box in cents.

    $$(x - 6)\left(\frac{2700}{x} + 5\right) = 2700$$

16. A group of students want to rent a house which rents for $360 a month. They find that if two more people rent the house with them, the monthly share of the rent for each will be reduced by $15. Find the original number of persons in the group.

17. A grocer bought some heads of lettuce for $16. After he had sold 30 heads at twice the price per head as he had paid, he reduced the price for the remaining lettuce by 15 cents a head. After he sold all of the lettuce, he found that he made a profit of $8.50 on all of the lettuce he sold. Find the original price he charged per head.

18. A realtor intends to subdivide a piece of land into a certain number of lots and then sell each lot for a total sale of $45,000. However, if he reduces the number of lots by 1 and increases the price per lot by $1000, the total sale would be $48,000. Find the original number of lots he intended to have in the subdivision.

19. An architect uses the formula $s^2 = 8rh - 4h^2$ to find the height h of a circular arch whose radius r is 30 feet and whose span s is 50

feet. Solve $2500 = 240h - 4h^2$ for h. The arch is less than a semi-circle ($h < 30$).

20. If an object is shot vertically upward in the air, then its height h in feet above the starting point is given by
$$h = rt - 16t^2$$
where r is the speed in feet per second with which the object is shot and t is the time in seconds after the object is shot. For a rocket fired at 96 feet per second, this equation becomes
$$h = 96t - 16t^2$$
    (a) Find the time t when the rocket is 128 feet above the starting point ($h = 128$). Explain the two solutions.
    (b) Find the time t when the rocket hits the ground ($h = 0$).

21. Use the equation below to find the hydrogen-ion concentration x in moles per liter for a 0.0010 molar solution of formic acid. The ionization constant of formic acid is 0.00018.
$$\frac{x^2}{0.0010 - x} = 0.00018$$

22. Find the current x in amps for a certain electronic circuit that has an output of 605 watts of power, a line voltage of 110 volts, and a resistance of 5 ohms. Use $605 = 110x - 5x^2$.

C. INEQUALITIES

1. When two resistors having resistances of x ohms and y ohms, respectively, are connected in parallel, the resulting resistance R is given by
$$R = \frac{xy}{x + y}$$
If $y = 6$ ohms, what are the possible values for x so that $R \geq 2$ ohms and $R \leq 5$ ohms? Solve
$$2 \leq \frac{6x}{x + 6} \leq 5$$

2. The temperature F in degrees Fahrenheit is related to the temperature C in degrees Celsius by
$$F = \frac{9C + 160}{5}$$
What is the range of Celsius temperatures when $50 \leq F \leq 95$? Solve
$$50 \leq \frac{9C + 160}{5} \leq 95$$

3. If the Fahrenheit temperature F, where $F = 1.8C + 32$, is approximated from the Celsius temperature C by the formula $F = 2C + 30$, for what values of C is the error less than $4°F$? Solve
$$|2C + 30 - (1.8C + 32)| < 4$$

4. Using the results of Problem 3 and the formula $C = \frac{5}{9}(F - 32)$, find the corresponding values of F for which the error is less than $4°F$.

5. How accurately must a 2-inch side of a square be measured so that its perimeter is correct to within $\frac{1}{4}$ inch? Let $x =$ error in the measurement of a side. Then

$|4(2 + x) - 8| \leq \frac{1}{4}$

Solve for x.

6. The grades of a student on 4 tests were 72, 85, 76, and 81. In what range must his grade on the 5th test be so his average will be 80 or higher? The maximum grade is 100. Solve

$$\frac{x + 72 + 85 + 76 + 81}{5} \geq 80$$

7. Each side of a rectangular 30×20-meter plot of land is to be lengthened by x meters so the resulting rectangular area is at least twice as large as the original area. Find the possible values for x. (Note $x > 0$.) Solve $(30 + x)(20 + x) \geq 1200$.

8. A nursery manager wants to buy x trees at \$8 each and three times as many shrubs at \$4 each and keep his cost between \$400 and \$450. How many trees can he buy? Solve $400 \leq 8x + 4(3x) \leq 450$.

9. A statistical study of traffic flow through a certain tunnel produced the empirical equation

$n = 6(50v - v^2)$

where n is the number of vehicles per hour and v is the average speed in mph maintained in the tunnel. Find the values of v so that n is at least 3150. Solve $6(50v - v^2) \geq 3150$.

3.7 CHAPTER REVIEW

(1–2) Solve for x. (3.1)

1. $9x - 5(x + 2) = 16 - (x - 4)$
2. $(x + 4)^2 + 15 = (x + 3)(x - 3)$
3. Solve $6x - 5y = 30$ for x and for y. (3.1)
4. Solve $T = mg - mf$ for f. (3.1)

(5–6) State the solution set. (3.1)

5. $8 - 5x \leq 2(x - 3)$ 6. $(x + 6)(x - 2) > x(x + 8)$

(7–12) Solve. (3.2)

7. $(x - 3)^2 = 16$ 8. $x^2 = x + 20$
9. $x^3 - 2x^2 - 49x + 98 = 0$ 10. $x^4 = 200(x^2 - 50)$
11. $\dfrac{1}{x + 5} - \dfrac{1}{4x - 20} = \dfrac{2}{x^2 - 25}$
12. $\dfrac{y}{4y - 24} + \dfrac{y - 18}{y^2 - 4y - 12} = \dfrac{3y + 14}{4y + 8}$

(13–20) Solve. (3.3)

13. $|6x - 10| = 20$ **14.** $18 - |3 - 2y| = 1$
15. $|3x + 7| \leq 2$ **16.** $|15 - 5t| > 20$
17. $4|-x - 2| \geq |-20|$ **18.** $12 - |x - 6| > 0$
19. $|4x - 9| < -2$ **20.** $|4x - 9| \geq -2$

(21–22) Solve by completing the square. Check. (3.4)

21. $x^2 + 12x + 26 = 0$ **22.** $x^2 - 8x + 20 = 0$

(23–24) Solve by using the quadratic formula. Check by using the sum and product of roots theorem. (3.4)

23. $3x^2 + 2x + 2 = 0$ **24.** $2x^2 = 8x - 5$

(25–26) Solve and check. (3.4)

25. $x^4 = 2(x^2 + 12)$ **26.** $x^6 - 16x^3 + 64 = 0$

(27–32) Solve and check. (3.5)

27. $x^2 < 3x + 10$ **28.** $x^2 + 10x + 24 \geq 0$

29. $\dfrac{x - 1}{x + 2} \leq 2$ **30.** $\dfrac{x + 3}{10} > \dfrac{1}{x}$

31. $\left|\dfrac{6}{x - 4}\right| > 1$ **32.** $\left|\dfrac{x}{x - 2}\right| \leq 1$

FUNCTIONS AND RELATIONS

The function and the relation are very important concepts in mathematics. They play a key role both in the development of mathematical topics and also in applications.

In this chapter we will study linear and quadratic relations and functions. We shall also see how these can be interpreted geometrically by using a rectangular coordinate system.

4.1   GENERAL TOPICS

DEFINITION

An **ordered pair** of real numbers a and b is an expression having the form (a, b) where a is called the **first component** and b is called the **second component**.

As examples, $(3, 5)$, $(-2, 1.5)$, and $(\sqrt{5}, 2)$ are ordered pairs of real numbers.

DEFINITION

A **solution of an equation in two variables,** x and y, is an ordered pair of numbers (a, b) such that the equation becomes a true statement for $x = a$ and $y = b$.

For example, $(4, 6)$ is a solution of $3x + y = 18$.
For $x = 4$ and $y = 6$, $3(4) + 6 = 12 + 6 = 18$.
Note that $(6, 4)$ is not a solution of $3x + y = 18$.
$3(6) + 4 = 22$ and $22 \neq 18$
In other words, $(4, 6) \neq (6, 4)$. The order in which the numbers are written is important.

DEFINITION

Two ordered pairs are equal if and only if their first components are equal and their second components are equal.
In symbols, $(a, b) = (c, d)$ if and only if $a = c$ and $b = \text{d}$.

A one-to-one correspondence between the set of ordered pairs of real numbers and the set of points on a plane can be established by constructing a **rectangular coordinate system.**

We select two perpendicular number lines, a horizontal line with its positive direction to the right and a vertical line with its positive direction upward, intersecting at their zero-points. The horizontal line is called the **x-axis,** the vertical line is called the **y-axis,** and the point of intersection of the two lines is called the **origin.**

The origin is assigned the ordered pair $(0, 0)$.

A point on the x-axis is assigned an ordered pair of the form $(a, 0)$ and a point on the y-axis is assigned an ordered pair of the form $(0, b)$.

In general, a point P is assigned the ordered pair (a, b) if and only if a vertical line through P intersects the x-axis at $(a, 0)$ and a horizontal line through P intersects the y-axis at $(0, b)$.

For a point designated as $P(a, b)$, the numbers a and b are called the **coordinates** of point P with a called the **abscissa** or **x-coordinate** and with b called the **ordinate** or **y-coordinate.** Frequently we say "the point (a, b)" when we mean "the point whose coordinates are (a, b)."

The x- and y-axes divide the plane into four regions called **quadrants.** The quadrants are numbered using Roman numerals in the counterclockwise direction. Quadrant 1 is the upper right region where the x- and y-coordinates are both positive.

Figure 4.1 shows a rectangular coordinate system with a point graphed in each quadrant.

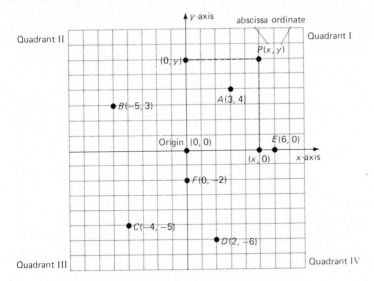

FIG. 4.1. Rectangular coordinate system.

DEFINITION

A **relation** is a set of ordered pairs.

A particular relation may be specified by a list of the ordered pairs that belong to the relation or by an equation or inequality that determines membership in the set. The following sets are examples of relations.

$A = \{(1, 4), (2, 9), (3, 16)\}$

$B = \{(0, 0), (1, 1), (1, -1), (4, 2), (4, -2), \ldots\}$

$C = \left\{(x, y)\, | \, y = \dfrac{1}{x - 2}\right\}$

$D = \{(x, y)\, | \, y = \sqrt{x - 3}\}$

DEFINITION

The **domain** of a relation is the set of its first components and the **range** of a relation is the set of its second components.

When the domain is not stated, it is understood to be the set of real numbers such that the defining statement of the relation produces a real number in the range for each number in the domain.

EXAMPLE 1 State the domain and range of relations A, B, C, and D.

Solution

A. The domain is $\{1, 2, 3\}$.
   The range is $\{4, 9, 16\}$.

B. The domain is $\{0, 1, 4, \ldots\}$.
   The range is $\{0, 1, -1, 2, -2, \ldots\}$.

C. The equation $y = \dfrac{1}{x - 2}$ defines a real number for all values of x except $x = 2$. The domain is the set of real numbers with 2 excluded; R, $x \neq 2$. To find the range, solve the equation $y = \dfrac{1}{x - 2}$ for x and note any restrictions on y.

$$xy - 2y = 1$$
$$xy = 2y + 1$$
$$x = \frac{2y + 1}{y}$$

All values are possible for y except $y = 0$. The range is the set of real numbers with 0 excluded; R, $y \neq 0$.

D. The expression $\sqrt{x - 3}$ designates a positive real number provided $x - 3$ is not negative. If $x - 3 \geq 0$, then $x \geq 3$. The domain is $\{x \mid x \geq 3\}$. The range is $\{y \mid y \geq 0\}$.

DEFINITION

A **function** is a relation with the property that each number in the domain is paired with exactly one number in the range.

EXAMPLE 2 Which of the following are functions?

(a) $\{(2, 6), (3, 8), (4, 10), (5, 12)\}$
(b) $\{(4, 2), (4, -2), (9, 3), (9, -3), (0, 0)\}$
(c) $\{(x, y) \mid y = 2x - 8\}$
(d) $\{(x, y) \mid x = y^2\}$

Solution

(a) A function since each first component is paired with exactly one second component.

(b) Not a function because there are at least two pairs that have the same first component and different second components. For example, $(4, 2)$ and $(4, -2)$ have the same first component 4.

(c) A function since the indicated operations on x (multiply by 2 and subtract 8) produce exactly one number y for each real value for x.

(d) Not a function since two pairs can be found

having the same first components and different second components; namely, $(9, 3)$ and $(9, -3)$.

Also, solving for y, $y = \pm\sqrt{x}$ and for each positive value of x, there are two values for y.

A letter such as f, or g, or h is often used as the name of a function. The notation $f(x)$, which is read "f of x," designates the unique value of y with which x is paired; and $y = f(x)$.

EXAMPLE 3   Given $f = \{(x, y) | y = x^2 - 2x\}$.
Find (a) $f(x)$, (b) $f(3)$, (c) $f(-4)$, (d) $f(t)$, (e) $f(t + 1)$.

Solution

(a) Since $y = f(x)$,
$$f(x) = x^2 - 2x$$
$$f(\ ) = (\ )^2 - 2(\ )$$
(b) $f(3) = (3)^2 - 2(3) = 9 - 6 = 3$
(c) $f(-4) = (-4)^2 - 2(-4) = 16 + 8 = 24$
(d) $f(t) = t^2 - 2t$
(e) $f(t + 1) = (t + 1)^2 - 2(t + 1) = t^2 - 1$

DEFINITION

The **graph of a relation** is the set of points that are the graphs of the ordered pairs belonging to the relation.

When a relation is defined by an equation or inequality, the graph is the graph of the solutions of the equation or inequality.

When a relation is a function, any vertical line will intersect the graph of the function in exactly one point or in no point. If a vertical line intersects the graph of a relation in two or more points, the relation is not a function.

EXAMPLE 4   Use the vertical line test to determine which of the graphs in Figure 4.2 is the graph of a function.

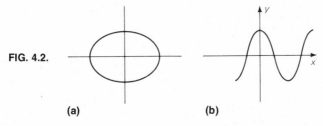

FIG. 4.2.

(a)   (b)

Solution

(a) Not a function because there are many vertical lines that intersect the graph in two points.

(b) A function because each vertical line intersects the graph in at most one point.

EXERCISES 4.1

(1–8) (a) Graph each of the following.
(b) Then join the points in alphabetical order with line segments. (Join A to B, B to C, C to D, and so on.)
(c) Use the vertical line test to determine whether the graph is the graph of a function. See Figure 4.1 and Example 4.

1. $A(-2, 10)$, $B(0, 6)$, $C(1, 4)$, $D(3, 0)$, $E(5, 4)$, $F(8, 10)$
2. $A(-4, 8)$, $B(-3, 4.5)$, $C(-2, 2)$, $D(0, 0)$, $E(2, 2)$, $F(3, 4.5)$, $G(4, 8)$
3. $A(5, 0)$, $B(4, 3)$, $C(0, 5)$, $D(-3, 4)$, $E(-5, 0)$, $F(-4, -3)$, $G(0, -5)$, $H(3, -4)$
4. $A(2, 0)$, $B(1, 1)$, $C(0, 2)$, $D(-1, 1)$, $E(-2, 0)$, $F(-1, -1)$, $G(0, -2)$, $H(1, -1)$
5. $A(\frac{1}{2}, -12)$, $B(1, -6)$, $C(2, -3)$, $D(3, -2)$, $E(6, -1)$, $F(12, -\frac{1}{2})$
6. $A(-7, 4)$, $B(0, 3)$, $C(5, 2)$, $D(9, 0)$, $E(5, -2)$, $F(0, -3)$, $G(-7, -4)$
7. $A(7, -7)$, $B(3, -5)$, $C(0, -3)$, $D(-2, 0)$, $E(0, 3)$, $F(3, 5)$, $G(7, 7)$
8. $A(-3, -9)$, $B(-2, -1)$, $C(-1, 1)$, $D(0, 0)$, $E(1, -1)$, $F(2, 1)$, $G(3, 9)$

(9–24) State the domain, d, and range, r, of each relation and determine whether the relation is a function. See Examples 1 and 2.

9. $\{(5, 2), (6, 2), (7, 1), (8, 1)\}$
10. $\{(2, 3), (2, 4), (3, 5), (4, 7)\}$
11. $\{(0, 0), (1, 1), (8, 2), (1, -1), (8, -2)\}$
12. $\{(1, 5), (2, 5), (3, 5), (4, 5), (5, 5)\}$
13. $\{x, y) \mid 2x + y = 8\}$ **14.** $\{(x, y) \mid y = |x|\}$
15. $\{(x, y) \mid y = x^2 - 1\}$ **16.** $\{(x, y) \mid y^2 = x - 1\}$
17. $\left\{(x, y) \mid y^2 = \dfrac{1}{x}\right\}$ **18.** $\left\{(x, y) \mid y = \dfrac{2}{x - 4}\right\}$
19. $\{(x, y) \mid y = \sqrt{6 - x}\}$ **20.** $\{(x, y) \mid x = \sqrt{y^2 - 9}\}$
21. $\{(x, y) \mid y = 4\}$ **22.** $\{(x, y) \mid x = 3\}$
23. $\{(x, y) \mid x = 0\}$ **24.** $\{(x, y) \mid y = 0\}$

(25–36) Each equation below defines a function f.
(a) Find $f(x)$.
(b) Find $f(5)$.
(c) Find $f(-3)$.
(d) Find $f(0)$.
(e) Find $f(t + 1)$.
(f) Find $f(2c)$.
See Example 3.

25. $y = 10 - 2x$ **26.** $y = x^2 - 5x$
27. $y = |2 - x|$ **28.** $y = \dfrac{6}{x - 3}$
29. $y = \dfrac{x}{x + 5}$ **30.** $y = \sqrt{25 - x^2}$
31. $y = 9 - x^2$ **32.** $y = 4 - |2x - 4|$

33. $y = 2\sqrt{x + 4}$ **34.** $y = x^3 - 3x^2$

35. $3x - y = 12$ **36.** $y = \sqrt[3]{x + 3}$

(37–46) Find the ordered pairs which are solutions of the given equation for the given conditions.

EXAMPLE 5 $x^2 - 2y^2 = 7$

(a) $x = 5$ (b) $y = -1$ (c) $y = 0$ (d) $x = 0$

Solution

(a) Replace x by 5 and solve for y.

$$5^2 - 2y^2 = 7$$
$$-2y^2 = -18$$
$$y^2 = 9 \quad \text{and} \quad y = \pm 3.$$

The solutions are $(5, 3)$ and $(5, -3)$.

(b) Replace y by -1 and solve for x.

$$x^2 - 2(1) = 7$$
$$x^2 = 9 \quad \text{and} \quad x = \pm 3$$

The solutions are $(3, -1)$ and $(-3, -1)$.

(c) For $y = 0$, $x^2 - 2(0) = 7$ and $x^2 = 7$,
$$x = \pm\sqrt{7}.$$

The solutions are $(\sqrt{7}, 0)$ and $(-\sqrt{7}, 0)$.

(d) For $x = 0$, $-2y^2 = 7$ and $y^2 = \dfrac{-7}{2}$ which

has no real solutions. Thus there are no solutions for $x = 0$.

37. $2x - 3y = 12$
  (a) $x = 3$ (b) $y = -4$ (c) $x = 0$
38. $5x + y = 20$
  (a) $x = -2$ (b) $y = 10$ (c) $y = 0$
39. $y = 2|8 - 2x|$
  (a) $x = 5$ (b) $y = 8$ (c) $y = 0$
40. $x = 2 + |y - 3|$
  (a) $x = 6$ (b) $y = 1$ (c) $x = 0$
41. $x^2 - 4x = y$
  (a) $x = -2$ (b) $y = 4$ (c) $y = -4$
42. $x^2 + y^2 = 10$
  (a) $x = -3$ (b) $y = 2$ (c) $y = 0$
43. $x^2 - 4y^2 = 9$
  (a) $x = 5$ (b) $y = -2$ (c) $x = -3$
44. $x = 4y - y^2 - 4$
  (a) $x = -4$ (b) $y = 2$ (c) $x = 0$
45. $xy = 8$
  (a) $x = -2$ (b) $y = 8$ (c) $y = -16$
46. $|x + y| = 5$
  (a) $x = 0$ (b) $y = 4$ (c) $x = -8$

4.2 LINEAR RELATIONS AND FUNCTIONS

DEFINITION

A **linear equation in two variables** x and y is an equation equivalent to $Ax + By = C$, where A, B, and C are real numbers and A and B are not both zero; that is, $A^2 + B^2 \neq 0$.

DEFINITION

A **linear relation** is one that can be expressed as $\{(x, y) \mid Ax + By = C,$ where $A^2 + B^2 \neq 0\}$

In other words, a linear relation is a relation that is defined by a linear equation. Some examples of linear equations are

$2x - 3y = 6$		(1)
$x + 2y = 0;$	note $C = 0$	(2)
$x = 5;$	note $B = 0$	(3)
$y = -6;$	note $A = 0$	(4)

In analytic geometry it is proved that the graph of a linear relation (or linear equation) is a straight line; thus the description "linear."

If $B \neq 0$, then the equation $Ax + By = C$ defines a **linear function.** Solving for y,

$$y = \frac{-A}{B}x + \frac{C}{B}$$

Equations (1), (2), and (4) define linear functions while Equation (3) does not.

To graph a linear equation, we find two or more solutions and graph these ordered pairs. Then we join the plotted points using a straight edge (or ruler) and extend the line through the points.

Convenient points to use are those for which $x = 0$ or $y = 0$. These are called **intercept points.**

DEFINITION

For the graph of an equation in two variables, a is an **x-intercept** if and only if $(a, 0)$ is on the graph, and b is a **y-intercept** if and only if $(0, b)$ is on the graph.

EXAMPLE 1 Graph $2x - 3y = 6$.

Solution   Replace x or y by any value and solve the equation for the value of the other variable. Selecting $x = 0$, $y = 0$, and $x = 6$, we obtain the following table of ordered pairs. The graph is shown in Figure 4.3.

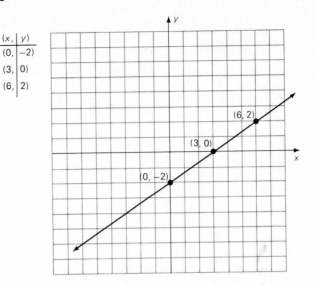

(x,	y)
(0,	−2)
(3,	0)
(6,	2)

FIG. 4.3. Graph of 2x − 3y = 6.

EXAMPLE 2 Graph x + 2y = 0.

Solution Selecting x = 0, we obtain y = 0 and the origin (0, 0) is on the graph. There is only one intercept point, so we select x = 4 and x = 6. The graph is shown in Figure 4.4.

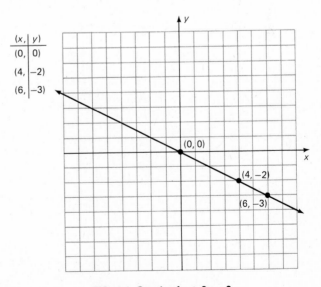

(x,	y)
(0,	0)
(4,	−2)
(6,	−3)

FIG. 4.4. Graph of x + 2y = 0.

EXAMPLE 3 Graph $x = 5$.

Solution Only y can be selected arbitrarily since x must be 5 for all ordered pairs. The table and graph are shown in Figure 4.5.

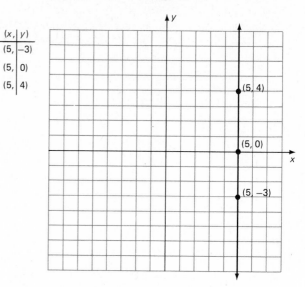

$(x,$	$y)$
$(5,$	$-3)$
$(5,$	$0)$
$(5,$	$4)$

FIG. 4.5. Graph of $x = 5$.

EXAMPLE 4 Graph $y = -6$.

Solution Only x can be selected arbitrarily since y must be -6 for all ordered pairs. The table and graph are shown in Figure 4.6.

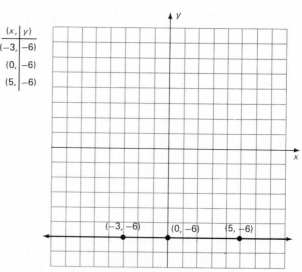

$(x,$	$y)$
$(-3,$	$-6)$
$(0,$	$-6)$
$(5,$	$-6)$

FIG. 4.6. Graph of $y = -6$.

In general, the graph of $x = a$ is a vertical line passing through $(a, 0)$ and $y = b$ is a horizontal line passing through $(0, b)$.

An important feature of a line that is the graph of a function is its **slope**. The slope indicates how steep the line is and whether the line rises or falls to the right. The slope of a line is defined as the difference in the y-coordinates of two points on the line (the rise) divided by the difference of the x-coordinates (the run). See Figure 4.7.

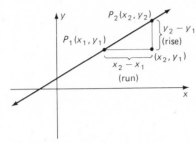

FIG. 4.7. Slope $m = \dfrac{y_2 - y_1}{x_2 - x_1}$.

DEFINITION

For $x_1 \neq x_2$ the **slope** m of a line passing through points $P_1(x_1, y_1)$ and $P_2(x_2, y_2)$ is defined by

$$m = \frac{y_2 - y_1}{x_2 - x_1}$$

Note that

$$\frac{y_2 - y_1}{x_2 - x_1} = \frac{y_1 - y_2}{x_1 - x_2}$$

and thus it makes no difference which point is called the first point P_1 and which point is called the second point P_2.

Furthermore the same value is obtained for m no matter what two points are selected on the line. This can be verified by considering properties of similar triangles.

EXAMPLE 5 Find the slope of the graph of $2x - 3y + 12 = 0$ by using:
(a) the points $A(3, 6)$ and $B(0, 4)$ on the line.
(b) the points $C(-3, 2)$ and $D(6, 8)$ on the line.

Solution

(a) Choose
$$P_1 = A(3, 6)$$
$$P_2 = B(0, 4)$$
$$m = \frac{4 - 6}{0 - 3} = \frac{-2}{-3} = \frac{2}{3}$$
$$m = \frac{2}{3}$$

(b) Choose
$$P_1 = C(-3, 2)$$
$$P_2 = D(6, 8)$$
$$m = \frac{8 - 2}{6 - (-3)} = \frac{6}{9} = \frac{2}{3}$$
$$m = \frac{2}{3}$$

Now let (x, y) be a general point on the line through distinct points $P_1(x_1, y_1)$ and $P_2(x_2, y_2)$. Since the slope of a line is independent of the choice of the points, equating slopes,

$$\frac{y - y_1}{x - x_1} = \frac{y_2 - y_1}{x_2 - x_1}, \quad \text{for } x_1 \neq x_2$$

and

$$\frac{y - y_1}{x - x_1} = m$$

Multiplying each side by $x - x_1$,

$$y - y_1 = m(x - x_1) \tag{5}$$

Equation (5) is called the **point-slope form** of a line. The form $Ax + By + C = 0$ is called the **standard form** or **general form.**

The point-slope form of a line is useful for determining the equation of a line as shown in Examples 6 and 7.

EXAMPLE 6 Find the standard form of the equation of the line through $(5, 2)$ and $(3, 6)$.

Solution Finding the slope,

$$m = \frac{6 - 2}{3 - 5} = \frac{4}{-2} = -2$$

Using the point-slope form with $(x_1, y_1) = (5, 2)$,

$$y - y_1 = m(x - x_1)$$
$$y - 2 = -2(x - 5)$$
$$y - 2 = -2x + 10$$
$$2x + y - 12 = 0, \quad \text{standard form}$$

EXAMPLE 7 Find the standard form of the equation of the line having slope $\frac{1}{3}$ and passing through the point $(-2, 4)$.

Solution Using the point-slope form with $m = \frac{1}{3}$ and $(x_1, y_1) = (-2, 4)$,

$$y - y_1 = m(x - x_1)$$
$$y - 4 = \frac{1}{3}[x - (-2)]$$
$$3y - 12 = x + 2$$
$$-x + 3y - 14 = 0$$

and

$$x - 3y + 14 = 0, \quad \text{preferred standard form}$$

While $-x + 3y - 14 = 0$ is in standard form, to avoid errors in later calculations, we prefer not to have the equation begin with a minus sign.

EXAMPLE 8 Find an equation of a line having a slope of 4 and a y-intercept of 6.

Solution Since the y-intercept is 6, the point $(0, 6)$ is on the line.

$$m = 4 \quad \text{and} \quad (x_1, y_1) = (0, 6)$$
$$y - y_1 = m(x - x_1)$$
$$y - 6 = 4(x - 0)$$
$$y = 4x + 6$$

Note in Example 8 that the coefficient of x is the slope and the constant term is the y-intercept. This property is true in general for a line with slope m and y-intercept b.

$$y - b = m(x - 0)$$
$$y = mx + b \qquad\qquad\qquad (6)$$

Equation (6) is called the **slope-intercept form** of a line. It is useful for finding the slope and y-intercept of a line when its equation is given, and for determining the functional value y when x is given.

EXAMPLE 9 Find the slope and y-intercept of the line
$$2x + 5y = 10$$

Solution   Solving for y,
$$5y = -2x + 10$$
$$y = \frac{-2}{5}x + 2$$

Comparing,
$$y = mx + b$$
$$m = -\tfrac{2}{5} \quad \text{and} \quad b = 2$$

The slope, the coefficient of x, is $-\tfrac{2}{5}$ and the y-intercept, the constant, is 2.

It should be noted that a vertical line such as $x = a$ has **no slope**; that is, the **slope is undefined.**

A horizontal line such as $y = b$ has a **zero slope,** $m = 0$. All other lines have nonzero slopes.

Two nonvertical lines are parallel if and only if they have the same slope.

All vertical lines are parallel and all horizontal lines are parallel.

Two lines that are not parallel intersect in exactly one point.

EXERCISES 4.2

(1–12)   Graph each of the following. See Examples 1, 2, 3, and 4.

1. $x + 4y = 8$ 2. $5x + 2y = 10$
3. $3x - 5y = 15$ 4. $2x - y = 12$
5. $x + y = 0$ 6. $x - y = 0$
7. $2x - y = 0$ 8. $3x + 2y = 0$
9. $x = 4$ 10. $y = 3$
11. $y = -2$ 12. $x = -6$

(13–22)   Find the slope of the line through the two given points. Sketch the line. See Example 5.

13. $(6, 5)$ and $(2, 8)$ 14. $(5, 3)$ and $(-5, 7)$
15. $(-4, 9)$ and $(-2, 3)$ 16. $(9, 7)$ and $(-6, 1)$
17. $(-5, -3)$ and $(4, 1)$ 18. $(1, -6)$ and $(-1, 4)$

19. $(3, 8)$ and $(-5, -2)$ **20.** $(-2, 2)$ and $(-3, 8)$
21. $(-6, -4)$ and $(6, -4)$ **22.** $(2, 0)$ and $(0, 1)$

(23–30) Find the standard form of the equation of the line passing through the given points. See Example 6.

23. $(3, 4)$ and $(0, 7)$ **24.** $(2, -3)$ and $(-2, 5)$
25. $(-4, 1)$ and $(2, -2)$ **26.** $(-3, 4)$ and $(5, 4)$
27. $(6, -2)$ and $(4, -2)$ **28.** $(3, 5)$ and $(0, 0)$
29. $(-2, -4)$ and $(3, 6)$ **30.** $(0, -5)$ and $(4, 0)$

(31–38) Find the standard form of the equation of the line passing through the given point and having the given slope. See Example 7.

31. $(5, 2)$, $m = 3$ **32.** $(4, -1)$, $m = -2$
33. $(-6, 2)$, $m = \frac{-1}{2}$ **34.** $(-3, -5)$, $m = \frac{2}{3}$
35. $(0, 0)$, $m = \frac{5}{4}$ **36.** $(-2, 0)$, $m = 1$
37. $(0, -3)$, $m = 0$ **38.** $(0, 0)$, $m = \frac{-3}{4}$

(39–50) Find the slope and y-intercept of the line whose equation is given. See Example 9.

39. $3x + 5y = 30$ **40.** $2x - 6y = 18$
41. $x - 4y = 16$ **42.** $2x + 3y + 6 = 0$
43. $5x + 2y + 20 = 0$ **44.** $x - 2y + 8 = 0$
45. $y + 8 = 0$ **46.** $y - 7 = 0$
47. $3x - y = 0$ **48.** $5x + 3y = 0$
49. $x + y + 1 = 0$ **50.** $x - y + 5 = 0$

(51–60) Determine if the given pair of lines are parallel or if they intersect.

EXAMPLE 10
(a) $x + 2y = 10$
   $x - 2y + 10 = 0$
(b) $2x - y = 6$
   $6x - 3y = 2$

Solution
(a) Writing each equation in the slope-intercept form,
$$x + 2y = 10 \rightarrow y = \frac{-1}{2}x + 5 \quad \text{and} \quad m_1 = \frac{-1}{2}$$
$$x - 2y + 10 = 0 \rightarrow y = \frac{1}{2}x + 5 \quad \text{and} \quad m_2 = \frac{1}{2}$$
Comparing slopes, $m_1 = \frac{-1}{2}$ and $m_2 = \frac{1}{2}$. Since the slopes are different, the lines intersect.
(b) Writing in slope-intercept form,
$$2x - y = 6 \rightarrow y = 2x - 6 \quad \text{and} \quad m_1 = 2$$
$$6x - 3y = 2 \rightarrow y = 2x - \frac{2}{3} \quad \text{and} \quad m_2 = 2$$
Since the slopes are equal, $m_1 = m_2 = 2$, the lines are parallel.

51. $x + 3y = 9$
  $3x - y = 9$

52. $2x = 3y + 6$
  $3y = 2x - 12$

53. $4x - 4y = 5$
  $3x - 3y = 2$

54. $x - 5y = 10$
  $5x + y = 10$

55. $2x = 3y$
  $3x + 2y = 0$

56. $x - y = 10$
    $y = x - 5$

57. $x + y = 10$
  $x - y = 4$

58. $5x = 2y$
  $5x - 2y = 10$

59. $2y = 5x - 10$
  $5x = 2y - 10$

60. $x + 4y = 8$
  $4x - y = 8$

(61–70) In analytic geometry, it is shown that two nonvertical lines are perpendicular if and only if the product of their slopes is -1; $m_1 m_2 = -1$. Using this fact, compare the slopes of each pair of lines in Exercises 51–60 and determine which lines are perpendicular.

(71–78) Write the standard form of the equation of each line meeting the stated conditions.

71. Through $(5, 2)$ and parallel to $2x - y = 4$.
72. Through $(-1, 3)$ and parallel to $x + 3y = 6$.
73. Through $(-5, 4)$ and perpendicular to $x + y = 6$.
74. Through $(2, -6)$ and perpendicular to $3x - 2y = 1$.
75. Having a y-intercept of 6 and parallel to $4x + y = 8$.
76. Having a y-intercept of -4 and perpendicular to $2x + 5y = 10$.
77. Having an x-intercept of 4 and a y-intercept of 5.
78. Having an x-intercept of a and a y-intercept of b.
79. Show that $\dfrac{x}{a} + \dfrac{y}{b} = 1$ is an equation of the line whose x-intercept is

a and whose y-intercept is b where $a \neq 0$ and $b \neq 0$. (This equation is called the "intercept form.")

80. Show that, for $a \neq 0$ and $b \neq 0$, the lines

$$\frac{x}{a} + \frac{y}{b} = 1 \quad \text{and} \quad \frac{x}{b} - \frac{y}{a} = 1$$

are perpendicular.

81. Show that $a_1 x + b_1 y = c_1$ and $a_2 x + b_2 y = c_2$ are parallel if and only if $a_1 b_2 = a_2 b_1$.
82. Show that $a_1 x + b_1 y = c_1$ and $a_2 x + b_2 y = c_2$ are perpendicular if and only if $a_1 a_2 = -b_1 b_2$.

4.3 QUADRATIC FUNCTIONS

DEFINITION

A **quadratic function** is a relation whose defining equation is equivalent to

$y = ax^2 + bx + c,$ where $a \neq 0$

Although a quadratic function f is the set of ordered pairs,
$$f = \{(x, y) \mid y = ax^2 + bx + c, \text{ where } a \neq 0\}$$
for convenience we say "the function $y = ax^2 + bx + c$." In general, the defining equation or $f(x)$ is referred to as the function.

The simplest quadratic functions are $y = x^2$ and $y = -x^2$. By selecting solutions of these equations, graphing the ordered pairs, and joining the points with a smooth curve, we obtain the curves shown in Figure 4.8.

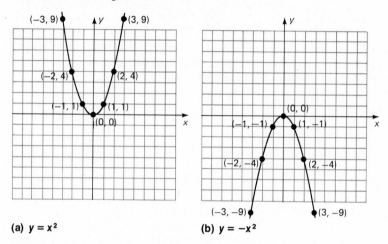

(a) $y = x^2$ **(b)** $y = -x^2$

FIG. 4.8. Parabolas, vertex at (0, 0).

The graph of a quadratic function is called a **parabola.** Note that for $y = x^2$, $a = 1$ and the parabola opens upward. For $y = -x^2$, $a = -1$ and the parabola opens downward. In general:

    If $a > 0$, the parabola opens upward.

    If $a < 0$, the parabola opens downward.

An important feature of a parabola is its turning point, called the **vertex.** If $a > 0$, the vertex is the **minimum point,** the point where y has its least value. If $a < 0$, the vertex is the **maximum point,** the point where y has its greatest value. A parabola also has a line of symmetry, called its **axis.** For the graph of a quadratic function, the axis is the vertical line passing through the vertex.

To graph a quadratic function it is convenient to locate the vertex of the parabola. This can be done by completing the square with respect to the x-terms.

$$y = ax^2 + bx + c$$
$$ax^2 + bx = y - c$$
$$a\left[x^2 + \frac{b}{a}x + \left(\frac{b}{2a}\right)^2\right] = y - c + \frac{b^2}{4a}$$
$$a\left(x + \frac{b}{2a}\right)^2 = y - \frac{4ac - b^2}{4a}$$

Now y will have its least value for a positive, and its greatest value for a negative, when the square is zero.

If $\left(x + \dfrac{b}{2a}\right)^2 = 0$, then $x = \dfrac{-b}{2a}$ and $y = \dfrac{4ac - b^2}{4a}$.

THEOREM

The coordinates of the vertex of the parabola $y = ax^2 + bx + c$ are
$$\left(\frac{-b}{2a}, \frac{4ac - b^2}{4a}\right)$$

THEOREM

The coordinates of the vertex of the parabola $a(x - h)^2 = y - k$ are (h, k).

EXAMPLE 1 Graph $y = 2x^2 - 6x - 1$.

Solution   First, find the vertex by (a) completing the square or (b) using the theorem.

(a) Completing the square,
$$2x^2 - 6x = y + 1$$
$$2\left(x^2 - 3x + \tfrac{9}{4}\right) = y + 1 + \tfrac{9}{2}$$
$$2\left(x - \tfrac{3}{2}\right)^2 = y + \tfrac{11}{2}$$
$$\left(x - \tfrac{3}{2}\right)^2 = 0 \quad \text{and} \quad y + \tfrac{11}{2} = 0$$
$$x = \tfrac{3}{2} \quad \text{and} \quad y = \tfrac{-11}{2}$$

The vertex is $\left(\tfrac{3}{2}, \tfrac{-11}{2}\right)$. The axis is $x = \tfrac{3}{2}$.

(b) Using the formula, for $y = 2x^2 - 6x - 1$,
$$a = 2,\ b = -6,\ \text{and}\ c = -1,$$
$$x = \frac{-b}{2a} = \frac{-(-6)}{2(2)} = \frac{3}{2} \quad \text{and} \quad y = \frac{4(2)(-1) - (36)}{4(2)} = \frac{-44}{8}$$
$$= \frac{-11}{2}$$

Now selecting values for x less than $\tfrac{3}{2}$ and greater than $\tfrac{3}{2}$, we form the following table of ordered pairs.

x	-1	0	1	$\tfrac{3}{2}$	2	3	4
y	7	-1	-5	$\tfrac{-11}{2}$	-5	-1	7

↑
vertex

Plotting these points and joining them with a smooth curve, we obtain the graph shown in Figure 4.9.

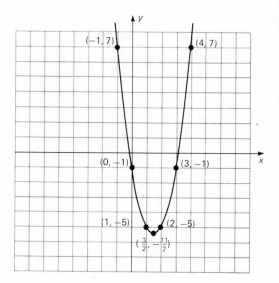

FIG. 4.9. Graph of $y = 2x^2 - 6x - 1$.

EXAMPLE 2 Graph $y = 4x - x^2$.

Solution Completing the square with respect to the x-terms,

$$-(x^2 - 4x\quad) = y$$
$$-(x^2 - 4x + 4) = y - 4$$
$$-(x - 2)^2 = y - 4$$
$$(x - 2)^2 = 0 \quad \text{and} \quad y - 4 = 0$$
$$x = 2 \quad \text{and} \quad y = 4$$

The vertex is $(2, 4)$ and the axis is $x = 2$.
The table below gives points near the vertex and the graph is shown in Figure 4.10.

x	-1	0	1	2	3	4	5
y	-5	0	3	4	3	0	-5

↑
vertex

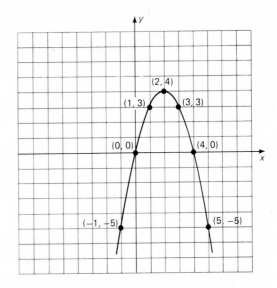

FIG. 4.10. Graph of $y = 4x - x^2$.

The graphs of some parabolas may be readily obtained by plotting the vertex and the intercept points, especially when these are integral. This is illustrated in Example 3.

EXAMPLE 3   Graph $y = x^2 - 2x - 8$ by plotting intercept points.
Solution

1. *Vertex.* Completing the square,
   $$x^2 - 2x + 1 = y + 8 + 1$$
   $$(x - 1)^2 = y + 9$$
   The vertex is $(1, -9)$.
2. *y-intercept.* Let $x = 0$; then $y = -8$.
   The y-intercept point is $(0, -8)$.
3. *x-intercept.* Let $y = 0$. Then
   $$x^2 - 2x - 8 = 0$$
   $$(x + 2)(x - 4) = 0$$
   $$x = -2 \quad \text{or} \quad x = 4$$

The x-intercept points are $(-2, 0)$ and $(4, 0)$. The graph is shown in Figure 4.11.

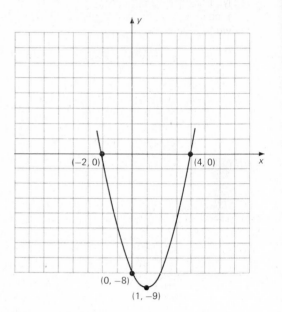

FIG. 4.11. Graph of $y = x^2 - 2x - 8$.

For $y = ax^2 + bx + c$, the x-intercepts are two distinct real numbers when $b^2 - 4ac > 0$. If $b^2 - 4ac = 0$, then there is only one x-intercept and the parabola is tangent to the x-axis. When $b^2 - 4ac < 0$, then the solutions of $ax^2 + bx + c = 0$ are imaginary and the graph of $y = ax^2 + bx + c$ has no x-intercepts. These cases are shown in Figure 4.12.

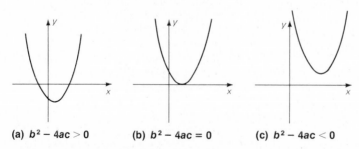

(a) $b^2 - 4ac > 0$ (b) $b^2 - 4ac = 0$ (c) $b^2 - 4ac < 0$

FIG. 4.12. Graphs of $y = ax^2 + bx + c$.

When the x-intercept points are not used to obtain the graph, they can be used to check that the graph has been drawn correctly. From the graph, the x-intercepts can be estimated and then compared with the solutions of $ax^2 + bx + c = 0$. This is shown in Example 4.

EXAMPLE 4 Check the x-intercepts of the graph in Figure 4.9.

Solution

1. From the graph, the x-intercepts are estimated as -0.2 and 3.2.
2. Solving $2x^2 - 6x - 1 = 0$ by using the quadratic formula,
$$x = \frac{6 \pm \sqrt{36 - 4(-2)}}{2(2)} = \frac{3 \pm \sqrt{11}}{2}$$
$$x = \frac{3 - \sqrt{11}}{2} \approx -0.16 \quad \text{or} \quad x = \frac{3 + \sqrt{11}}{2} \approx 3.16$$

The results agree to the nearest tenth.

EXAMPLE 5 An apartment owner has 100 apartments, all rented, at $150 per unit per month. He finds for each $5 monthly increase in the rent, he has two vacant units. How much should the owner charge per month to obtain the maximum income?

Solution   Let
$$x = \text{number of } \$5 \text{ increases in rent}$$
$$150 + 5x = \text{rental price per unit}$$
$$100 - 2x = \text{number of units rented}$$
$$y = \text{monthly income}$$
Then
$$y = (100 - 2x)(150 + 5x)$$
$$y = -10x^2 + 200x + 15{,}000$$
Since the graph of this equation is a parabola opening downward, the maximum value of y will occur at the vertex. Completing the square,
$$-10(x^2 - 20x) = y - 15{,}000$$
$$-10(x^2 - 20x + 10^2) = y - 15{,}000 - 1000$$
$$-10(x - 10)^2 = y - 16{,}000$$
The maximum occurs at $x = 10$. The monthly rental per unit should be
$$150 + 5(10) = 200 \text{ dollars.}$$

EXERCISES 4.3

(1–20) Graph each of the following quadratic functions. State the coordinates of the vertex and the equation of the axis. See Examples 1, 2, and 3.

1. $y = \frac{1}{2}x^2$	**2.** $y = 2x^2$
3. $y = -2x^2$	**4.** $3y = -2x^2$
5. $y = x^2 - 4$	**6.** $y = x^2 + 4$
7. $y = x^2 - 2x$	**8.** $y = x - x^2$
9. $y = 8x - x^2$	**10.** $y = x^2 - 6x$
11. $y = x^2 - 6x + 5$	**12.** $y = x^2 + 3x - 10$
13. $y = 4 - 2x - x^2$	**14.** $y = 3 + 4x - x^2$
15. $y = 2x^2 + 8x - 3$	**16.** $y = 3x^2 - 6x + 2$
17. $2y = 4 + 10x - 5x^2$	**18.** $2y = 3 + 10x - 2x^2$
19. $2y = x^2 - 4x + 4$	**20.** $2y = x^2 + 2x + 1$

(21–26) Check the x-intercepts of the graphs of each of the following. See Example 4.

21. Exercise 13	**22.** Exercise 14
23. Exercise 15	**24.** Exercise 16
25. Exercise 17	**26.** Exercise 18

(27–36) Solve each of the following. See Example 5.

27. A farmer has 480 feet of fencing he wants to use to fence 3 sides of a rectangular area. (The fourth side is along a stream.) Find the dimensions of the rectangle so that the area will be a maximum.

Let
x = width
$480 - 2x$ = length
y = the area
$y = x(480 - 2x)$

Find x for the maximum value of y. Then find the value of $480 - 2x$.

28. An airline offers a group a price of \$600 per person for a certain flight if at least 200 persons take the flight. For every passenger over 200 it will reduce the cost per person by \$2. How many persons should take this flight for the airline to receive its maximum amount? What will be the corresponding cost per person?

Let
x = number of passengers over 200
$200 + x$ = total numbers of passengers
$600 - 2x$ = cost per person
y = amount the airline receives
$y = (200 + x)(600 - 2x)$

Complete the solution.

29. An object is shot vertically upward at an initial speed of 800 feet per second. Its height h at the end to t seconds is given by $h = 800t - 16t^2$. Find the maximum height the object reaches.

30. An artist wants to place metallic stripping along the 4 sides of a rectangular piece of plywood and also down the middle of the area parallel to a width. He wants to use exactly 24 feet of stripping. What dimensions should he use for the rectangle so that the area is a maximum?

Let

$x = $ number of feet in one width
$L = $ number of feet in one length

$3x + 2L = 24$

$$L = \frac{24 - 3x}{2}$$

$y = $ the area

$$y = \frac{x(24 - 3x)}{2}$$

Complete the problem.

31. A manufacturing company finds that its profit, y, in dollars, is related to the number of parts x that it produces by the equation $y = 1600x - 2x^2$. How many parts should the company produce for a maximum profit?

32. A machinist wants to make the T-shape shown in the figure using exactly 20 inches of metal tubing. The T should be formed in such a way that the length c, the hypotenuse of a right triangle, will be a minimum. What lengths should be used for a and b? Note if c^2 is a minimum, c will also be a minimum.

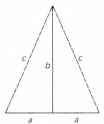

Let

$2a + b = 20$

$b = 20 - 2a$

$y = c^2$

$y = a^2 + (20 - 2a)^2$

Find a and b so y is a minimum.

33. When a boat is at A, another boat is at B, 25 miles due west of A. The boat at A is traveling due west at 15 mph and the boat at B is traveling due south at 20 mph. Find the time when the distance between the boats is the least. Also find this minimum distance. If the square of the distance is a minimum, then so is the distance. Let $y = $ the square of the distance. Then, using the figure,

$y = (20t)^2 + (25 - 15t)^2$

Find t so y is a minimum. Then find the distance.

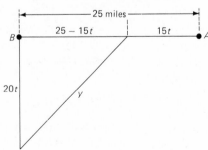

34. A manufacturer finds that the cost C for producing x units of a certain commodity is given by

$C = x^2 - 100x + 2900$

How many units should be produced so the cost is a minimum?

35. Find two numbers x and y so that $x + y = 24$ and $3x^2 + y^2$ is a minimum.

36. A piece of wire 48 inches long is to be formed into a square and into the two equal legs of an isosceles triangle. Find the length of a side of the square and the length of a leg of the triangle so that the sum of the areas of the square and the triangle is a minimum.

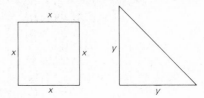

Using the figure,
$4x + 2y = 48$
and
$$y = 24 - 2x$$
Let
A = sum of the areas
Then
$$A = x^2 + \tfrac{1}{2}(24 - 2x)^2$$

4.4 QUADRATIC RELATIONS

DEFINITION

A **quadratic equation in two variables,** x and y, is an equation equivalent to
$$Ax^2 + Bxy + Cy^2 + Dx + Ey + F = 0$$
where $A \neq 0$ or $B \neq 0$ or $C \neq 0$; that is, $A^2 + B^2 + C^2 \neq 0$.

DEFINITION

A **quadratic relation** is a relation defined by a quadratic equation in two variables.

If $P(x, y) = 0$ represents a quadratic equation in two variables, then the corresponding quadratic relation is
$$\{(x, y) \mid P(x, y) = 0\}$$
In analytic geometry it is shown that the graph of a quadratic equation is a parabola, a circle, an ellipse, a hyperbola; or the

graph is degenerate. When it is **degenerate,** the graph is a point, no point, one straight line, or two straight lines.

For example, the graph of $x^2 + y^2 = 0$ is the origin $(0, 0)$.

The graph of $(x - 3)^2 + (y - 5)^2 = 0$ is the point $(3, 5)$.

The graph of $x^2 + y^2 = -4$ is no point (the solution set of $x^2 + y^2 = -4$ is the empty set).

The graph of $x^2 - y^2 = 0$ is the pair of lines $x - y = 0$ and $x + y = 0$.

The graph of $x^2 + 2xy + y^2 = 0$ is the graph of one line $x + y = 0$.

In this section we consider quadratic relations defined by special types of quadratic equations.

In the last section we saw that the graph of $y = ax^2 + bx + c$ was a parabola, opening upward for $a > 0$ and downward for $a < 0$.

The graph of $x = ay^2 + by + c$ is also a parabola, opening to the right for $a > 0$ and to the left for $a < 0$. Its axis (line of symmetry) is a horizontal line with equation $y = \frac{-b}{2a}$. The vertex occurs at the point on the parabola for which the ordinate $y = \frac{-b}{2a}$ and can also be found by completing the square with respect to the y-terms.

EXAMPLE 1   Graph $x = 2y^2 - 2y - 3$.

Solution   First find the vertex by (a) using $y = \frac{-b}{2a}$ or (b) completing the square.

(a) Using $y = \frac{-b}{2a}$, $a = 2$, $b = -2$, $y = \frac{-(-2)}{2(2)} = \frac{1}{2}$.

For $y = \frac{1}{2}$, $x = 2\left(\frac{1}{4}\right) - 2\left(\frac{1}{2}\right) - 3 = -\frac{7}{2}$.

The vertex is $\left(\frac{-7}{2}, \frac{1}{2}\right)$.

(b) Completing the square,
$$2(y^2 - y) = x + 3$$
$$2\left(y^2 - y + \frac{1}{4}\right) = x + 3 + \frac{1}{2}$$
$$2\left(y - \frac{1}{2}\right)^2 = x + \frac{7}{2}$$
$$y - \frac{1}{2} = 0 \quad \text{and} \quad x + \frac{7}{2} = 0$$
$$y = \frac{1}{2} \quad \text{and} \quad x = \frac{-7}{2}$$

The vertex is $\left(\frac{-7}{2}, \frac{1}{2}\right)$.

Now make a table of ordered pairs of solutions, selecting values for y. Graph the points and join them with a smooth curve to form the parabola.

The table is shown below and the graph is shown in Figure 4.13. Note since $a = 2$ and $2 > 0$, the graph opens to the right.

Computed values →	x	9	1	-3	$\frac{-7}{2}$	-3	1	9
Selected values →	y	-2	-1	0	$\frac{1}{2}$	1	2	3

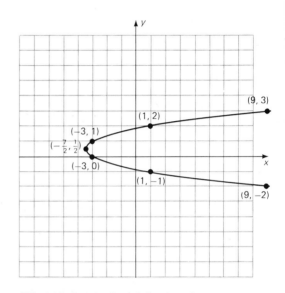

FIG. 4.13. Graph of $x = 2y^2 - 2y - 3$.

EXAMPLE 2 Graph $y^2 + x + 6y + 1 = 0$.

Solution

$x = -y^2 - 6y - 1$

Since $a = -1$ and $-1 < 0$, the parabola opens to the left. Completing the square,

$$y^2 + 6y = -x - 1$$
$$y^2 + 6y + 9 = -x - 1 + 9$$
$$(y + 3)^2 = -x + 8$$
$$y + 3 = 0 \quad \text{and} \quad -x + 8 = 0$$
$$y = -3 \quad \text{and} \qquad x = 8$$

The vertex is $(8, -3)$.

The table is shown below and the graph is shown in Figure 4.14.

Computed values →	x	-8	-1	4	7	8	7	4	-1	-8
Selected values →	y	-7	-6	-5	-4	-3	-2	-1	0	1

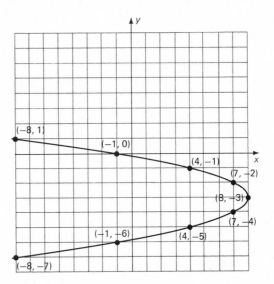

FIG. 4.14. Graph of $y^2 + x + 6y + 1 = 0$.

When the general shape of a curve can be identified from its equation, the graph can be easily sketched by graphing a few ordered pairs that are solutions of the equation. The general graphs of some special types of quadratic equations are shown in Figures 4.15 and 4.16.

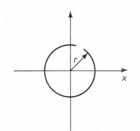

(a) $x^2 + y^2 = r^2$. Circle; center $(0, 0)$, radius r.

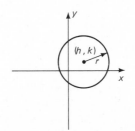

(b) $(x - h)^2 + (y - k)^2 = r^2$. Circle; center (h, k), radius r.

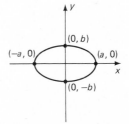

(c) $b^2x^2 + a^2y^2 = a^2b^2$. Ellipse; $a \neq b, ab \neq 0$.

FIG. 4.15. Circles and ellipses.

(a) $xy = k, k > 0$.

(b) $xy = k, k < 0$.

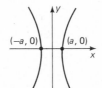

(c) $b^2x^2 - a^2y^2 = a^2b^2$.

(d) $a^2y^2 - b^2x^2 = a^2b^2$.

FIG. 4.16. Hyperbolas.

EXAMPLE 3   Graph $x^2 + y^2 - 6x + 4y - 3 = 0$.

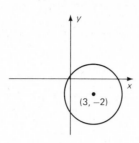

FIG. 4.17

Solution   Comparing with Figures 4.15 and 4.16, the only possibility is Figure 4.15(b). Completing the squares with respect to the x-terms and the y-terms,

$$(x^2 - 6x \quad) + (y^2 + 4y \quad) = 3$$
$$(x^2 - 6x + 9) + (y^2 + 4y + 4) = 3 + 9 + 4$$
$$(x - 3)^2 + (y + 2)^2 = 16$$
$$(x - h)^2 + (y - k)^2 = r^2$$
$$h = 3 \quad \text{and} \quad k = -2 \quad \text{and} \quad r = 4$$

This is a circle with center $(3, -2)$ and with radius $r = 4$. The circle shown in Figure 4.17 is drawn by using a compass with point at $(3, -2)$ and open to a radius of 4.

EXAMPLE 4   Graph $4x^2 + 25y^2 = 100$.

Solution   Comparing with Figures 4.15 and 4.16, the graph can be recognized as an ellipse, with intercepts at $(5, 0)$, $(-5, 0)$, $(0, 2)$, and $(0, -2)$. Solving for y, other ordered pairs shown in the table can be obtained by selecting values for x and computing the corresponding y-values. The graph is shown in Figure 4.18. Note that for y to be a real number, $25 - x^2 \geq 0$ and thus $-5 \leq x \leq 5$.

x	0	± 3	± 4	± 5
y	± 2	$\pm \dfrac{8}{5}$	$\pm \dfrac{6}{5}$	0

$$y = \pm \frac{2}{5}\sqrt{25 - x^2}$$

FIG. 4.18. $4x^2 + 25y^2 = 100$.

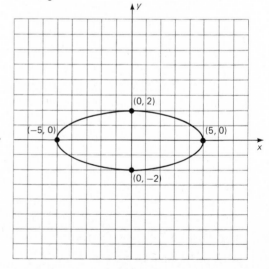

EXAMPLE 5   Graph $4x^2 - y^2 = 36$.

Solution   Noting the minus sign between the x^2- and y^2-terms, the graph can be identified as a hyperbola of the type shown in Figure 4.16(c). Solving for y,

$$y^2 = 4x^2 - 36$$
$$y = \pm 2\sqrt{x^2 - 9}$$

Note that $|x| \geq 3$ for y to be a real number. A table of ordered pairs is shown below and the graph is shown in Figure 4.19.

x	± 3	± 4	± 5	± 6
y	0	$\pm 2\sqrt{7}$	± 8	$\pm 6\sqrt{3}$

$2\sqrt{7} \approx 5.3$ and $6\sqrt{3} \approx 10.4$.

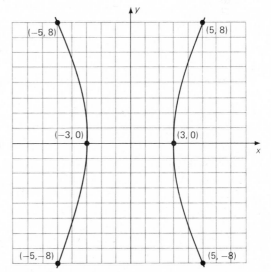

FIG. 4.19. $4x^2 - y^2 = 36$.

EXAMPLE 6   Graph $y^2 - 4x^2 = 36$.

Solution   The curve is identified as a hyperbola having the general shape shown in Figure 4.16(d).

$$y^2 = 4x^2 + 36 \quad \text{and} \quad y = \pm 2\sqrt{x^2 + 9}$$

A table of ordered pairs is shown below and the graph is shown in Figure 4.20.

$(x,$	$y)$
0	± 6
± 1	$\pm 2\sqrt{10}$, where $2\sqrt{10} \approx 6.3$
± 2	$\pm 2\sqrt{13}$, where $2\sqrt{13} \approx 7.2$
± 3	$\pm 6\sqrt{2}$, where $6\sqrt{2} \approx 8.5$
± 4	± 10

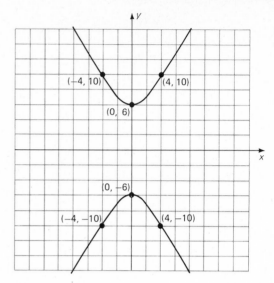

FIG. 4.20. $y^2 - 4x^2 = 36$.

ASYMPTOTES OF A HYPERBOLA

The lines obtained by replacing the constant term by 0 in the equation of a hyperbola are called **asymptotes** of the hyperbola.

For $xy = k$, the equation $xy = 0$ yields the asymptotes $x = 0$ and $y = 0$.

For $b^2x^2 - a^2y^2 = a^2b^2$, the asymptotes are obtained from $b^2x^2 - a^2y^2 = 0$ which yields the lines

$$bx - ay = 0 \quad \text{and} \quad bx + ay = 0$$

These are also the asymptotes for $a^2y^2 - b^2x^2 = a^2b^2$.

The two branches of a hyperbola lie entirely within two of the regions formed by its asymptotes. For larger and larger absolute values of x, or y, a branch of the hyperbola comes closer and closer to its asymptote. This property can be used as an aid in graphing a hyperbola, as shown in Example 7.

EXAMPLE 7 For the hyperbola defined by $4x^2 - 9y^2 = 36$, (a) state the asymptotes, (b) graph the asymptotes, and (c) graph the hyperbola using the asymptotes as guide lines.

Solution

(a) Replacing the constant by 0,
$$4x^2 - 9y^2 = 0$$
$$(2x - 3y)(2x + 3y) = 0$$
$2x - 3y = 0 \quad \text{and} \quad 2x + 3y = 0$ are the asymptotes

(b) and (c) See Figure 4.21.

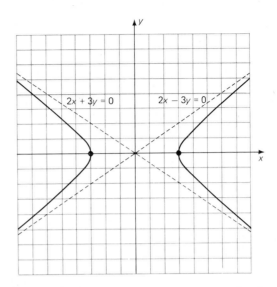

FIG. 4.21. Graph of $4x^2 - 9y^2 = 36$ with asymptotes (dotted lines).

EXERCISES 4.4

(1–14) Graph each parabola. See Examples 1 and 2.

1. $x = y^2 - 2y - 8$

2. $x = y^2 + 4y - 5$

3. $x = 6y - y^2 - 8$

4. $x = 6 - 2y - y^2$

5. $x = 5y - y^2$

6. $x = y^2 - 4y$

7. $x = y^2 + 1$

8. $x = 9 - y^2$

9. $x = y^2 + 3y - 6$

10. $x = 5 + 4y - 2y^2$

11. $x = 8y - y^2 - 16$

12. $x = 1 + 2y + y^2$

13. $x = 2y^2 - 2y + 1$

14. $x = -y^2$

(15–26) State the center and radius and graph each circle. See Example 3.

15. $x^2 + y^2 = 16$

16. $x^2 + y^2 = 25$

17. $4x^2 + 4y^2 = 9$

18. $9x^2 + 9y^2 - 100 = 0$

19. $(x - 5)^2 + (y + 3)^2 = 36$

20. $(x + 1)^2 + (y - 4)^2 = 4$

21. $x^2 + y^2 - 4x - 8y = 5$

22. $x^2 + y^2 - 6x + 4y + 4 = 0$

23. $x^2 + y^2 + 6x + 2y = 6$

24. $x^2 + y^2 - 6x - 8y = 0$

25. $x^2 + y^2 = 6x$

26. $x^2 + y^2 = 8y$

(27–40) Graph each of the following and state whether it is an ellipse or a hyperbola. See Examples 4–7. State the asymptotes of each hyperbola.

27. $4x^2 + y^2 = 100$

28. $4x^2 - y^2 = 100$

29. $x^2 - y^2 = 9$

30. $x^2 + 4y^2 = 100$

31. $xy - 6 = 0$

32. $xy + 8 = 0$

33. $x^2 + 9y^2 = 36$

34. $25y^2 - 4x^2 = 100$

35. $9x^2 - 4y^2 = 36$

36. $25x^2 + 4y^2 = 100$

37. $xy + 12 = 0$

38. $xy - 12 = 0$

39. $x^2 - 16y^2 + 16 = 0$

40. $x^2 - y^2 + 1 = 0$

(41–50) Identify the graph of each of the following as a parabola, a circle, an ellipse, or a hyperbola.

41. $x^2 + 9y^2 = 9$

42. $4x^2 - y^2 + 4 = 0$

43. $x^2 + y^2 = 9$

44. $4x - y^2 + 4 = 0$

45. $x + 9y^2 = 9$

46. $4x^2 + 4y^2 - 4 = 0$

47. $x^2 - 9y^2 = 9$

48. $xy + 4 = 0$

49. $xy = 9$

50. $4x^2 + y^2 - 4 = 0$

4.5 INEQUALITY RELATIONS

The graph of a linear or quadratic function defined by $y = f(x)$ divides the plane into three mutually distinct sets of points:

1. The points on the curve; $y = f(x)$.
2. The points above the curve; $y > f(x)$.
3. The points below the curve; $y < f(x)$.

Figure 4.22 illustrates these sets for the linear function and the quadratic function.

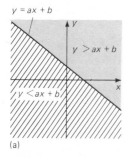
(a)

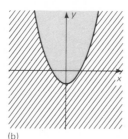
(b)

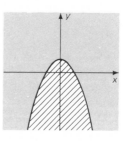

FIG. 4.22

(a) $y = ax + b$, solid line   (b) $y = ax^2 + bx + c$, solid curve
$y > ax + b$, shaded area $y > ax^2 + bx + c$, shaded area
$y < ax + b$, hatched area $y < ax^2 + bx + c$, hatched area

EXAMPLE 1   Graph $2x - y < 8$.

Solution   First solve for y:

$2x < y + 8$
$y > 2x - 8$

Graph the equality, $y = 2x - 8$, using a dotted line to indicate the line is not part of the graph. Then shade the region above the line as shown in Figure 4.23.

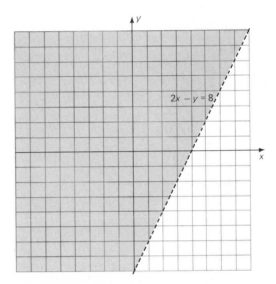

FIG. 4.23. $2x - y < 8$.

EXAMPLE 2   Graph $x^2 \geq 4x + y + 5$.

Solution   First solve for y:

$y \leq x^2 - 4x - 5$

Graph the equality, $y = x^2 - 4x - 5$, a parabola, using a solid line to indicate the parabola is part of the curve. Then shade the region below the parabola as shown in Figure 4.24.

159

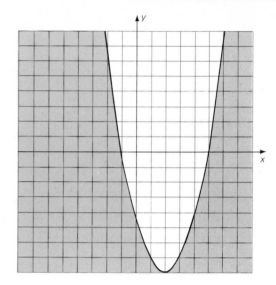

FIG. 4.24. $x^2 \geq 4x + y + 5$.

If a line or a parabola is defined by an equation having the form $x = g(y)$, then the graph of this relation also divides the plane into three mutually distinct sets of points:

1. The points on the curve (or line); $x = g(y)$.
2. The points to the right of the curve; $x > g(y)$.
3. The points to the left of the curve; $x < g(y)$.

Figure 4.25 illustrates these sets for general lines and parabolas.

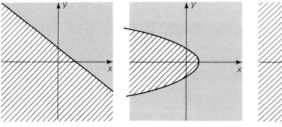
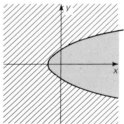

(a)

a. $x = ay + b$, solid line
$x > ay + b$, shaded area
$x < ay + b$, hatched area

(b)

b. $x = ay^2 + by + c$, solid curve
$x > ay^2 + by + c$, shaded area
$x < ay^2 + by + c$, hatched area

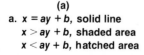

FIG. 4.25

EXAMPLE 3 Graph $x + 2y \geq 10$.

Solution Solve for x:

$x \geq 10 - 2y$

Graph the equality, $x = 10 - 2y$, using a solid line and shade the area to the right of the line as shown in Figure 4.26.

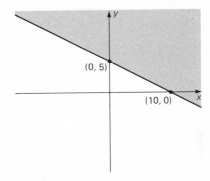

FIG. 4.26. $x + 2y \geq 10$.

EXAMPLE 4 Graph $y^2 < 4y - x$.

Solution Solve for x:

$x < 4y - y^2$

Graph the equality, the parabola $x = 4y - y^2$, using a dotted curved line and shade the area to the left of the curve as shown in Figure 4.27.

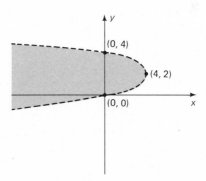

FIG. 4.27. $y^2 < 4y - x$.

The graph of a circle or an ellipse divides the plane into three mutually distinct sets of points; those on the curve, those inside the curve, and those outside the curve. Figure 4.28 illustrates these cases for a general circle and a general ellipse.

FIG. 4.28

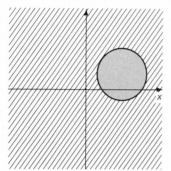
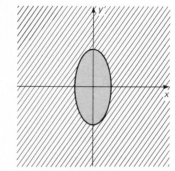

(a) $(x - h)^2 + (y - k)^2 = r^2$,
solid curve;
$(x - h)^2 + (y - k)^2 < r^2$,
shaded area;
$(x - h)^2 + (y - k)^2 > r^2$,
hatched area.

(b) $a^2 x^2 + b^2 y^2 = a^2 b^2$,
solid curve;
$a^2 x^2 + b^2 y^2 < a^2 b^2$,
shaded area;
$a^2 x^2 + b^2 y^2 > a^2 b^2$,
hatched area.

EXAMPLE 5 Graph $x^2 + y^2 - 4x + 2y > 4$.

Solution Rewrite the corresponding equation in standard form:

$(x - 2)^2 + (y + 1)^2 = 9$

Graph the circle using a dotted curve.

For the graph of $(x - 2)^2 + (y + 1)^2 > 9$, shade the region outside the circle as shown in Figure 4.29.

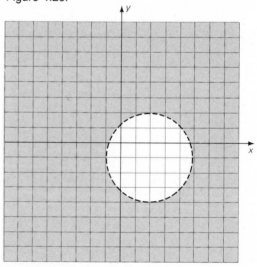

FIG. 4.29. $x^2 + y^2 - 4x + 2y > 4$.

EXAMPLE 6   Graph $4x^2 + y^2 \leq 16$.

Solution   Graph the ellipse, $4x^2 + y^2 = 16$, using a solid curved line. Then shade the region inside the ellipse as shown in Figure 4.30.

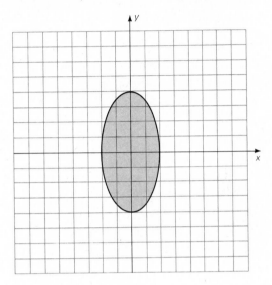

FIG. 4.30. $4x^2 + y^2 \leq 16$.

EXERCISES 4.5

(1–54)   Graph each of the following.

1. $y \leq 2x - 6$	**2.** $y \geq 3x - 9$
3. $y > 10 - 5x$	**4.** $y < 6 - x$
5. $y \geq 3$	**6.** $y \leq -2$
7. $3x + y < 12$	**8.** $2x + y > 10$
9. $x - y \geq 8$	**10.** $4x - y \leq 8$
11. $2x - y + 6 \leq 0$	**12.** $3x - 2y - 12 > 0$
13. $x + 2y > 14$	**14.** $5y > 15 + 3x$
15. $6 - x \geq 3y$	**16.** $2y + x + 4 \leq 0$
17. $x < 4$	**18.** $x > -5$
19. $x \geq 2y$	**20.** $3x \leq 2y$
21. $y \leq x^2$	**22.** $y \leq -x^2$
23. $y \geq 4 - x^2$	**24.** $y \geq x^2 - 9$
25. $y < 6x - x^2$	**26.** $y < x^2 + 2x$
27. $x^2 + 4x < y - 4$	**28.** $x^2 + 9 < 6x + 2y$
29. $4x + y + 1 \geq 2x^2$	**30.** $8x + 2y - 3 \leq 2x^2$

31. $x + y^2 \geq 0$

32. $x - y^2 \geq 0$

33. $x < y^2 - 1$

34. $x \leq y^2 - 4y$

35. $x \geq 2y - y^2$

36. $2x < 16 - y^2$

37. $x + y + 2 > y^2$

38. $2x - 2y + 5 > 2y^2$

39. $2x \leq 9 - y^2$

40. $2y^2 + 5 \geq 6y - 2x$

41. $x^2 + y^2 \leq 4$

42. $x^2 + y^2 < 9$

43. $x^2 + y^2 > 25$

44. $x^2 + y^2 \geq 16$

45. $x^2 + 9y^2 \geq 36$

46. $4x^2 + y^2 \leq 16$

47. $25x^2 + 4y^2 < 100$

48. $9x^2 + 16y^2 > 144$

49. $x^2 + y^2 \geq 6x - 8y$

50. $x^2 + y^2 \leq 4x + 2y$

51. $x^2 + y^2 < 8y$

52. $x^2 + y^2 > 12x$

53. $2x - 2y - 1 \geq x^2 + y^2$

54. $9 - x^2 - y^2 \leq 13 + 4x - 8y$

(55–60) Determine whether or not each given ordered pair is a solution of the inequality.

EXAMPLE 7

$x^2 + y^2 \leq 8x$.

(a) $(3, 2)$ (b) $(2, -5)$ (c) $(4, -4)$

Solution

(a) For $(3, 2)$, $x = 3$ and $y = 2$.

$3^2 + 2^2 \leq 8(3)$, $13 \leq 24$ *true*

Since $x^2 + y^2 \leq 8x$ is true for $(3, 2)$, $(3, 2)$ is a solution.

(b) For $(2, -5)$, $x = 2$ and $y = -5$.

$2^2 + (-5)^2 \leq 8(2)$, $29 \leq 16$ *false*

$(2, -5)$ is not a solution.

(c) For $(4, -4)$, $x = 4$ and $y = -4$.

$4^2 + (-4)^2 \leq 8(4)$, $32 \leq 32$ *true*

$(4, -4)$ is a solution.

55. $x^2 + y^2 \leq 25$

(a) $(-3, 4)$

(b) $(4, -2)$

(c) $(2, 5)$

56. $x^2 + 4y^2 \geq 16$

(a) $(2, 3)$

(b) $(4, 0)$

(c) $(2, -1)$

57. $4x^2 + y^2 > 100$

(a) $(4, -6)$

(b) $(-6, 4)$

(c) $(3, 5)$

58. $x^2 + y^2 < 6y$

(a) $(-3, 3)$

(b) $(-2, 2)$

(c) $(-4, 4)$

59. $y^2 \geq 6 - x$

(a) $(-4, 5)$

(b) $(-4, 3)$

(c) $(-3, -3)$

60. $x^2 \leq y + 2x$

(a) $(4, 8)$

(b) $(-3, 4)$

(c) $(3, 4)$

(61–70) State an inequality whose graph is described.

61. The region inside the circle $x^2 + y^2 = 1$.

62. The region inside and on the ellipse $x^2 + 4y^2 = 16$.

63. The region above the line $2x + 3y = 12$.

64. The region below the line $3x - y = 15$.
65. The region below and on the parabola $x^2 = 4 - y$.
66. The region above and on the parabola $x^2 = y + 2x$.
67. The region outside and on the ellipse $9x^2 + y^2 = 9$.
68. The region outside the circle $x^2 - 2x = 2y - y^2$.
69. The region to the right of the parabola $y^2 = x + 4y$.
70. The region to the left of the parabola $4y^2 = 9 - x$.

4.6 VARIATION

Variation is an important application of a relationship between two or more variables. There are many practical problems where the value of one variable depends on the value (or values) of one or more other variables.

DEFINITION

The variable y **varies directly** (or is **directly proportional**) as the variable x if and only if $y = kx$ where k is a constant.

DEFINITION

The variable y **varies inversely** as the variable x if and only if $y = \dfrac{k}{x}$ where k is a constant. The constant k is called the **constant of variation**.

EXAMPLE 1   The cost C of a turkey varies directly as the weight w of the turkey. A turkey weighing $5\frac{1}{2}$ pounds costs \$3.85.

(a) Find an equation expressing this variation.
(b) Find the cost of a turkey weighing 7.2 pounds.

Solution

(a) $C = kw$

For $C = 385$ cents and $w = 5.5$ pounds,

$385 = k\,5.5$

$k = \dfrac{385}{5.5} = 70$

In this case, k is the unit price, 70 cents per pound.

$C = 70w$

(b) Using $C = 70w$ with $w = 7.2$,

$C = 70(7.2) = 504$ cents

The cost is \$5.04.

EXAMPLE 2 If the voltage of an electrical circuit is constant, then the resistance R in ohms varies inversely as the amperage I. When $I = 22$ amps and $R = 5$ ohms:
(a) Find an equation expressing this variation.
(b) Find the resistance when $I = 10$ amps.

Solution

(a) For inverse variation, $R = \dfrac{k}{I}$ or $RI = k$.

Using $R = 5$, $I = 22$; $k = 5(22) = 110$.

Thus $RI = 110$ and $R = \dfrac{110}{I}$.

(b) When $I = 10$, $R = \dfrac{110}{10} = 11$ *ohms*.

When one variable varies directly as another, the ratio of the two variables is a constant.

If $y = kx$, then $\dfrac{y}{x} = k$.

An equation stating that two ratios are equal is called a **proportion;** in symbols,

$$\frac{a}{b} = \frac{c}{d}$$

is a proportion.

For many problems involving direct variation, it is not necessary to find the constant of variation. A proportion can be used instead.

EXAMPLE 3 How many grams of sodium are in 100 grams of sodium hydroxide if the ratio of the weight of sodium to the weight of sodium hydroxide is 23 to 40?

Solution Let $x = $ number of grams of sodium.

Equating ratios of the form $\dfrac{\text{sodium}}{\text{sodium hydroxide}}$,

$$\frac{x}{100} = \frac{23}{40}$$

$$x = \frac{2300}{40} = 57.5 \text{ grams}$$

DEFINITION

The variable z **varies jointly** as the variables x and y if and only if
$z = kxy$
where k is a constant.

A **combined variation** is one in which one variable varies directly or jointly as one or more variables and varies inversely as one or more variables.

For example, if w varies jointly as x and y and inversely as the square of z, then

$$w = \frac{kxy}{z^2}$$

where k is a constant.

The gravitational force F between two objects varies directly as the product of their masses m_1 and m_2 and inversely as the square of the distance r between them;

$$F = \frac{km_1 m_2}{r^2}$$

EXAMPLE 4 The amount of illumination L varies directly as the wattage w of a lamp and inversely as the square of the distance d in feet from the lamp. A 60-watt lamp provides adequate light on an object 2 feet away. What wattage is needed to provide adequate light on an object 3 feet away?

Solution   Writing the equation,

$$L = \frac{kw}{d^2}$$

For $w = 60$ and $d = 2$,

$$L = \frac{k60}{2^2}$$

For $w = 60$ and $d = 3$,

$$L = \frac{kx}{3^2}$$

Equating the values of L,

$$\frac{kx}{9} = \frac{k60}{4}$$

Solving for x,

$$x = \frac{60(9)}{4}$$

$$= 135 \text{ watts}$$

EXERCISES 4.6

(1–8) Solve by using a proportion. See Example 3.

1. How many pounds of iron are in 150 pounds of iron oxide if the ratio of iron to iron oxide by weight is 7 to 10?
2. The ratio of miles to kilometers is 5 to 8. How many miles are equal to 900 kilometers?
3. If 3 kilograms of apricots cost $2.94, what is the cost of 5 kilograms?
4. The scale on a map reads 1 inch = 25 miles. How far apart, in miles, are two towns if the distance between them on the map measures 4.6 inches?

5. A tree casts a shadow 64 feet long at the same time that a 3-foot yardstick casts a shadow $4\frac{1}{2}$ feet long. How tall is the tree?

6. If 10 meters of wire weigh 850 grams, how many grams does 35 meters weigh?

7. A car used 9 gallons of gas to travel 162 miles traveling at a constant speed. How many gallons of gas will the car use to travel 300 miles if it travels at the same speed?

8. If 150 million gallons of water are used daily by the people of a town whose population is 200,000, how much water will be needed 10 years from now if the population then is 300,000?

(9–22) (a) Write an equation expressing the variation described.
   (b) Find the value as indicated.

9. The distance d traveled at a constant speed varies directly as the time t. When $t = 3$ hours, $d = 135$ miles. Find d when $t = 4$ hours.

10. Hooke's law states that the force F required to stretch a spring varies directly as the distance s the spring is stretched. If 5 kilograms of force stretch a spring 12 decimeters, how much force would be required to stretch the spring 20 decimeters?

11. The time it takes to do a certain job varies inversely as the number of workers on the job. If it takes 15 workers to do a certain job in 8 days, how many workers would be needed to do the job in 5 days?

12. For a certain horizontal beam, the weight it can hold varies inversely as the length between its supports. If a 12-foot beam can hold 1500 pounds, how many pounds can an 18-foot beam hold?

13. For two meshed gears, the number of rotations of a gear is inversely proportional to the number of teeth the gear has. The larger gear has 24 teeth and rotates at 45 rpm. If the smaller gear has 18 teeth, how fast does it rotate?

14. The cost of producing certain articles varies inversely as the number of articles produced per day. The cost of producing 80 articles per day is $480. What would be the cost if 120 articles were produced per day?

15. The area A of a triangle varies jointly as its base b and altitude h. If $A = 600$ square inches when $b = 40$ inches and $h = 30$ inches, find A when $b = 15$ meters and $h = 20$ meters.

16. The volume V of a cylinder varies jointly as its height h and the square of its radius r. When $h = 10$ centimeters and $r = 6$ centimeters, $V = 1131$ cubic centimeters. Find V when $h = 5$ inches and $r = 4$ inches.

17. The pressure of a gas varies directly as its absolute temperature and inversely as its volume. When the absolute temperature is 297° and the volume is 22 cubic feet, the pressure is 12 pounds per square inch. What is the pressure when the absolute temperature is 288° and the volume is 20 cubic feet?

18. The horsepower H needed to drive an airplane varies directly as

the cube of its speed. When the speed is 500 mph, $H = 3750$. Find H for a speed of 600 mph.

19. For a certain copper wire, its electrical resistance R varies directly as its length L and inversely as the square of its diameter d. When $L = 60$ feet, $d = 0.05$ inch and $R = 0.25$ ohm. Find R when $L = 100$ feet and $d = 0.08$ inch.

20. The braking distance d of a car varies directly as the square of the speed v of the car. When $v = 20$ mph, $d = 22$ feet. Find d when $v = 60$ mph. Find v when $d = 30$ feet.

21. The rate r of the flow of water from the bottom of a tank varies directly as the square root of the depth d of the water in the tank. When $d = 16$ feet, $r = 30$ gallons per minute. Find r when $d = 9$ feet. Find d when $r = 15$ gallons per minute.

22. The safe load W a certain circular column can support varies directly as the fourth power of its radius r and inversely as the square of its length L. When $r = 6$ inches and $L = 12$ feet, $W = 7.2$ tons. Find W when $r = 8$ inches and $L = 8$ feet. Find r when $L = 10$ feet and $W = 5$ tons.

4.7 INVERSE FUNCTIONS AND RELATIONS

Every relation R has a companion relation obtained by interchanging the components of the ordered pairs belonging to R. This relation is called the inverse of R, R^{-1}, and is read "R inverse."

DEFINITION OF INVERSE RELATION

$$R^{-1} = \{(x, y) | (y, x) \text{ is in } R\}$$

Since the points (a, b) and (b, a) are symmetric to the line $y = x$, the graphs of R and R^{-1} are symmetric to the line $y = x$. This symmetry can be used to readily obtain the graph of R^{-1} from the graph of R.

When a relation R is defined by an equation or statement in the variables x and y, the inverse relation is defined by the statement obtained by replacing x by y and y by x. This has the effect of interchanging the ordered pairs.

In graphing parabolas, we were dealing with the concept of inverse relations. Namely, the relation defined by $x = ay^2 + by + c$ is the inverse of the relation defined by

$$y = ax^2 + bx + c$$

EXAMPLE 1   Given $R = \{(x, y)|y = 2x + 6\}$.

(a) Find R^{-1}.

(b) Graph R and R^{-1} on the same set of axes.

(c) Determine if R and R^{-1} are functions.

Solution

(a) $R^{-1} = \{(x, y)|x = 2y + 6\}$

Solving $x = 2y + 6$ for y, $y = \dfrac{x - 6}{2}$ and thus

$$R^{-1} = \left\{(x, y)|y = \frac{x - 6}{2}\right\}$$

(b) The graphs are shown in Figure 4.31. The $y = x$ line is drawn as a dotted line to show the symmetry of the graphs of R and R^{-1} about this line.

(c) R is a function and R^{-1} is a function. From the equations we can note that exactly one value can be computed for each value of x.

Also we can note from the graphs that every vertical line intersects each graph in no more than one point.

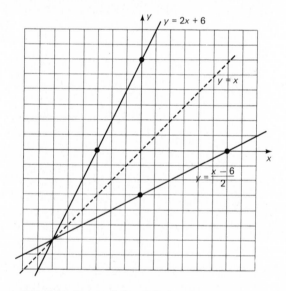

FIG. 4.31 Graph of R and R^{-1}.

EXAMPLE 2   Given R defined by $y = \frac{1}{2}\sqrt{25 - x^2}$.

(a) Find an equation defining R^{-1}.
(b) State the domain and range of R and R^{-1}.
(c) Graph R and R^{-1} on the same set of axes.
(d) Determine if R and R^{-1} are functions.

Solution

(a) For R^{-1}, $x = \frac{1}{2}\sqrt{25 - y^2}$

Solving for y,
$$(2x)^2 = 25 - y^2$$
$$y^2 = 25 - 4x^2$$
$$y = \pm\sqrt{25 - 4x^2}$$

(b) To find the domain of R, solve $25 - x^2 \geq 0$ and $-5 \leq x \leq 5$. For the range of R, note that the radical indicates a positive number or 0, and the greatest value for y is $\frac{5}{2}$. Thus $0 \leq y \leq \frac{5}{2}$.

   For the domain and range of R^{-1}, replace x by y and y by x in the domain and range statements for R. Thus:

Domain of R^{-1}: $0 \leq x \leq \frac{5}{2}$

Range of R^{-1}: $-5 \leq y \leq 5$

(c) The graphs are shown in Figure 4.32.
(d) R is a function but R^{-1} is not a function.

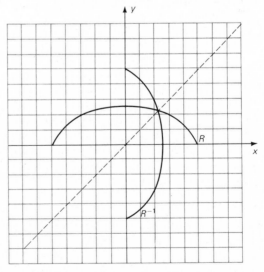

FIG. 4.32. Graph of R and R^{-1}: R, $y = \frac{1}{2}\sqrt{25 - x^2}$; R^{-1}, $x = \frac{1}{2}\sqrt{25 - y^2}$.

Examples 1 and 2 show that the inverse of a function is not always a function. For R^{-1} to be a function, its graph must also pass the vertical line test. This is equivalent to saying the graph of R must pass the horizontal line test; that is, every horizontal line must intersect R in at most one point. In other words, the function R must not have two different ordered pairs with the same second component.

When a relation R passes both the horizontal line test and the vertical line test, the relation is a function and its inverse is also a function. Such a relation is called a **one-to-one function.**

DEFINITION

A **one-to-one function** is a relation such that:

1. There is exactly one y-value for each x-value in its domain.
2. There is exactly one x-value for each y-value in its range.

Recall that when a relation is a function, the notation $f(x)$ is used to indicate the unique value of y paired with x; that is, $y = f(x)$.

Similarly, when the inverse of a function is also a function, the notation $f^{-1}(x)$ is used to indicate the unique value of y; that is, $y = f^{-1}(x)$.

Procedure for Finding $f^{-1}(x)$

1. Interchange x and y in the equation that defines f.
2. Solve the resulting equation for y.
3. Replace y by $f^{-1}(x)$ in the final equation of Step 2.

EXAMPLE 3   Given $f = \{(x, y)| y = 2\sqrt{x - 3}\}$.
(a) Find $f(x)$ and state its domain and range.
(b) Find $f^{-1}(x)$ and state its domain and range.
Solution
(a) Since $y = f(x)$, $f(x) = 2\sqrt{x - 3}$.
Domain: $x \geq 3$, Range: $y \geq 0$.
(b) 1. f^{-1} is defined by $x = 2\sqrt{y - 3}$, interchanging x and y in the equation for f.

2. Solving for y, $y = \dfrac{x^2}{4} + 3$.

3. Since $y = f^{-1}(x)$, $f^{-1}(x) = \dfrac{x^2}{4} + 3$.

Domain: $x \geq 0$, Range: $y \geq 3$.
The domain and range statements for f^{-1} are obtained by replacing x by y and y by x in the domain and range statements for f.

EXAMPLE 4   For $f(x) = 2\sqrt{x - 3}$ and $f^{-1}(x) = \dfrac{x^2}{4} + 3$, find $f^{-1}(f(x))$
and $f(f^{-1}(x))$.

Solution

(a) 1. To find $f^{-1}(f(x))$, rewrite the equation for
        $f^{-1}(x)$ but remove x and hold its place
        with open parentheses.

$$f^{-1}(\ \ ) = \frac{(\ \ )^2}{4} + 3$$

  2. Insert $f(x)$ on the left side of the equation
        and the expression to which $f(x)$ is equal
        on the right side.

$$f^{-1}(f(x)) = \frac{(2\sqrt{x - 3})^2}{4} + 3$$

  3. Simplify the right side.

$$f^{-1}(f(x)) = \frac{4(x - 3)}{4} + 3$$
$$= x$$

(b) 1. To find $f(f^{-1}(x))$, rewrite the equation for
        $f(x)$, removing x and holding its place with
        open parentheses.

$$f(\ \ ) = 2\sqrt{(\ \ )} - 3$$

  2. Insert $f^{-1}(x)$ on the left side and the ex-
        pression to which $f^{-1}(x)$ is equal on the
        right side.

$$f(f^{-1}(x)) = 2\sqrt{\left(\frac{x^2}{4} + 3\right)} - 3$$

  3. Simplify the right side.

$$f(f^{-1}(x)) = \sqrt{x^2}$$
$$= x \qquad \text{since } x \geq 0$$

Note that the domain of f^{-1} is $x \geq 0$ since
the range of f is $y \geq 0$.

Note also in Example 4 that
$f^{-1}(f(x)) = x$ and $f(f^{-1}(x)) = x$.
These statements are true in general whenever
f and f^{-1} are functions.

EXERCISES 4.7

(1–20) For each relation R defined by the given statement, graph R
and R^{-1} on the same set of axes and determine if R and R^{-1} are functions. See Examples 1 and 2.

1. $2x + y = 10$ **2.** $x - 3y = 9$
3. $y = x + 5$ **4.** $y = x - 4$
5. $y = 3x$ **6.** $x = 2y$

7. $y = x^2$

8. $x = -y^2$

9. $y = 4 - x^2$

10. $y = 6x - x^2$

11. $y = \sqrt{x - 2}$

12. $y = \sqrt{4 - x}$ \

13. $x^2 + 9y^2 = 9$

14. $4x^2 + y^2 = 4$

15. $y = 2\sqrt{4 - x^2}$

16. $y = 3\sqrt{x^2 - 1}$

17. $y = \frac{1}{2}\sqrt{x^2 + 9}$

18. $xy = 6$

19. $x^2 + y^2 = 25$

20. $y = 1 + \sqrt{x + 1}$

(21–32) For the function defined by the given equation:
(a) Find $f(x)$ and the domain and range of f.
(b) Find $f^{-1}(x)$ and the domain and range of f^{-1}.
(c) Find $f^{-1}(f(x))$ and $f(f^{-1}(x))$.
See Examples 3 and 4.

21. $y = 2\sqrt{x + 4}$

22. $y = \sqrt{9 - x}$

23. $y = x^3$

24. $y = \sqrt[3]{x - 1}$

25. $xy = 10$

26. $x = y^2 - 1$ and $y \geq 0$

27. $y = x^2$ and $x \geq 0$

28. $y = \sqrt{4 - x^2}$ and $x \geq 0$

29. $4x^2 + y^2 = 16$ and $x \geq 0$ and $y \geq 0$

30. $4x^2 - y^2 = 4$ and $x \geq 0$ and $y \geq 0$

31. $xy = x + 2y$

32. $xy = x - 3y - 3$

4.8 CHAPTER REVIEW

(1–2) State the domain and range of each relation and determine if it is a function. (4.1)

1. $\{(x, y) | y = \sqrt{x - 4}\}$

2. $\{(x, y) | x^2 = y + 9\}$

(3–4) For the function f defined by the given equation, find
(a) $f(x)$,   (b) $f(0)$,   (c) $f(4)$,   (d) $f(-2)$,   (e) $f(t - 3)$ (4.1)

3. $y = x^2 - 4x - 12$

4. $y + 1 = \sqrt{4x + 9}$

5. Find the ordered pairs that are solutions of $4x^2 + y^2 = 100$ for each of the following. (4.1)
(a) $x = 0$,   (b) $y = 0$,   (c) $x = -4$,   (d) $y = 8$

(6–9) Graph each of the following. (4.2)

6. $4x + 5y = 20$

7. $7x - 2y = 14$

8. $y = 4$

9. $x = -5$

10. Find the slope of the line through the points $(6, -4)$ and $(2, 3)$. (4.2)

(11–12) Find the standard form of the equation of the line satisfying the given conditions. (4.2)

11. Passing through $(10, 6)$ and $(0, -6)$.
12. Passing through $(-4, 2)$ and with slope -3.
13. Find the slope and y-intercept of the line whose equation is $4x - 3y = 24$. (4.2)

(14–15) Determine if the given pair of lines are parallel or if they intersect. (4.2)

14. $6x - y = 12$ and $x - 6y = 12$
15. $x + 4y = 8$ and $x = 4 - 4y$

(16–19) Graph each of the following. State the coordinates of the vertex and the equation of the axis. (4.3)

16. $y = x^2 - 4x$ **17.** $y = 8 - 2x^2$
18. $2y = x^2 - 6x - 8$ **19.** $2y = 4 - 3x - x^2$
20. The profit y in dollars is related to the number of units x produced by a certain manufacturer by the equation

$$y = 60x - x^2 - 100$$

Find the number of units that should be produced for a maximum profit. (4.3)

(21–28) Graph each of the following and state whether the graph is a parabola, a circle, an ellipse, or a hyperbola. (4.4)

21. $x^2 + y^2 = 8x - 6y$ **22.** $y^2 = 2x + 2y$
23. $y^2 - 4x^2 = 9$ **24.** $x^2 = 36 - y^2$
25. $x^2 = 36 - 4y^2$ **26.** $x^2 = y^2 + 16$
27. $x = 6y - y^2$ **28.** $xy = 10$

(29–38) Graph each of the following. (4.5)

29. $2x + y \leq 8$ **30.** $x - 3y > 9$ **31.** $x < 3$
32. $y \geq -4$ **33.** $y > 2x - x^2$ **34.** $y \leq x^2 - 1$
35. $x \geq y^2 - 4y + 4$ **36.** $x^2 + y^2 \leq 16$ **37.** $9x^2 + 4y^2 \geq 36$
38. $x < 4 + 3y - y^2$

(39–42) Answer each question. (4.6)

39. Orange juice is made by mixing 1 can of liquid concentrate with 3 cans of water. How many liters of concentrate are needed to make 24 liters of orange juice?
40. The pressure P due to the ocean water on a diver varies directly as the depth d below the surface of the water. When $d = 50$ feet, $P = 3120$ pounds per square foot. What is the pressure when $d = 250$ feet?
41. The time T to do a certain job varies inversely as the number n of persons working on the job. If 15 persons can do the job in 10 days, how long will it take 25 persons to do the same job?

42. The cost C to manufacture one unit of a certain product varies directly as the salary s of the worker and inversely as the square of the number n of units the worker makes each day. When $s = \$40$ and $n = 6$, then $C = \$60$. What is the value of C when $s = \$48$ and $n = 9$?

(43–44) For the relation R defined by the given equation, graph R and R^{-1} and determine if R and R^{-1} are functions. (4.7)

43. $y = 2\sqrt{9 - x^2}$ **44.** $y = \sqrt{2x - 6}$

(45–46) For the function f defined by the given equation:
(a) Find $f(x)$ and the domain and range of f.
(b) Find $f^{-1}(x)$ and the domain and range of f^{-1}.
(c) Find $f^{-1}(f(x))$ and $f(f^{-1}(x))$.

45. $2y = \sqrt{16 - 2x}$ **46.** $y = \dfrac{12}{x - 2}$

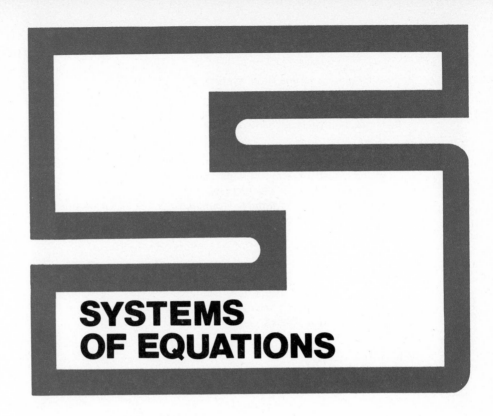

SYSTEMS OF EQUATIONS

There are many applications in which two or more quantities must satisfy two or more equations. The equations are said to form a system. In this chapter we will study methods for solving systems whose equations are either linear or quadratic or both.

5.1 LINEAR SYSTEMS, TWO VARIABLES

The set of equations $2x + y = 10$ and $x - 3y + 9 = 0$ form a system of two linear equations in the variables x and y. The solution of this system is the ordered pair $(3, 4)$ because $(3, 4)$ is the only solution of *both* equations.

DEFINITION

The **solution set of a system of equations** is the intersection of the solution sets of the equations in the system.

In general, for the linear system

$$A_1 x + B_1 y = C_1$$
$$A_2 x + B_2 y = C_2$$

there are three possibilities for the solution set of the system.

1. The **solution set contains exactly one ordered pair.** The equations are called **consistent and independent.** The graphs of the equations are two lines that intersect in exactly one point. The lines have different slopes.
2. The **solution set is empty.** The equations are called **inconsistent.** The graphs of the equations are two parallel lines that do not intersect. The lines have the same slope but different y-intercepts.
3. The **solution set** is the solution set of either equation and **contains infinitely many ordered pairs.** The equations are called **dependent.** The graphs are the same line. The slopes are equal and the y-intercepts are equal. Figure 5.1 shows these three possibilities.

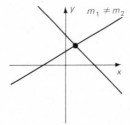 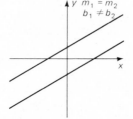 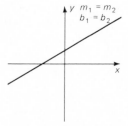

1. Exactly one solution; consistent, independent equations.
2. No solution; inconsistent equations.
3. Infinitely many solutions; dependent equations.

FIG. 5.1. Three possibilities for the system, $A_1 x + B_1 y = C_1$ and $A_2 x + B_2 y = C_2$.

Graphing the two equations of a system helps us understand the nature of the solution set. However, as a general rule, the coordinates of a point of intersection can only be approximated. To obtain exact values we solve the system algebraically using either the substitution method or the addition method. Both of these methods involve procedures for obtaining a simpler equivalent system.

DEFINITION

Two systems of equations are equivalent if and only if they have the same solution set.

SUBSTITUTION METHOD

The **substitution method** involves the use of the substitution axiom of equality which produces an equivalent system.

THEOREM 1 SUBSTITUTION THEOREM

Let $P(x, y)$ be a polynomial in the variables x and y.

1. The system $P(x, y) = 0$ and $y = f(x)$ is equivalent to the system $P(x, f(x)) = 0$ and $y = f(x)$.
2. The system $P(x, y) = 0$ and $x = g(y)$ is equivalent to the system $P(g(y), y) = 0$ and $x = g(y)$.

Substitution Method Procedure

1. Solve one of the equations for one of the variables obtaining $y = f(x)$ [or $x = g(y)$].
2. In the other equation, replace y by $f(x)$ [or x by $g(y)$].
3. Solve the resulting equation in one variable.
4. Using $y = f(x)$ [or $x = g(y)$], substitute the value of the variable found in Step 3 and determine the value of the other variable.
5. State each solution as an ordered pair.
6. Check each solution in each equation of the original system.

The substitution method is especially useful when the coefficient of either x or y in one of the equations is 1.

Examples 1 and 2 illustrate the substitution method for an independent system to obtain an equivalent system having the form $x = a$ and $y = b$.

Examples 3 and 4 illustrate the substitution method for systems that are dependent or inconsistent.

EXAMPLE 1 Solve
$$2x + y = 22$$
$$3x - 2y = 12$$
using the substitution method.

Solution

1. Solving $2x + y = 22$ for y,
$$y = 22 - 2x$$
2. Replacing y in the other equation by $22 - 2x$,
$$3x - 2(22 - 2x) = 12$$
3. Solving for x,
$$7x - 44 = 12$$
$$7x = 56$$
$$x = 8$$
4. Replacing x by 8 in the first equation and solving for y,
$$y = 22 - 2(8)$$
$$= 6$$
The equivalent system is $x = 8$ and $y = 6$.
5. The solution is $(8, 6)$.
6. *Check* For $2x + y = 22$,
$$2(8) + 6 = 16 + 6 = 22.$$
For $3x - 2y = 12$,
$$3(8) - 2(6) = 24 - 12 = 12.$$

Figure 5.2 shows that the systems are equivalent. Both systems have the same solution set and the graphs have the same point of intersection.

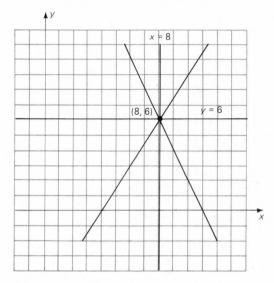

FIG. 5.2. Equivalent systems,
$\begin{cases} 2x + y = 22 \\ 3x - 2y = 12 \end{cases}$ **and** $\begin{cases} x = 8 \\ y = 6 \end{cases}$.

EXAMPLE 2 Solve
$$x - 2y = 9$$
$$2x - 5y = 20$$
using the substitution method.

Solution The first equation is solved for x, since the coefficient of x is 1.

1. $\qquad\qquad x = 2y + 9$
2. $2(2y + 9) - 5y = 20$
3. $\qquad -y + 18 = 20$
   $$y = -2$$
4. $\qquad\qquad x = 2(-2) + 9$
   $$= 5$$
5. The solution is $(5, -2)$.
6. **Check** For $x - 2y = 9$,
   $$5 - 2(-2) = 5 + 4 = 9.$$
   For $2x - 5y = 20$,
   $$2(5) - 5(-2) = 10 + 10 = 20.$$

EXAMPLE 3   Solve
$2x - y = 6$
$2y = 4x + 5$

Solution   Solving the first equation for y,
$-y = -2x + 6$
$y = 2x - 6$
Then
$2(2x - 6) = 4x + 5$
$4x - 12 = 4x + 5$
$-12 = 5$ *false*
Since a false statement is obtained, the equations are inconsistent and there is no solution. The solution set is the empty set.
Check   Compare the slopes and y-intercepts of each equation. For $2x - y = 6$; $y = 2x - 6$ and $m = 2$ and $b = -6$. For $2y = 4x + 5$; $y = 2x + \frac{5}{2}$ and $m = 2$ and $b = \frac{5}{2}$.

Since the slopes are equal and the y-intercepts are not equal, the lines are parallel. Thus there is no solution.

EXAMPLE 4   Solve
$x - 2y = 8$
$2y = x - 8$

Solution   Solving the first equation for x,
$x = 2y + 8$
Then
$2y = (2y + 8) - 8$
$2y = 2y$
$0 = 0$ *true*
These last two equations are true for all values of y. The value for x, however, depends on the value for y. The solution set can be written in any of the following forms:
$\{(x, y) \mid x - 2y = 8\}$
$\{(2y + 8, y)\}$
$\left\{ \left(x, \dfrac{x - 8}{2} \right) \right\}$
Check
Method 1   Replace x by $2y + 8$ in each equation.
For $x - 2y = 8$, $(2y + 8) - 2y = 8$ and $8 = 8$.
For $2y = x - 8$, $2y = (2y + 8) - 8$ and $2y = 2y$.

Method 2 Replace y by $\dfrac{x-8}{2}$ in each equation.

For $x - 2y = 8$, $x - 2\left(\dfrac{x-8}{2}\right) = x - x + 8 = 8$.

For $2y = x - 8$, $2\left(\dfrac{x-8}{2}\right) = x - 8 = x - 8$.

Method 3 Compare the slopes and y-intercepts.

For $x - 2y = 8$, $y = \frac{1}{2}x - 4$; $m = \frac{1}{2}$ and $b = -4$.

For $2y = x - 8$, $y = \frac{1}{2}x - 4$; $m = \frac{1}{2}$ and $b = -4$.

Since the slopes are equal *and* the y-intercepts are equal, the graphs of the equations are the same line. The lines are also said to be *coincident*.

ADDITION METHOD

The **addition method** involves adding a constant multiple of one equation to a constant multiple of the second equation in order to eliminate one of the variables.

Suppose (a, b) is a solution of
$$A_1x + B_1y - C_1 = 0 \quad \text{and} \quad A_2x + B_2y - C_2 = 0.$$
Then for the constants k and n,
$$k(A_1a + B_1b - C_1) + n(A_2a + B_2b - C_2) = k(0) + n(0) = 0$$
Thus (a, b) is also a solution of
$$k(A_1x + B_1y - C_1) + n(A_2x + B_2y - C_2) = 0$$
This shows that the solution of the system is a solution of the above equation and the addition method procedure produces an equivalent system.

THEOREM 2 ADDITION METHOD THEOREM

Let $P(x, y)$ and $Q(x, y)$ be polynomials in the variables x and y. Let k and n be constants, with $n \neq 0$. Then the system $P(x, y) = 0$ and $Q(x, y) = 0$ is equivalent to the system $P(x, y) = 0$ and $kP(x, y) + nQ(x, y) = 0$.

Addition Method Procedure

1. Multiply one equation by a constant k and the other equation by a constant n so that the sum of the resulting terms in one of the variables is zero.
2. Add the two equations of Step 1.
3. Solve the resulting equation in one variable.
4. Substitute the value found in Step 3 in either one of the original equations and solve for the other variable, or repeat Steps 1–3.
5. State each solution as an ordered pair.
6. Check each solution in each equation of the original system.

Examples 5 and 6 illustrate the addition method.

EXAMPLE 5 Solve
$$2x + 3y = 14$$
$$3x - 5y = 2$$
using the addition method.

Solution

1. To eliminate y, multiply the first equation by 5 and the second equation by 3. Then add the resulting equations.

$$5(2x + 3y = 14) \rightarrow 10x + 15y = 70$$
$$3(3x - 5y = 2) \ \rightarrow \ \underline{9x - 15y = 6}$$

2. $\qquad\qquad\qquad\qquad 19x \qquad\quad = 76$
3. $\qquad\qquad\qquad\qquad\qquad x = 4$

1. To eliminate x, multiply the first equation by 3 and the second equation by -2. Then add the resulting equations.

$$3(2x + 3y = 14) \rightarrow \ \ 6x + \ \ 9y = 42$$
$$-2(3x - 5y = 2) \ \rightarrow \ \underline{-6x + 10y = -4}$$

2. $\qquad\qquad\qquad\qquad\qquad 19y = 38$
3. $\qquad\qquad\qquad\qquad\qquad\quad y = 2$
5. The solution is $(4, 2)$.
6. *Check*
   For $2x + 3y = 14$, $2(4) + 3(2) = 8 + 6 = 14$.
   For $3x - 5y = 2$, $3(4) - 5(2) = 12 - 10 = 2$.

Example 6 illustrates the use of the addition method to eliminate one variable and then the use of the substitution axiom to find the value of the other variable.

EXAMPLE 6 Solve
$$3x + 4y = 10$$
$$2x + 5y = 23$$

Solution First, eliminating y,

1. $\quad 5(3x + 4y = 10) \rightarrow \ \ 15x + 20y = 50$
   $\quad -4(2x + 5y = 23) \rightarrow - \ \underline{8x - 20y = -92}$
2. $\qquad\qquad\qquad\qquad\qquad 7x \qquad\quad = -42$
3. $\qquad\qquad\qquad\qquad\qquad\qquad x = -6$

Now, replacing x by -6 in either equation, the first for example,

4. $3(-6) + 4y = 10$
   $\qquad\quad 4y = 28$
   $\qquad\quad\ \ y = 7$
5. The solution is $(-6, 7)$.
6. *Check*
   For $3x + 4y = 10$, $3(-6) + 4(7) = -18 + 28 = 10$.
   For $2x + 5y = 23$, $2(-6) + 5(7) = -12 + 35 = 23$.

Some systems of equations that are not linear may become linear by substitutions. Then the resulting system can be solved by the methods discussed in this section. Example 7 illustrates this technique.

EXAMPLE 7 Solve

$$\frac{3}{x} + \frac{2}{y} = \frac{1}{2}$$

$$\frac{6}{x} - \frac{1}{y} = \frac{3}{8}$$

Solution Let $r = \frac{1}{x}$ and $s = \frac{1}{y}$. Then the system becomes

$$3r + 2s = \frac{1}{2}$$

$$6r - s = \frac{3}{8}$$

Solving by the addition method for r and s,

$$3r + 2s = \frac{1}{2}$$

$$\underline{12r - 2s = \frac{3}{4}}$$

$$15r \quad = \frac{5}{4}$$

$$r = \frac{5}{60} = \frac{1}{12}$$

Using the first equation,

$$3\left(\frac{1}{12}\right) + 2s = \frac{1}{2}$$

$$2s = \frac{1}{2} - \frac{1}{4} = \frac{1}{4}$$

$$s = \frac{1}{8}$$

Now replacing r by $\frac{1}{x}$ and s by $\frac{1}{y}$,

$$r = \frac{1}{12} \quad \text{and} \quad s = \frac{1}{8}$$

$$\frac{1}{x} = \frac{1}{12} \quad \text{and} \quad \frac{1}{y} = \frac{1}{8}$$

$x = 12$ and $y = 8$
The solution is $(12, 8)$.

Check For $\frac{3}{x} + \frac{2}{y} = \frac{1}{2}, \frac{3}{12} + \frac{2}{8} = \frac{1}{4} + \frac{1}{4} = \frac{1}{2}.$

For $\frac{6}{x} - \frac{1}{y} = \frac{3}{8}, \frac{6}{12} - \frac{1}{8} = \frac{1}{2} - \frac{1}{8} = \frac{3}{8}.$

EXERCISES 5.1

(1–20) Solve and check. See Examples 1–6.

1. $2x + y = 16$
$\quad x - y = 2$

2. $x + 2y = 100$
$\quad 3x - 5y = 25$

3. $4x - 5y = 20$
   $x - y = 8$

4. $2x + 3y = 7$
   $3x + y = 21$

5. $x = y + 6$
   $y = 6 - x$

6. $x = y + 10$
   $y = x - 10$

7. $2y = x + 6$
   $3x - 6y = 10$

8. $2x - 3y = 4$
   $3x + 4y = 23$

9. $2x - 3y = 26$
   $3x + 2y = 13$

10. $x = 3y + 6$
    $2x - 6y + 3 = 0$

11. $2x - y = 5$
    $y + 5 = 2x$

12. $3x - 4y = 0$
    $4x - 3y = 21$

13. $4x + 3y = 1$
    $3x + 4y = 6$

14. $2y - 5x = 16$
    $3y + 4x = 24$

15. $x = 3 - 2y$
    $2x + 4y = 6$

16. $2x - 2y = 5$
    $3x - 3y = 10$

17. $4x - 3y = 50$
    $2x + 5y = 90$

18. $4x - 3y = 25$
    $3x + 4y = 50$

19. $2x = 5y - 10$
    $3x = 5y + 5$

20. $2y = 2 - 3x$
    $4y = 10 - 5x$

(21–28) Solve and check. See Example 7.

21. $\dfrac{1}{x} + \dfrac{1}{y} = \dfrac{1}{2}$

$\dfrac{1}{x} - \dfrac{1}{y} = \dfrac{1}{6}$

22. $\dfrac{1}{x} + \dfrac{1}{y} = \dfrac{1}{6}$

$\dfrac{6}{x} - \dfrac{2}{y} = \dfrac{1}{5}$

23. $\dfrac{2}{x} + \dfrac{5}{y} = 1$

$\dfrac{10}{x} - \dfrac{5}{y} = 2$

24. $\dfrac{2}{x} + \dfrac{3}{y} = 1$

$\dfrac{4}{x} + \dfrac{2}{y} = 1$

25. $\dfrac{3}{x} - \dfrac{4}{y} = 7$

$\dfrac{2}{x} - \dfrac{1}{y} = 8$

26. $\dfrac{2}{x} + \dfrac{3}{y} = 5$

$\dfrac{1}{x} - \dfrac{2}{y} = 20$

27. $\dfrac{5}{x} + \dfrac{7}{y} = 1$

$\dfrac{2}{x} + \dfrac{3}{y} = 1$

28. $\dfrac{5}{x} - \dfrac{3}{y} = 8$

$\dfrac{3}{x} + \dfrac{5}{y} = 15$

(29–36) Solve for x and y.

29. $x + y = 2a$
    $x - y = 2b$

30. $x + y = a + \sqrt{a^2 - b^2}$
    $x - y = a - \sqrt{a^2 - b^2}$

31. $ax + by = ab$
    $bx + ay = ab$

32. $ax + by = c$
    $bx + ay = c$

33. $ax + by = a$
    $cx + dy = c$

34. $ax + by = 1$
    $cx + dy = 1$

35. $ax - by = a^2 - b^2$
    $bx + ay = 2ab$

36. $ax + by = a^3$
    $bx + ay = b^3$

(37–45) Solve algebraically, by using two variables.

37. Six gallons of gas and 1 quart of oil cost $5.50. Five gallons of gas and 2 quarts of oil cost $5.75. Find the cost of 1 gallon of gas and of 1 quart of oil.

   Let x = cost of 1 gallon of gas and y = cost of 1 quart of oil. Then $6x + y = \$5.50$ and $5x + 2y = \$5.75$. Complete the solution.

38. If 4 trees and 5 flats of ground cover cost $50 while 3 trees and 2 flats of ground cover cost $27, find the price of each tree and each flat.

39. Traveling with a tailwind, an airplane makes a 1260-mile trip in 2 hours. Traveling against the same wind, now a head wind, the same airplane makes a 1680-mile trip in 3 hours. Find the speed of the airplane in still air and the speed of the wind.

40. A certain taxi company charges a fixed amount for the first mile and then a uniform charge for each additional $\frac{1}{4}$th of a mile. A 5-mile trip costs $4.85 while an $8\frac{1}{4}$-mile trip costs $8.10. Find the fixed fee for the first mile and the charge for each $\frac{1}{4}$ mile.

41. A woman has a total income of $1360 from $8000 invested in mortgages and $7000 invested in bonds. Her brother has a total income of $1300 from $5000 invested in the same kind of mortgages and $10,000 invested in the same kind of bonds. Find the rate of interest for the mortgages and for the bonds.

42. One week an electrician worked for 50 hours and his apprentice worked for 30 hours. Their combined wages for the week was $750. The next week the electrician worked 40 hours and his apprentice worked 40 hours. Their combined wages for the second week were $680. Find how much each is paid per hour.

43. On one job two machines worked together and completed the job in 2 hours. Later one machine worked for 4 hours alone on the same type of job and then the other machine completed the job in 1 hour. Find how long it would have taken each machine to do the job working alone.

   Let x = time for one and let y = time for the other. Then
$$\frac{2}{x} + \frac{2}{y} = 1 \quad \text{and} \quad \frac{4}{x} + \frac{1}{y} = 1$$
Find x and y.

44. When resistor A and resistor B are joined in parallel, the total resistance in the electric circuit is 6 ohms. When 7 of the A resistors and 2 of the B resistors are joined in parallel, the total resistance is 1.2 ohms. Find how many ohms are in each resistor.

   Let
x = number of ohms for A
y = number of ohms for B
Then
$$\frac{1}{x} + \frac{1}{y} = \frac{1}{6} \quad \text{and} \quad \frac{7}{x} + \frac{2}{y} = \frac{1}{1.2}$$
Solve for x and y.

45. A nutritionist wants to prepare a combination of foods to obtain 230 grams of protein and 130 grams of fiber. Food A contains 30% protein and 20% fiber. Food B contains 40% protein and 15% fiber. How many grams of each food should the nutritionist use?

Let

x = number of grams of food A

y = number of grams of food B

Then

$$0.30x + 0.40y = 230$$

and

$$0.20x + 0.15y = 130$$

Solve for x and y.

5.2 LINEAR SYSTEMS, THREE VARIABLES

The equation $2x + 3y - 4z = 2$ is an example of a linear equation in three variables, x, y, and z. The ordered triple $(4, 2, 3)$ is a solution of this equation. For $x = 4$, $y = 2$, and $z = 3$ the equation $2x + 3y - 4z = 2$ becomes

$$2(4) + 3(2) - 4(3) = 8 + 6 - 12 = 2$$

a true statement. In general, we have the following definitions.

DEFINITION

A **linear equation in the three variables** x, y, and z is an equation equivalent to

$$Ax + By + Cz = D$$

where A, B, C, and D are constants such that $A \neq 0$ or $B \neq 0$ or $C \neq 0$ ($A^2 + B^2 + C^2 \neq 0$).

DEFINITION

A **solution of a linear equation in three variables** x, y, and z is an **ordered triple** of numbers (a, b, c) such that the equation becomes a true statement when x is replaced by a, y is replaced by b, and z is replaced by c.

Note that if $A \neq 0$ and if y and z are replaced by any real numbers, then the value of x can be obtained by solving the resulting equation for x.

For $2x + 3y - 4z = 2$, let $y = 4$ and $z = 5$. Then

$$2x + 3(4) - 4(5) = 2$$

$$2x - 8 = 2 \quad \text{and} \quad x = 5$$

Thus $(5, 4, 5)$ is a solution.

Also, we can solve for y when $B \neq 0$ and any two values are assigned to x and z. Also, we can solve for z when $C \neq 0$ and any two values are assigned to x and y. Thus a linear equation in three variables has infinitely many solutions.

DEFINITION

The **solution set of a system of linear equations in three variables** is the intersection of the solution sets of the equations in the system.

Geometrically, the graph of a linear equation in three variables is a plane. Two distinct planes are either parallel or they intersect in exactly one line.

For a system having the form
$$A_1x + B_1y + C_1z = D_1$$
$$A_2x + B_2y + C_2z = D_2$$
$$A_3x + B_3y + C_3z = D_3$$
there are **three possibilities** for the solution set:

1. There is **exactly one solution.** The equations are called **consistent and independent.** The three planes intersect in exactly one point as shown in Figure 5.3(a).
2. There is **no solution.** The solution set is the empty set. The equations are called **inconsistent.** The three planes do not have a point in common. See Figure 5.3(b).
3. There are **infinitely many solutions.** The equations are called **dependent.** The three planes intersect in a line or the graphs of the three equations are the same plane. See Figure 5.3(c).

ALGEBRAIC STATEMENT	GEOMETRIC STATEMENT	ILLUSTRATION
(a) There is exactly one solution.	The three planes intersect in exactly one point, P.	(a) 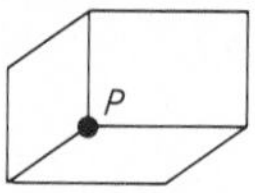
(b) The solution set is empty.	The three planes do not have a point in common.	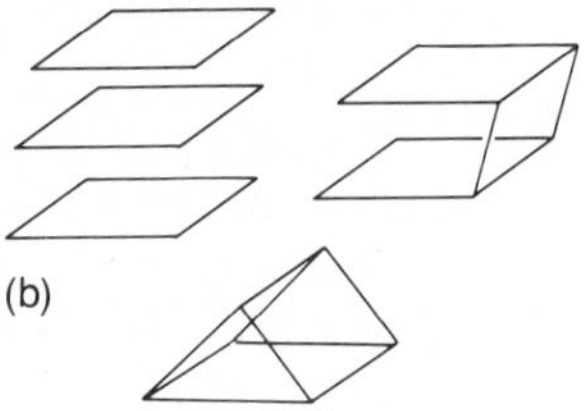 (b)
(c) There are infinitely many solutions.	The three planes intersect in a line, or the three planes are coincident.	(c) 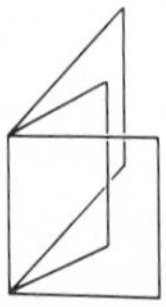

FIG. 5.3

The solution set of a system of three linear equations in three variables can be obtained by using methods similar to those used for systems of two linear equations in two variables.

One of the variables can be eliminated by adding multiples of two of the equations. Eliminating the same variable using a different pair of equations, we obtain two equations in two variables. These can be solved by the methods of the previous section. By substituting the values of these two variables in one of the original equations of the system, the value of the third variable is obtained.

EXAMPLE 1   Solve the system

$$x + 2y - 3z = -11$$
$$x - y - z = 2$$
$$x + 3y + 2z = -4$$

Solution

1. Eliminate z by using the first and second equations:

$$-1(x + 2y - 3z = -11) \rightarrow -x - 2y + 3z = 11$$
$$3(x - y - z = 2) \rightarrow \underline{3x - 3y - 3z = 6}$$
$$2x - 5y \quad\quad = 17$$

2. Eliminate z by using the second and third equations:

$$2(x - y - z = 2) \rightarrow 2x - 2y - 2z = 4$$
$$1(x + 3y + 2z = -4) \rightarrow \underline{x + 3y + 2z = -4}$$
$$3x + y \quad\quad = 0$$

3. Now solve the system $2x - 5y = 17$, $3x + y = 0$.

$$1(2x - 5y = 17) \rightarrow 2x - 5y = 17$$
$$5(3x + y = 0) \rightarrow \underline{15x + 5y = 0}$$
$$17x \quad\quad = 17 \quad \text{and} \quad x = 1$$

$$3(2x - 5y = 17) \rightarrow 6x - 15y = 51$$
$$-2(3x + y = 0) \rightarrow \underline{-6x - 2y = 0}$$
$$-17y = 51 \quad \text{and} \quad y = -3$$

4. Use any one of the three original equations to find z. Using the third equation, $x + 3y + 2z = -4$, with $x = 1$ and $y = -3$.

$$1 + 3(-3) + 2z = -4$$
$$-8 + 2z = -4$$
$$2z = 4 \quad \text{and} \quad z = 2$$

5. State the solution set.
   The solution set is $(1, -3, 2)$.

6. Check the solution in each equation.

$$x + 2y - 3z = 1 + 2(-3) - 3(2) = 1 - 6 - 6 = 1 - 12 = -11$$
$$x - y - z = 1 - (-3) - 2 = 1 + 3 - 2 = 4 - 2 = 2$$
$$x + 3y + 2z = 1 + 3(-3) + 2(2) = 1 - 9 + 4 = -8 + 4 = -4$$

189

EXAMPLE 2 Solve

$2x + 2y - 2z = 1$
$5x - 2y + z = 2$
$3x + 3y - 3z = 5$

Solution   Eliminating z using the first two equations,

$1(2x + 2y - 2z = 1) \rightarrow$ $2x + 2y - 2z = 1$
$2(5x - 2y + z = 2) \rightarrow$ $\underline{10x - 4y + 2z = 4}$
$12x - 2y\ \ = 5$

Eliminating z using the last two equations,

$3(5x - 2y + z = 2) \rightarrow 15x - 6y + 3z =\ \ 6$
$1(3x + 3y - 3z = 5) \rightarrow \underline{3x + 3y - 3z =\ \ 5}$
$18x - 3y\ \ = 11$

Solving the system,

$12x - 2y = 5$ and $18x - 3y = 11$
$3(12x - 2y = 5) \rightarrow 36x - 6y = 15$
$-2(18x - 3y = 11) \rightarrow \underline{-36x + 6y = -22}$
$0 = -7, \textit{false}$

Since a false statement is obtained, the equations are inconsistent and the solution set is the empty set, $\varnothing$. A check may be obtained by repeating the solution process, eliminating either x or y as the initial step.

EXAMPLE 3   Solve the system

$x - 2y + 3z = 4$
$2x + y - z = 1$
$3x - y + 2z = 5$

Solution   Eliminating z using the first two equations,

1. $1(x - 2y + 3z = 4) \rightarrow$ $x - 2y + 3z = 4$
$3(2x + y - z = 1) \rightarrow \underline{6x + 3y - 3z = 3}$
$7x +\ \ y\ \ = 7$

Eliminating z using the last two equations,

2. $2(2x + y - z = 1) \rightarrow 4x + 2y - 2z = 2$
$1(3x - y + 2z = 5) \rightarrow \underline{3x -\ \ y + 2z = 5}$
$7x +\ \ y\ \ = 7$

Since the same equation is obtained in both cases, the solution set is infinite. Solving this equation for y, $y = 7 - 7x$.

Now, replacing y by $7 - 7x$ in the second equation of the system,

$2x + (7 - 7x) - z = 1$
$-5x - z = -6$
$z = 6 - 5x$

The solution set may be written as $\{(x, 7 - 7x, 6 - 5x)\}$

Check

For $x - 2y + 3z = 4$,
$x - 2(7 - 7x) + 3(6 - 5x) = x - 14 + 14x + 18 - 15x = 4$
For $2x + y - z = 1$,
$2x + 7 - 7x - 6 + 5x = 1$

For $3x - y + 2z = 5$,
$$3x - 7 + 7x + 2(6 - 5x) = 10x - 7 + 12 - 10x = 5$$

There are alternate forms for the solution set of Example 3. The components of the ordered triple may each be expressed as functions of y, or of z, or of a third variable, say t.

For example, solving $7x + y = 7$ for x,
$$x = \frac{7 - y}{7} = 1 - \frac{y}{7}$$

Now using $z = 6 - 5x$ and substituting,
$$z = 6 - 5\left(1 - \frac{y}{7}\right) = 1 + \frac{5y}{7}$$

Therefore another form for the solution set is $\left\{\left(1 - \frac{y}{7}, y, 1 + \frac{5y}{7}\right)\right\}$

If we let $y = 7t$ in order to get rid of the fractions, then we obtain another form for the solution set:
$$\{(1 - t, 7t, 1 + 5t)\}$$

Similarly, by solving $z = 6 - 5x$ for x and substituting in $y = 7 - 7x$, the solution set can be expressed in terms of z:
$$\left\{\left(\frac{6 - z}{5}, \frac{7z - 7}{5}, z\right)\right\}$$

Actually many forms are possible; however, each ordered triple of the solution set must satisfy each of the original equations of the system. This can be verified by checking.

EXAMPLE 4 Solve
$$x - 2y - 3z = 3$$
$$x + y - z = 2$$
$$2x - 3y - 5z = 5$$

Solution Eliminating x using the first two equations,
$$\begin{array}{r} -x + 2y + 3z = -3 \\ \underline{x + y - z = 2} \\ 3y + 2z = -1 \end{array}$$

Eliminating x using the first and third equations,
$$\begin{array}{r} 2(x - 2y - 3z = 3) \rightarrow \quad 2x - 4y - 6z = 6 \\ -1(2x - 3y - 5z = 5) \rightarrow \underline{-2x + 3y + 5z = -5} \\ -y - z = 1 \\ y + z = -1 \end{array}$$

Solving, $y + z = -1$ and
$$\begin{array}{r} 3y + 2z = -1 \\ \underline{-2y - 2z = +2} \\ y = 1 \end{array}$$

Since $y + z = -1$, $1 + z = -1$ and $z = -2$.

Using the first equation with $y = 1$ and $z = -2$, $x - 2y - 3z = 3$ becomes $x - 2 + 6 = 3$ and $x = -1$.

The solution is $(-1, 1, -2)$.

EXERCISES 5.2

(1–16) Solve and check.

1. $x + y - z = 8$
$x - y + z = 2$
$3x - y - z = 4$

2. $2x - y - z = 6$
$3x + y - 2z = 4$
$2x - 4y + z = 12$

3. $x + y = 3$
$y + z = 4$
$x + z = 11$

4. $x - y = 5$
$y - z = 7$
$x - z = 12$

5. $x + y + z = 6$
$2x - y + z = 3$
$3x - y - z = 2$

6. $5x + 3y - 8 = 0$
$3x - 4z - 10 = 0$
$x - 6y + 2z = 14$

7. $2x + y = 3 - 2z$
$2x - y = 3z + 22$
$3x + 2z = 2y - 1$

8. $3x + 2y = 4z - 10$
$4x + 2z = 3y - 14$
$4y + 3z = 21 - 2x$

9. $x - y - z = 4$
$3x - 3y - 3z = 7$
$6x + y - 2z = 5$

10. $a + b + c = 4$
$a + 2b + 2c = 1$
$2a + 2b + 5c = 2$

11. $a - 3b = 6c - 1$
$a + b = 1 - 2c$
$4a - b = 2 - 2c$

12. $2a + b = 9 - 3c$
$3a + 2c = b + 16$
$a - 3b = 2c + 8$

13. $a + b - c = 0$
$2a - b + c = 0$
$a - 2b + 3c = 0$

14. $x + 3y + z = 6$
$2x - y + 2z = 5$
$3x - 2y + 3z = 6$

15. $x - y + z = 0$
$3x + y - 3z = 0$
$9x - y - 3z = 0$

16. $x + 2y - z = 0$
$2x - y + 2z = 0$
$x - 2y + 2z = 0$

(17–20) Find the solution set of each system.

EXAMPLE 5

$x + 3y - 5z = 8$
$2x - y - 3z + 5 = 0$

Solution   Treat the system as a set of linear equations in two variables; for example, x and y. Solve for these two variables in terms of the third variable.

$x + 3y = 5z + 8$
$2x - y = 3z - 5$

Eliminating y,

$$x + 3y = 5z + 8$$
$$\underline{6x - 3y = 9z - 15}$$
$$7x \qquad = 14z - 7$$
$$x \qquad = 2z - 1$$

Substituting in the first equation,

$(2z - 1) + 3y = 5z + 8$
$3y = 3z + 9$
$y = z + 3$

The solution set is

$\{(2z - 1, z + 3, z) | z \text{ is any real number}\}$

17. $2x + y - 5z + 6 = 0$
$x - 3y + 8z + 10 = 0$

18. $3x + 2y = z + 12$
$2x - y = z + 1$

19. $2x + 3y = z + 17$
$3x + 2y = 4z + 13$

20. $2x - y - z = 3$
$3x - y - 3z = 9$

(21–28) Find the solution set of the system. First solve the system consisting of the first three equations. Then determine whether this solution satisfies the fourth equation. If it does, the solution set has been found. If it does not, the solution set is the empty set.

21. $x + y + z = 12$
$2x - y + z = 6$
$3x - y - z = 4$
$5x + 2y - 2z = 24$

22. $x + y + z = 4$
$2x - y + 3z = 14$
$3x - y - 2z = 1$
$5x - 2y + z = 15$

23. $x + 3y - z = 2$
$3x - 2y + 3z = 4$
$2x - y + 2z = 3$
$3x + 2y + z = 5$

24. $x - 3y + 2z = 1$
$2x - 5y + 3z = 1$
$3x + 2y - 4z = 4$
$6x - 6y + z = 6$

25. $x + y - 2z = 3$
$2x - y - z = 3$
$3x - 3y + z = 7$
$x + 2y + 2z = 10$

26. $x - 2y + z = 3$
$2x - y - z = 0$
$2x + y - 3z = -4$
$5x - 2y - 2z = -1$

27. $x + y - 2z = 10$
$x - 2y - z = 8$
$2x - y - 3z = 18$
$3x - 3y - 4z = 26$

28. $x - y - z = 0$
$2x + y + z = 9$
$2x - 2y + 3z = 5$
$4x + y + 5z = 12$

(29–36) Solve, using three variables.

29. To manufacture three different products, a company uses 2400 pounds of ingredient *A,* 310 pounds of ingredient *B,* and 28 pounds of ingredient *C.* The number of pounds needed for each one of the three products is shown in the table.

Ingredient \ Product	P	Q	R
A	25	20	150
B	5	2	10
C	0	1	$\frac{1}{2}$

How many items of each product should be produced in order to use all of the raw material?

Solve
$$25x + 20y + 150z = 2400$$
$$5x + 2y + 10z = 310$$
$$y + \tfrac{1}{2}z = 28$$

30. When Kirchhoff's laws are applied to the electrical circuit shown in Figure 5.4, the following equations are obtained.

$$I_1 - I_2 - I_3 = 0$$
$$5I_1 + 20I_3 = 100$$
$$15I_2 - 20I_3 = 15$$
$$5I_1 + 15I_2 = 115$$

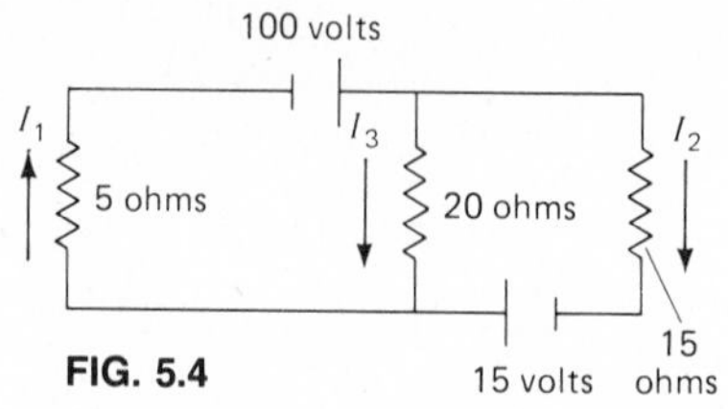

FIG. 5.4

Solve this system by showing that the solution of the first three equations is also a solution of the fourth equation.

31. Three machines, working together, require 20 minutes to complete a certain job. The first two machines require 30 minutes to do the job working together. The first and the third machines require 36 minutes to do the job working together. Find the time it would take for each machine working alone to do the job.

32. The array below shows the amount of nutrient in each food per unit of food. Find how many units of each food will provide the total amounts of nutrients indicated in the last column.

Nutrient	Food			Total
	P	Q	R	
A	1	3	2	32
B	2	2	0	20
C	2	1	5	39

33. When copper is dissolved in nitric acid, the chemical equation for the reaction is
$$xCu + 2yHNO_3 \rightarrow xCu(NO_3)_2 + yH_2O + zNO$$
To balance the oxygen, O,
$$6y = 6x + y + z$$
To balance the nitrogen, N,
$$2y = 2x + z$$
Find a solution for these equations so that x, y, and z are the smallest positive integers possible.

34. An anhydride $C_8H_4O_3$ is made in the plastics industry by the oxidation of naphthalene according to the equation
$$xC_{10}H_8 + yO_2 \rightarrow xC_8H_4O_3 + zCO_2 + zH_2O$$
To balance the carbon C,
$$10x = 8x + z$$
To balance the hydrogen H,
$$8x = 4x + 2z$$
To balance the oxygen O,
$$2y = 3x + 3z$$
Find the least positive integers for x, y, and z.

35. When the three resistors A, B, and C are joined in parallel, the total resistance in the electrical circuit is 6 ohms. When 1 of the A, 1 of the B, and 3 of the C are used, the total resistance is 4.5 ohms. When 3 of A, none of B, and 9 of C are used, the total resistance is 2 ohms. Find the number of ohms for each type of resistor.

$$\frac{1}{x} + \frac{1}{y} + \frac{1}{z} = \frac{1}{6}$$

$$\frac{1}{x} + \frac{1}{y} + \frac{3}{z} = \frac{1}{4.5} = \frac{2}{9}$$

$$\frac{3}{x} + \frac{9}{z} = \frac{1}{2}$$

36. In genetics, when a hybrid is bred with a hybrid, the offspring may be dominant, recessive, or hybrid. If d, r, and h represent the probabilities for each of these types, then in future generations,

$$d + r + h = 1$$

$$\tfrac{1}{2}d + \tfrac{1}{4}h = d$$

$$\tfrac{1}{2}d + \tfrac{1}{2}r + \tfrac{1}{2}h = h$$

$$\tfrac{1}{2}r + \tfrac{1}{4}h = r$$

Solve this system for d, r, and h.

5.3 NONLINEAR SYSTEMS, SUBSTITUTION METHOD

A **nonlinear system of equations** is a system in which one or more of the equations in the system is not linear. In this section we consider systems of two equations in two variables where one equation is linear and the other quadratic or where both equations are quadratic.

When a quadratic equation is involved, the values for x or y may be either real or imaginary numbers. Graphing the two equations provides insight as to the nature and number of the real solutions. However, as a general rule, only approximations to the real solutions can be obtained from the graphs. Again an algebraic solution process is necessary to obtain the exact values of both the real and imaginary solutions.

For a system consisting of one linear and one quadratic equation, the solution set may contain two real solutions, one real solution, or two imaginary solutions. These possibilities are illustrated in Figure 5.5 for a line and an ellipse.

 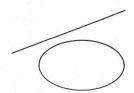

(a) Two real solutions.

(b) One real solution (a point of tangency.).

(c) No real solutions; two imaginary solutions.

FIG. 5.5. One linear equation and one quadratic equation.

For a system consisting of two quadratic equations, the possibilities for the solutions are as follows:

1. 4 real solutions
2. 3 real solutions
3. 2 real solutions and 2 imaginary solutions
4. 1 real solution and 2 imaginary solutions
5. 0 real solution and 4 imaginary solutions

These possibilities are illustrated in Figure 5.6.

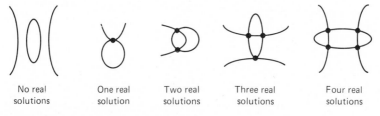

No real solutions

One real solution

Two real solutions

Three real solutions

Four real solutions

FIG. 5.6 Two quadratic equations.

SUBSTITUTION METHOD

The substitution method used to solve linear systems can also be used to solve nonlinear systems. This method is especially useful when it is easy to solve for one of the variables in terms of the other variable.

EXAMPLE 1 Solve
$$4x^2 + y^2 = 100$$
$$2x - y = 2$$
by the substitution method.

Solution Solving the linear equation for y,
$y = 2x - 2.$

Substituting in the quadratic equation and solving for x,

$$4x^2 + (2x - 2)^2 = 100$$
$$4x^2 + 4x^2 - 8x + 4 = 100$$
$$8x^2 - 8x - 96 = 0$$
$$x^2 - x - 12 = 0$$
$$(x - 4)(x + 3) = 0$$
$$x = 4 \quad \text{or} \quad x = -3$$

The original system is now replaced by the equivalent system.

$$y = 2x - 2 \quad \text{and} \quad (x = 4 \text{ or } x = -3)$$

which in turn is equivalent to

$$(y = 2x - 2 \text{ and } x = 4) \quad \text{or} \quad (y = 2x - 2 \text{ and } x = -3)$$

Solving, for $x = 4$, $y = 2(4) - 2 = 6$;

for $x = -3$, $y = 2(-3) - 2 = -8$.

The solutions are $(4, 6)$ and $(-3, -8)$.

The solution set is $\{(4, 6), (-3, -8)\}$.

Check

$$4x^2 + y^2 = 100; \qquad 4(4^2) + 6^2 = 64 + 36 = 100$$
$$4(-3)^2 + (-8)^2 = 36 + 64 = 100$$
$$2x - y = 2; \qquad 2(4) - 6 = 8 - 6 = 2$$
$$2(-3) - (-8) = -6 + 8 = 2$$

Figure 5.7 shows the geometric interpretation of this problem.

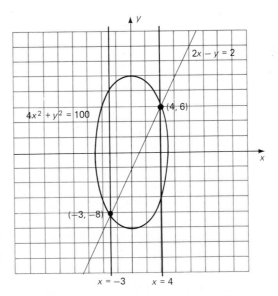

FIG. 5.7. Equivalent systems, $\begin{Bmatrix} 4x^2 + y^2 = 100 \\ 2x - y = 2 \end{Bmatrix}$ **and** $\begin{Bmatrix} 2x - y = 2 \\ x = 4 \text{ or } x = -3 \end{Bmatrix}$.

EXAMPLE 2 Solve
$$x^2 - y^2 = 16$$
$$y^2 = x + 4$$
by the substitution method.

Solution Solving the second equation for x, $x = y^2 - 4$. Substituting in the first equation,
$$(y^2 - 4)^2 - y^2 = 16$$
$$y^4 - 9y^2 = 0$$
$$y^2(y^2 - 9) = 0$$
$$y = 0 \quad \text{or} \quad y = 3 \quad \text{or} \quad y = -3$$
Using the second equation with these three linear equations, we can readily solve this equivalent system:

$x = y^2 - 4$ or	$x = y^2 - 4$ or	$x = y^2 - 4$
$y = 0$	$y = 3$	$y = -3$
$x = -4$	$x = 9 - 4$	$x = 9 - 4$
	$x = 5$	$x = 5$
$(-4, 0)$	$(5, 3)$	$(5, -3)$

The solution set is $\{(-4, 0), (5, 3), (5, -3)\}$.

Each of these solutions is checked by substituting into both equations of the original system.

Alternate Solution Note that y^2 appears in both equations. One can also replace y^2 in the first equation by its equal expression in the second equation.

Using $y^2 = x + 4$, then $x^2 - y^2 = 16$ becomes
$$x^2 - (x + 4) = 16$$
and
$$x^2 - x - 20 = 0$$
$$(x - 5)(x + 4) = 0$$
$$x = 5 \quad \text{or} \quad x = -4$$
Forming the equivalent system,
$$y^2 = x + 4 \quad \text{and} \quad (x = 5 \text{ or } x = -4)$$
when $x = 5$, $y^2 = 5 + 4 = 9$ and $y = \pm 3$;
when $x = -4$, $y^2 = -4 + 4 = 0$ and $y = 0$.

Figure 5.8 provides the geometric interpretation of this problem.

EXAMPLE 3 Solve
$$xy = 12$$
$$y^2 - 4x^2 = 20$$
by the substitution method.

Solution The first equation can easily be solved for either x or y. Selecting y,
$$y = \frac{12}{x}$$
Substituting in the second equation,
$$\left(\frac{12}{x}\right)^2 - 4x^2 = 20$$

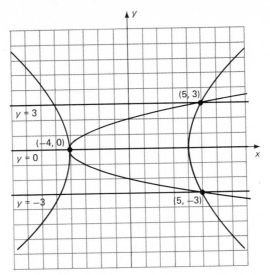

FIG. 5.8. Equivalent systems, $\begin{Bmatrix} x^2 - y^2 = 16 \\ y^2 = x + 4 \end{Bmatrix}$ and $\begin{Bmatrix} y^2 = x + 4 \\ y = 0 \text{ or } y = 3 \text{ or } y = -3 \end{Bmatrix}$.

$$\frac{144}{x^2} - 4x^2 = 20$$
$$144 - 4x^4 = 20x^2$$
$$4x^4 + 20x^2 - 144 = 0$$
$$x^4 + 5x^2 - 36 = 0$$
$$(x^2 + 9)(x^2 - 4) = 0$$
$$x^2 + 9 = 0 \quad \text{or} \quad x^2 - 4 = 0$$
$$x^2 = -9 \quad \text{or} \quad x^2 = 4$$
$$x = \pm 3i \quad \text{or} \quad x = \pm 2$$

Using the first equation to obtain the corresponding y-values,

$$x = 3i \quad \text{and} \quad y = \frac{12}{x} = \frac{12}{3i} = \frac{4i}{ii} = -4i$$

$$x = -3i \quad \text{and} \quad y = \frac{12}{x} = \frac{12}{-3i} = \frac{4i}{-ii} = 4i$$

$$x = 2 \quad \text{and} \quad y = \frac{12}{x} = \frac{12}{2} = 6$$

$$x = -2 \quad \text{and} \quad y = \frac{12}{x} = \frac{12}{-2} = -6$$

The solution set is
$\{(2, 6), (-2, -6), (3i, -4i), (-3i, 4i)\}$
Checking in $y^2 - 4x^2 = 20$,
For $(2, 6)$, $36 - 4(2^2) = 36 - 16 = 20$.
For $(-2, -6)$, $36 - 4(-2)^2 = 36 - 16 = 20$.
For $(3i, -4i)$, $(-4i)^2 - 4(3i)^2 = 16i^2 - 36i^2$
$$= -16 + 36 = 20.$$
For $(-3i, 4i)$, $(4i)^2 - 4(-3i)^2 = 16i^2 - 36i^2 = 20$.

Figure 5.9 shows the geometric interpretation of this problem. Note that the graphic illustration does not exhibit the imaginary solutions.

199

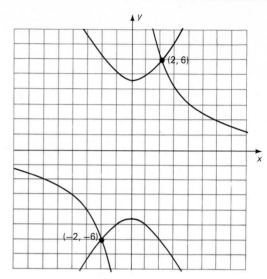

FIG. 5.9. $xy = 12$ and $y^2 - 4x^2 = 20$.

EXERCISES 5.3

(1–10) Solve each system algebraically. Sketch the graph of each system, and approximate, to the nearest tenth, the real solutions.

1. $y = 9 - x^2$
$y = x + 7$

2. $y = 2x - 1$
$y^2 - x^2 = 16$

3. $x^2 - y^2 = 9$
$x = 2y + 3$

4. $5x + 4y = 20$
$xy = 5$

5. $xy = 6$
$x^2 + 4y^2 = 25$

6. $x^2 - y^2 = 5$
$3y^2 - x^2 = 3$

7. $y = 4x - 5$
$9x^2 + y^2 = 9$

8. $x^2 = 2y + 20$
$x^2 + y^2 = 100$

9. $x^2 + y^2 = 25$
$y^2 = 10 - 2x$

10. $x^2 + y^2 = 4x + 2y - 1$
$y^2 = x + 2y - 1$

(11–20) Solve by the substitution method. Check each solution.

11. $3x + y = 8$
$x^2 - y^2 = 16$

12. $3x - y = 6$
$y^2 = 8y - 3x$

13. $y = x^2 + 2x + 4$
$x = 4y - 16$

14. $2x^2 + y^2 = 12$
$x - 2y = 2$

15. $2x - y = 1$
$4x^2 + y^2 = 25$

16. $2x = y - 2$
$2x^2 - 2xy + y^2 = 10$

17. $xy + 12 = 0$
$2x + 3y = 6$

18. $xy = 4$
$x + 3y = 10$

19. $y^2 = 4x$
$x^2 = y^2 - 3$

20. $xy + y = 1$
$xy - x = 4$

5.4 NONLINEAR SYSTEMS, ADDITION METHOD

ELIMINATING ONE OF THE VARIABLES

When both equations of a system are quadratic, the substitution method generally yields a complicated equation involving radicals, and, upon simplification, a fourth degree equation which is not always easy to solve.

For certain special cases, it is possible to eliminate one of the variables by the **addition method** which is similar to that used for linear systems. Each equation of the system is multiplied by an appropriate constant and the resulting equations are added. The procedure is justified by reasoning similar to that used for the linear systems.

If (a, b) is a solution of both

$$A_1 x^2 + B_1 xy + C_1 y^2 + D_1 x + E_1 y + F_1 = 0$$

and

$$A_2 x^2 + B_2 xy + C_2 y^2 + D_2 x + E_2 y + F_2 = 0$$

then it is also a solution of any linear combination of the two quadratic equations. Thus (a, b) is a solution of

$$k(A_1 x^2 + B_1 xy + C_1 y^2 + D_1 x + E_1 y + F_1)$$
$$+ n(A_2 x^2 + B_2 xy + C_2 y^2 + D_2 x + E_2 y + F_2) = 0$$

since

$$k(A_1 a^2 + B_1 ab + C_1 b^2 + D_1 a + E_1 b + F_1) +$$
$$n(A_2 a^2 + B_2 ab + C_2 b^2 + D_2 a + E_2 b + F_2) = k(0) + n(0) = 0$$

EXAMPLE 1 Using the addition method, solve the system

$$3x^2 + 2y^2 = 23$$
$$x^2 - 3y^2 = -7$$

Solution

1. To eliminate y^2, multiply the first equation by 3:
   $9x^2 + 6y^2 = 69$
   Multiply the second equation by 2:
   $2x^2 - 6y^2 = -14$
2. Add the new equations:
   $11x^2 = 55$
3. Solve for x:
   $x^2 = 5.$
   $x = \pm\sqrt{5}.$
4. To eliminate x^2, multiply the first equation by 1:
   $3x^2 + 2y^2 = 23$
   Multiply the second equation by -3:
   $-3x^2 + 9y^2 = 21$

5. Add the new equations:
$$11y^2 = 44$$
6. Solve for y:
$$y^2 = 4$$
$$y = \pm 2$$
7. List the solution set:
$$\{(\sqrt{5}, 2), (\sqrt{5}, -2), (-\sqrt{5}, 2), (-\sqrt{5}, -2)\}$$
8. Check each solution in both equations of the original system. (The check is left for the student.)

ELIMINATING THE SECOND-DEGREE TERMS

When it is not possible to eliminate one of the variables by adding multiples of the two equations, it may be possible to eliminate the second-degree terms and obtain a linear equation. Since each solution of the quadratic system must also be a solution of the linear equation, the linear equation may be solved simultaneously with either equation of the quadratic system.

EXAMPLE 2 Solve
$$x^2 + y^2 - 2x = 9$$
$$x^2 + y^2 - 4y = 1$$

Solution

1. Eliminating the second-degree terms,
$$(x^2 + y^2 - 2x = 9)(-1) \rightarrow -x^2 - y^2 + 2x = -9$$
$$(x^2 + y^2 - 4y = 1)(1) \quad \rightarrow \quad x^2 + y^2 - 4y = 1$$
Adding, $\qquad\qquad\qquad\qquad\qquad 2x - 4y = -8$
Simplifying $\qquad\qquad\qquad\qquad\quad\; x - 2y = -4$

2. Solving the equivalent system,
$$x^2 + y^2 - 4y = 1$$
$$x - 2y = -4$$
Using the substitution method,
$$x = 2y - 4$$
$$(2y - 4)^2 + y^2 - 4y = 1$$
$$5y^2 - 20y + 15 = 0$$
$$y^2 - 4y + 3 = 0$$
$$(y - 3)(y - 1) = 0$$
Thus $y = 3$ or $y = 1$.
If $y = 3$, then $x = 2y - 4 = 2(3) - 4 = 2$.
If $y = 1$, then $x = 2y - 4 = 2(1) - 4 = -2$.
Thus the solution set is $\{(2, 3), (-2, 1)\}$.

ELIMINATING THE CONSTANTS

Sometimes the solution set of a quadratic system may be easily obtained by using the addition method to eliminate the constants.

EXAMPLE 3 Solve

$x^2 + xy = 4$
$y^2 - xy = 6$

Solution Eliminating the constants,

$(x^2 + xy = 4)(3) \quad \rightarrow \quad 3x^2 + 3xy = 12$
$(y^2 - xy = 6)(-2) \rightarrow \quad \underline{-2y^2 + 2xy = -12}$

Adding, $3x^2 + 5xy - 2y^2 = 0$

Factoring, $(3x - y)(x + 2y) = 0$

$3x - y = 0 \quad$ or $\quad x + 2y = 0$
$\quad y = 3x \quad$ or $\quad\quad x = -2y$

Now either equation of the original system can be replaced by this set of two linear equations. Thus the original system is equivalent to the union of the two systems:

$$\begin{cases} x^2 + xy = 4 \\ \quad\quad y = 3x \end{cases} \quad \text{or} \quad \begin{cases} x^2 + xy = 4 \\ \quad\quad x = -2y \end{cases}$$

Each of these systems can now be solved by the substitution method.

$x^2 + x(3x) = 4 \quad\quad\quad\quad (-2y)^2 + (-2y)y = 4$
$\quad\quad 4x^2 = 4 \quad\quad\quad\quad\quad\quad\quad\quad 2y^2 = 4$
$\quad\quad\quad x^2 = 1 \quad\quad\quad\quad\quad\quad\quad\quad\quad y^2 = 2$
$\quad\quad\quad\quad x = \pm 1 \quad\quad\quad\quad\quad\quad\quad\quad y = \pm\sqrt{2}$

If $x = 1$, then $y = 3x = 3$. If $y = \sqrt{2}$, then $x = -2y = -2\sqrt{2}$.
If $x = -1$, then $y = -3$. If $y = -\sqrt{2}$, then $x = 2\sqrt{2}$.

Thus the solution set is

$\{(1, 3), (-1, -3), (-2\sqrt{2}, \sqrt{2}), (2\sqrt{2}, -\sqrt{2})\}$

Each of these solutions should be checked in both equations of the original system.

EXERCISES 5.4

(1–12) Solve by the addition method, by eliminating a variable. Check each solution.

1. $x^2 + y^2 = 25$
$x^2 - y^2 = 7$

2. $x^2 + 4y^2 = 13$
$x^2 - 4y^2 = 5$

3. $2x^2 + y^2 = 13$
$3x^2 + y^2 = 17$

4. $2x^2 + y^2 - 5 = 0$
$x^2 - 5y^2 + 3 = 0$

5. $x^2 - y^2 + 7 = 0$
$3x^2 + 2y^2 = 24$

6. $3x^2 - 2y^2 = 6$
$5x^2 - 3y^2 = 7$

7. $4x^2 + xy = 11$
$8x^2 - 5xy = 8$

8. $9x^2 + 2xy = 15$
$9x^2 + 3xy = 10$

9. $xy + y^2 = 50$
$2y^2 - xy = 10$

10. $y^2 - 2xy = 9$
$10xy - y^2 = 3$

11. $x^2 + 2xy - y^2 = 28$
$2x^2 - 2xy + y^2 = 20$

12. $2x^2 - y^2 + 2x - 3y = 42$
$x^2 + 2y^2 = 4x + 6y = 41$

(13–18) Solve by eliminating either the second-degree terms or the constant. Check each solution.

13. $x^2 - xy = 3$
$2y^2 - xy = 2$

14. $x^2 + xy = 3$
$y^2 + xy = 6$

15. $\qquad xy = 24$
$xy + x - 2y = 22$

16. $\qquad xy = 120$
$xy + x - 4y = 124$

17. $x^2 + y^2 = 2x$
$x^2 + y^2 = 6y$

18. $x^2 - y^2 + 2x - y + 12 = 0$
$x^2 - y^2 + 4x - 2y + 12 = 0$

19. A hotel manager bought a supply of towels for $180. Later when the price increased by $1, she found she received 9 less towels for the same amount of money. Find the original price per towel.

Let

x = original price per towel
y = number of towels in first purchase

Then

$$xy = 180 \quad \text{and} \quad (x + 1)(y - 9) = 180$$

Complete the solution.

20. An open box is to be made from a rectangular piece of cardboard by cutting off 3-inch squares from the corners and then bending up the edges. If the volume is to be 288 cubic inches and the area of the original rectangular cardboard is 252 square inches, find the dimensions of the original piece of cardboard.

21. In economics, a supply law is an equation that relates the price y and the number of units x of a product the manufacturer can supply. A demand law is an equation relating the price y and the number of units x that the consumers demand. The values of x and y at a point of intersection of the demand curve and the supply curve are called the equilibrium quantity and price, respectively. Find these values for the following laws. Note that x and y cannot be negative.

Supply law: $\qquad y = x^2 + 20$
Demand law: $\qquad y = 90 - 4x - x^2$

22. The path of a projectile fired at a 45° angle with an initial velocity of 640 feet per second is given by the equation

$$y = \frac{12,800x - x^2}{12,800}$$

How high up does it hit a hill whose slope is given by the equation

$3y = 2x - 12,000$

where x and y are measured in feet? (Find the y-value at the point of intersection.)

23. When 2 resistors are joined in series, the total resistance is 25 ohms. When they are joined in parallel, the total resistance is 4 ohms. Find the number of ohms for each resistor.

$$x + y = 25 \quad \text{and} \quad \frac{1}{x} + \frac{1}{y} = \frac{1}{4}$$

24. Find the dimensions of a rectangle if its diagonal is 25 meters and its area is 300 square meters.

25. A motorist has a choice of two roads, A and B, to go from one town to another. His rate on road B is 20 mph faster and his time is 1 hour less. However, road B is 165 miles long while road A is 140 miles long. Find his rate r and time t on road A.

26. A stone is heard hitting the bottom of a ravine 7.7 seconds after it is dropped. If the speed of sound is 1120 feet per second and if the stone falls $16t^2$ feet in t seconds, find how deep the ravine is.

Let d = the depth of the ravine. Then

$$d = 16t^2 \quad \text{and} \quad t + \frac{d}{1120} = 7.7$$

Solve for d.

5.5 SYSTEMS OF INEQUALITIES, LINEAR PROGRAMMING

There are many problems in science, business, economics, military tactics, and other areas that require the solution of a system of inequalities. Usually these involve many variables and many inequalities and are solved by using an electronic computer.

In this section, as an introduction to this area of application, we consider systems of linear inequalities in two variables.

DEFINITION

The **solution set of a system of inequalities** is the intersection of the solution sets of the inequalities in the system.

For example, the solution set of the system

$$3y \leq 2x + 12 \quad \text{and} \quad 2x + 3y \leq 24$$

is the set

$$\{(x, y) \mid 3y \leq 2x + 12\} \cap \{(x, y) \mid 2x + 3y \leq 24\}$$

A graphical representation provides a better understanding of the solution set, as shown in Examples 1 and 2.

EXAMPLE 1 Graph the solution set of the system

$$x \geq 0, \quad y \geq 0, \quad 3y \leq 2x + 12, \quad \text{and} \quad 2x + 3y \leq 24$$

Solution Each inequality is graphed on the same set of axes and the region common to all of the inequalities is indicated by a shading. See Figure 5.10.

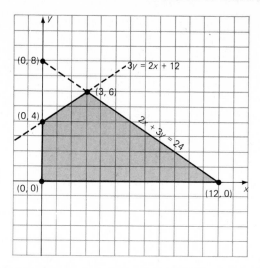

FIG. 5.10. System of Example 1, shaded convex region.

EXAMPLE 2 Graph the solution set of the system
$$0 \le x \le 11, \quad y \ge 0, \quad \text{and} \quad y \le |x - 5| + 2$$

Solution The shaded region in Figure 5.11 is the graph of the solution set.

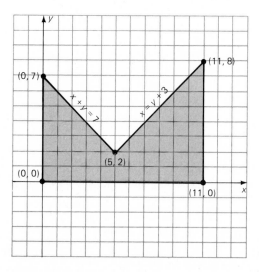

FIG. 5.11. Graph of system of Example 2, shaded region, not convex.

Linear programming, an application that has become increasingly more important since its introduction in the 1940s, is concerned with finding a solution of a system of linear inequalities so that the solution produces a maximum or a minimum of a specified function of the variables.

The inequalities of the system are called **constraints,** and the solution set of the system is called the set of **feasible solutions.** The linear function that is to be maximized or minimized is called the **objective function.**

A linear programming problem in two variables can be solved geometrically. If the graph of the solution set is a convex polygonal region, then the maximum or minimum value of the function will occur at one of the vertices.

Figure 5.10 shows a convex polygonal region while Figure 5.11 shows a polygonal region that is not convex.

DEFINITION

A **convex region** is a region that has the following property. For every two points A and B in the region, the line segment AB lies entirely in the region.

The following theorem, stated without proof, provides a method for solving a linear programming problem in two variables.

THEOREM

The **maximum and minimum values** of $ax + by + c$ over a closed convex polygonal region occur at one of the vertices of the region.

Linear Programming Procedure

1. Graph the solution set of the system of inequalities.
2. State the vertices of the solution set.
3. Evaluate the expression that is to be maximized or minimized at each of the vertices.
4. Select the maximum or minimum value from those evaluated in Step 3.

EXAMPLE 3 To produce each of two different products, A and B, a manufacturer must use each of two different machines. The number of hours required on each machine to produce the two products is given in the table, along with the maximum number of hours each machine can be run per week.

Machine	Product A	Product B	Maximum hours for machine
I	1	4	44
II	3	2	42

If the profit on product A is $4 per item and the profit on product B is $5 per item, how many of each product should be manufactured for maximum profit? Find the maximum profit.

Solution   Let
x = number of items of A
y = number of items of B
Then $x \geq 0$, $y \geq 0$, $x + 4y \leq 44$, and
$$3x + 2y \leq 42.$$
The profit $P = 4x + 5y$.

1. Graph the solution set of the inequalities. See Figure 5.12.
2. Find the vertices.

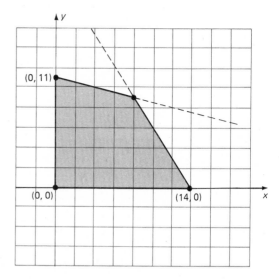

FIG. 5.12

Equations	Vertex
$x = 0, y = 0$	$(0, 0)$
$x = 0, x + 4y = 44$	$(0, 11)$
$y = 0, 3x + 2y = 42$	$(14, 0)$
$3x + 2y = 42, x + 4y = 44$ $3(44 - 4y) + 2y = 42$ $-10y = -90, y = 9,$ and $x = 8$	$(8, 9)$

3. Evaluate $P = 4x + 5y$ at each vertex.

Vertex	$P = 4x + 5y$
$(0, 0)$	0
$(0, 11)$	55
$(14, 0)$	56
$(8, 9)$	$32 + 45 = 77$

4. The maximum profit is $77 for 8 items of A and 9 items of B.

EXERCISES 5.5

(1–8) Graph the solution set of each of the following systems, and find the coordinates of the vertices.

1. $0 \leq x \leq 8, y \geq 0,$ and $2y \leq 3x + 6$
2. $2 \leq x \leq 10, 3 \leq y \leq 12,$ and $3x + 4y \leq 54$
3. $x \geq 0, y \geq 0, y \geq x, y \leq 2x,$ and $5y \leq 2x + 24$
4. $x \geq 0, y \geq 0, 2x + y \leq 24,$ and $x + 2y \leq 18$
5. $3 \leq x \leq 9, 2x + 3y \leq 24,$ and $x - y \geq 2$
6. $0 \leq x \leq 8, 0 \leq y \leq 12, 3x + 5y \geq 45, 2x - y \leq 4$
7. $0 \leq x \leq 10, 0 \leq y \leq 15, y + 4x \geq 10,$ and $2y + 3x \geq 15$
8. $x \leq 2y, y \leq 3x, 6 \leq x + y \leq 12$

9. A manufacturer can produce two different products A and B. He has 40 units of material available and 60 units of labor available. Each item of product A requires 4 units of material and 2 units of labor. Each item of product B requires 2 units of material and 6 units of labor. If the profit on product A is \$5 per item and the profit on product B is \$4 per item, how many items of each product should he manufacture for a maximum profit? Find the maximum profit.

10. A buyer wants to spend no more than \$900 on the purchase of blouses costing \$10 each and slacks costing \$15 a pair. The buyer wants at least 10 pair of slacks and at least three times as many blouses as slacks. Find the number of blouses and the number of pairs of slacks that should be purchased to obtain the maximum profit on their sale if:
    (a) the profit on each blouse is \$5 and the profit on each pair of slacks is \$8.
    (b) the profit on each blouse is \$6 and the profit on each pair of slacks is \$5.

11. Two foods A and B are to be combined in such a way that the combination supplies at least 120 units of vitamins and at least 80 units of minerals. Food A supplies 2 units of vitamins and 2 units of minerals per ounce. Food B supplies 3 units of vitamins and 1 unit of minerals per ounce. If the cost of food A is 10 cents per ounce and the cost of food B is 6 cents per ounce, how should the foods be combined for minimum cost?

12. A promoter wants to combine two entertainment groups for a 60-minute special. Group I provides 1 minute of music for every 3 minutes of comedy. Group II provides 2 minutes of music for each minute of comedy. At least 20 minutes of music and at least 30 minutes of comedy are wanted for maximum viewer appeal. How many minutes of each group should he have if he wants to minimize the cost? Assume the cost is the same for each group.

13. A grower can ship three kinds of citrus fruits from California to Chicago. He intends to ship 800 boxes of fruit on each truck. He must ship at least 200 boxes of oranges, at least 100 boxes of limes, and at least 200 boxes of lemons. If the profit on oranges is \$2 a box, on lemons \$3 a box, and on limes \$1 a box, how should his truck be loaded for maximum profit? (*Hint:* Let $x =$ number of boxes of oranges, $y =$ number of boxes of lemons, $800 - x - y =$ number of boxes of limes.)

14. A student publication containing 80 pages is to contain at least 40 pages of literary articles and at most 20 pages of advertising. It also contains art work limited to one-fourth the number of pages for the literary articles. If the cost per page is \$50 for literary pages, \$100 for art pages, and $-$\$50 for advertisement pages (the advertiser pays for the page), how should the publication be composed for minimum cost? (*Hint:* Let $x =$ number of literary pages, $y =$ number of art pages, and $80 - x - y =$ number of pages of advertisements.) What is the minimum cost?

15. A furniture dealer has space to display 40 sets of furniture consisting of living room sets, bedroom sets, and dining room sets. He wants to display at least 10 of each set but not more than 20 dining room sets. How many of each set should be displayed to maximize the profit if the profit on each living room set is $50, on each bedroom set is $40, and on each dining room set is $30?

16. One company owns two different textile mills that produce three different types of material daily as shown in the table below. Also shown is the minimum number of units of each material the company needs in order to fill its orders.

Material	Mill A	Mill B	Minimum units needed
Percale	4 units	2 units	20
Muslin	3 units	2 units	18
Terry	3 units	4 units	24

The daily cost of running Mill *A* is $5000. The daily cost of running Mill *B* is $3000. Find the number of days each mill should be operated in order to minimize the cost of production.

5.6 CHAPTER REVIEW

(1–5) Solve and check. (5.1)

1. $x + 2y = 24$
   $3x - 4y = 2$

2. $4x + 3y = 3$
   $5x + 4y = 6$

3. $y = 8 - 6x$
   $12x + 2y = 5$

4. $x + 3y = 4$
   $2x + 6y = 8$

5. $\dfrac{6}{x} + \dfrac{4}{y} = 1$
   $\dfrac{4}{x} + \dfrac{8}{y} = 1$

(6–9) Solve and check. (5.2)

6. $x + y - 2z = 0$
   $x - 2y + z = 3$
   $2x - y - z = 3$

7. $a + 2b + c = 6$
   $2a - 2b - 3c = 5$
   $3a + 3b - 2c = 2$

8. $x - 2y + z = 6$
   $2x + y - 3z = 2$

9. $x + y + z = 15$
   $3x - 2y - z = 4$
   $4x + 2y - 3z = 22$
   $x + 4y - 2z = 18$

(10–11) Solve each system and check. (5.3)

10. $2x = y^2 - 5y + 2$
  $2y = x + 8$

11. $x^2 - 3y^2 = 8$
  $xy = 4$

(12–14) Solve each system and check. (5.4)

12. $2x^2 + y^2 = 45$
  $y^2 - 2x^2 = 5$

13. $2x^2 + xy = 2$
  $y^2 + xy = 3$

14. $xy = 150$
  $xy + 2x - 6y = 120$

(15–16) Solve graphically. (5.5)

15. Find the maximum value of $P = 2x + 5y$ over the convex polygonal set defined by the system
  $y \geq 0, \quad 0 \leq x \leq 12, \quad x + y \leq 16, \quad$ and $\quad x \geq 2y - 14$

16. The table below gives the number of units produced daily by two different factories in the manufacture of three different products. The last column gives the minimum number of units of each product that are needed. The daily cost of operating factory A is \$2000. The daily cost of operating factory B is \$4000. Find the number of days each factory should be operated so that the cost will be a minimum.

	Factory A	Factory B	Minimum units needed
Product 1	3	3	15
Product 2	2	7	20
Product 3	8	2	16

MATRICES, DETERMINANTS, APPLICATIONS TO SYSTEMS

In 1750 the Swiss mathematician Gabriel Cramer (1704–1752) introduced a method for solving linear systems using determinants. This method is called **Cramer's Rule.** In this section we learn what determinants are and how to use them to solve linear systems; that is, how to use Cramer's Rule.

     The quadratic formula provides a rule for finding the solutions of a quadratic equation by performing calculations on the coefficients of the equation. A formula can also be found for the solution of a linear system of two equations in two variables as follows.

EXAMPLE 1   Solve

$$a_1 x + b_1 y = c_1$$
$$a_2 x + b_2 y = c_2$$

Solution   Finding x by eliminating y,

$$b_2(a_1 x + b_1 y = c_1) \rightarrow \quad a_1 b_2 x + b_1 b_2 y = b_2 c_1$$
$$-b_1(a_2 x + b_2 y = c_2) \rightarrow \quad \underline{-a_2 b_1 x - b_1 b_2 y = -b_1 c_2}$$
$$(a_1 b_2 - a_2 b_1)x = b_2 c_1 - b_1 c_2$$

If $a_1 b_2 - a_2 b_1 \neq 0$, then

$$x = \frac{b_2 c_1 - b_1 c_2}{a_1 b_2 - a_2 b_1}$$

Finding y by eliminating x,

$$-a_2(a_1 x + b_1 y = c_1) \rightarrow -a_1 a_2 x - a_2 b_1 y = -a_2 c_1$$
$$a_1(a_2 x + b_2 y = c_2) \rightarrow \quad \underline{a_1 a_2 x + a_1 b_2 y = a_1 c_2}$$
$$(a_1 b_2 - a_2 b_1)y = a_1 c_2 - a_2 c_1$$

If $a_1 b_2 - a_2 b_1 \neq 0$, then

$$y = \frac{a_1 c_2 - a_2 c_1}{a_1 b_2 - a_2 b_1}$$

While a formula has been found for x and y, it is not one that is easy to remember in this form. By introducing the expression

$$\begin{vmatrix} a & c \\ b & d \end{vmatrix}$$

and requiring it to have the value $ad - bc$, the equations for x and y can be written in a form that is much easier to remember; namely,

$$x = \frac{\begin{vmatrix} c_1 & b_1 \\ c_2 & b_2 \end{vmatrix}}{\begin{vmatrix} a_1 & b_1 \\ a_2 & b_2 \end{vmatrix}} \quad \text{and} \quad y = \frac{\begin{vmatrix} a_1 & c_1 \\ a_2 & c_2 \end{vmatrix}}{\begin{vmatrix} a_1 & b_1 \\ a_2 & b_2 \end{vmatrix}}$$

DEFINITION

A **determinant of order 2** is an expression having the form $\begin{vmatrix} a_1 & b_1 \\ a_2 & b_2 \end{vmatrix}$ and having the value $a_1 b_2 - a_2 b_1$. The numbers a_1, a_2, b_1, and b_2 are called the **elements** of the determinant.

The pair of equations in Example 1 that express x and y in terms of determinants is called **Cramer's Rule** for solving a system of two linear equations in two variables.

Now let

$$D = \begin{vmatrix} a_1 & b_1 \\ a_2 & b_2 \end{vmatrix} \quad \text{and} \quad N_x = \begin{vmatrix} c_1 & b_1 \\ c_2 & b_2 \end{vmatrix} \quad \text{and} \quad N_y = \begin{vmatrix} a_1 & c_1 \\ a_2 & c_2 \end{vmatrix}$$

Then

$$x = \frac{N_x}{D} \quad \text{and} \quad y = \frac{N_y}{D}$$

and D is called the **determinant of the coefficients.**

Note that N_x, the numerator determinant for x, can be obtained from D by replacing the first column, the coefficients of x, by the constants c_1 and c_2.

Similarly N_y, the numerator determinant for y, can be obtained from D by replacing the second column, the coefficients of y, by c_1 and c_2.

EXAMPLE 2 Solve
$$3x - 2y = 16$$
$$5x + 4y = 12$$
by using Cramer's Rule.

Solution

$$D = \begin{vmatrix} 3 & -2 \\ 5 & 4 \end{vmatrix} = 3(4) - (5)(-2) = 12 + 10 = 22$$

$$N_x = \begin{vmatrix} 16 & -2 \\ 12 & 4 \end{vmatrix} = 16(4) - (12)(-2) = 88$$

$$N_y = \begin{vmatrix} 3 & 16 \\ 5 & 12 \end{vmatrix} = 3(12) - 5(16) = -44$$

$$x = \frac{N_x}{D} = \frac{88}{22} = 4$$

$$y = \frac{N_y}{D} = \frac{-44}{22} = -2$$

The solution is $(4, -2)$.

If $D = 0$ and $N_x \neq 0$ or $N_y \neq 0$, then the solution set is the empty set.

If $D = 0$, $N_x = 0$, and $N_y = 0$, then the solution set contains infinitely many solutions.

When the general system
$$a_1 x + b_1 y + c_1 z = d_1$$
$$a_2 x + b_2 y + c_2 z = d_2$$
$$a_3 x + b_3 y + c_3 z = d_3$$
is solved for x by eliminating y and z, we obtain
$$(a_1 b_2 c_3 - a_1 b_3 c_2 - a_2 b_1 c_3 + a_2 b_3 c_1 + a_3 b_1 c_2 - a_3 b_2 c_1)x$$
$$= d_1 b_2 c_3 - d_1 b_3 c_2 - d_2 b_1 c_3 + d_2 b_3 c_1 + d_3 b_1 c_2 - d_3 b_2 c_1$$
Similar expressions can be obtained for y and z. In this way we can obtain a formula for the solution of the system but it is extremely difficult to remember. To remove this difficulty, we form arrays of numbers whose values are specified so that calculations involving them produce the desired solution of the linear system.

DEFINITION

A **determinant of order 3** is an expression having the form
$$\begin{vmatrix} a_1 & b_1 & c_1 \\ a_2 & b_2 & c_2 \\ a_3 & b_3 & c_3 \end{vmatrix}$$
and having the value
$$a_1 b_2 c_3 - a_1 b_3 c_2 - a_2 b_1 c_3 + a_2 b_3 c_1 + a_3 b_1 c_2 - a_3 b_2 c_1$$

Now, letting

$$D = \begin{vmatrix} a_1 & b_1 & c_1 \\ a_2 & b_2 & c_2 \\ a_3 & b_3 & c_3 \end{vmatrix}$$

and

$$N_x = \begin{vmatrix} d_1 & b_1 & c_1 \\ d_2 & b_2 & c_2 \\ d_3 & b_3 & c_3 \end{vmatrix}$$

$$N_y = \begin{vmatrix} a_1 & d_1 & c_1 \\ a_2 & d_2 & c_2 \\ a_3 & d_3 & c_3 \end{vmatrix}$$

$$N_z = \begin{vmatrix} a_1 & b_1 & d_1 \\ a_2 & b_2 & d_2 \\ a_3 & b_3 & d_3 \end{vmatrix}$$

the general system of three linear equations in three variables can be written in the following equivalent form:

$$Dx = N_x \quad \text{and} \quad Dy = N_y \quad \text{and} \quad Dz = N_z$$

If $D \neq 0$, then

$$x = \frac{N_x}{D} \quad \text{and} \quad y = \frac{N_y}{D} \quad \text{and} \quad z = \frac{N_z}{D}$$

This last statement is called **Cramer's Rule** for solving a system of three linear equations in three variables. The determinant D is called the **determinant of the coefficients.**

However, there is still one difficulty. While the values for x, y, and z are expressed in a form that is easy to remember, it is not as easy to remember how to find the value of a determinant by using the definition. Therefore we develop properties of determinants to make this evaluation easier.

First we can rewrite the value of the determinant in the definition, as follows:

$$a_1(b_2 c_3 - b_3 c_2) - a_2(b_1 c_3 - b_3 c_1) + a_3(b_1 c_2 - b_2 c_1)$$

$$= a_1 \begin{vmatrix} b_2 & c_2 \\ b_3 & c_3 \end{vmatrix} - a_2 \begin{vmatrix} b_1 & c_1 \\ b_3 & c_3 \end{vmatrix} + a_3 \begin{vmatrix} b_1 & c_1 \\ b_2 & c_2 \end{vmatrix}$$

This last expression is called an **expansion of the determinant** using column 1. Note that each determinant of order 2 can be obtained by "striking out" the row and column in which its factor from column 1 lies.

An expansion can be obtained by using any row or column but first we need a definition.

DEFINITION

A **cofactor** of an element in a determinant is the product of the positional sign of the element and the value of the determinant obtained by striking out the row and column in which the element lies. The positional sign for a determinant of order 3 is determined by the following array of signs:

$$
\begin{array}{ccc}
+ & - & + \\
- & + & - \\
+ & - & +
\end{array}
$$

THEOREM 1

The value of a determinant of order 3 is the sum of the products obtained by multiplying each element in a row (or column) by its cofactor.

This theorem can be proved by a direct but lengthy checking that each of the six possible expansions is equal to the value stated in the definition.

EXAMPLE 3 Evaluate

$$
\begin{vmatrix}
2 & 4 & 1 \\
1 & -2 & 2 \\
-3 & 1 & -1
\end{vmatrix}
$$

(a) by an expansion using column 2

(b) by an expansion using row 3

Solution

(a) $\begin{vmatrix} 2 & 4 & 1 \\ 1 & -2 & 2 \\ -3 & 1 & -1 \end{vmatrix} = (-1)4\begin{vmatrix} 1 & 2 \\ -3 & -1 \end{vmatrix} + (-2)\begin{vmatrix} 2 & 1 \\ -3 & -1 \end{vmatrix} - 1\begin{vmatrix} 2 & 1 \\ 1 & 2 \end{vmatrix}$

$$= -4(5) - 2(1) - (3) = -25$$

(b) $\begin{vmatrix} 2 & 4 & 1 \\ 1 & -2 & 2 \\ -3 & 1 & -1 \end{vmatrix} = -3\begin{vmatrix} 4 & 1 \\ -2 & 2 \end{vmatrix} - 1\begin{vmatrix} 2 & 1 \\ 1 & 2 \end{vmatrix} + (-1)\begin{vmatrix} 2 & 4 \\ 1 & -2 \end{vmatrix}$

$$= -3(10) - (3) - (-8) = -25$$

The following theorems are also useful for evaluating a determinant. They can be proved by expanding the determinants and checking the results.

THEOREM 2

If two rows or two columns of a determinant are identical, the value of the determinant is zero.

THEOREM 3

If each element of a row or column is multiplied by a constant, then the determinant is multiplied by this constant.

THEOREM 4

If each element of a row (or column) is multiplied by a constant and then added to the corresponding element of another row (or column), the value of the determinant is unchanged.

As examples:

Using Theorem 2,

$$\begin{vmatrix} 2 & 4 & 2 \\ 1 & 6 & 1 \\ 3 & 1 & 3 \end{vmatrix} = 0 \qquad \text{because column 1 is identical to column 3}$$

Using Theorem 3,

$$5\begin{vmatrix} 2 & 4 & 3 \\ 1 & 6 & 6 \\ 3 & 1 & 3 \end{vmatrix} = \begin{vmatrix} 5 \cdot 2 & 5 \cdot 4 & 5 \cdot 3 \\ 1 & 6 & 6 \\ 3 & 1 & 3 \end{vmatrix} = \begin{vmatrix} 10 & 20 & 15 \\ 1 & 6 & 6 \\ 3 & 1 & 3 \end{vmatrix}$$

and

$$\begin{vmatrix} 2 & 4 & 3 \\ 1 & 6 & 6 \\ 3 & 1 & 9 \end{vmatrix} = 3\begin{vmatrix} 2 & 4 & 1 \\ 1 & 6 & 2 \\ 3 & 1 & 3 \end{vmatrix}$$

Using Theorem 4, adding -3 times row 2 to row 3,

$$\begin{vmatrix} 2 & 4 & 1 \\ 1 & 6 & 2 \\ 3 & 1 & 2 \end{vmatrix} = \begin{vmatrix} 2 & 4 & 1 \\ 1 & 6 & 2 \\ 3 + 1(-3) & 1 + 6(-3) & 2 + 2(-3) \end{vmatrix}$$

$$= \begin{vmatrix} 2 & 4 & 1 \\ 1 & 6 & 2 \\ 0 & -17 & -4 \end{vmatrix}$$

EXAMPLE 4 Solve

$$x + y + z = 4$$
$$x - 2y - z = 1$$
$$2x - y - 2z = -1$$

using Cramer's Rule.

Solution

1. Finding D, the determinant of the coefficients,

$$D = \begin{vmatrix} 1 & 1 & 1 \\ 1 & -2 & -1 \\ 2 & -1 & -2 \end{vmatrix} = \begin{vmatrix} 1 & 1 & 2 \\ 1 & -2 & 0 \\ 2 & -1 & 0 \end{vmatrix}$$

Note that by adding column 1 to column 3, we obtain a determinant that is much easier to evaluate. Now expanding using the simplified column 3,

$$D = 2\begin{vmatrix} 1 & -2 \\ 2 & -1 \end{vmatrix} + 0 + 0 = 2(-1 + 4) = 6$$

2. Finding N_x, the numerator determinant for x, by replacing the coefficients of x by the constant terms,

$$N_x = \begin{vmatrix} 4 & 1 & 1 \\ 1 & -2 & -1 \\ -1 & -1 & -2 \end{vmatrix}$$

$$= \begin{vmatrix} 4 & 1 & 5 \\ 1 & -2 & 0 \\ -1 & -1 & -3 \end{vmatrix} \qquad \text{by adding column 1 to column 3}$$

$$= \begin{vmatrix} 4 & 9 & 5 \\ 1 & 0 & 0 \\ -1 & -3 & -3 \end{vmatrix} \qquad \text{by adding 2 times column 1 to column 2}$$

Now expanding using the simplified row 2,

$$N_x = -1 \begin{vmatrix} 9 & 5 \\ -3 & -3 \end{vmatrix} = -1(-27 + 15) = 12$$

3. Finding N_y, the numerator determinant for y, by replacing the coefficients of y by the constant terms,

$$N_y = \begin{vmatrix} 1 & 4 & 1 \\ 1 & 1 & -1 \\ 2 & -1 & -2 \end{vmatrix}$$

$$= \begin{vmatrix} 1 & 4 & 2 \\ 1 & 1 & 0 \\ 2 & -1 & 0 \end{vmatrix} \quad \text{by adding column 1} \\ \text{to column 3}$$

Now expanding using the simplified column 3,

$$N_y = 2 \begin{vmatrix} 1 & 1 \\ 2 & -1 \end{vmatrix} = 2(-1 - 2) = -6$$

4. Finding N_z, the numerator determinant for z, by replacing the coefficients of z by the constant terms,

$$N_z = \begin{vmatrix} 1 & 1 & 4 \\ 1 & -2 & 1 \\ 2 & -1 & -1 \end{vmatrix}$$

$$= \begin{vmatrix} 1 & 2 & 4 \\ 1 & -1 & 1 \\ 2 & 1 & -1 \end{vmatrix} \quad \text{by adding column 1} \\ \text{to column 2}$$

$$= \begin{vmatrix} 1 & 2 & 6 \\ 1 & -1 & 0 \\ 2 & 1 & 0 \end{vmatrix} \quad \text{by adding column 2} \\ \text{to column 3}$$

Now expanding using the simplified column 3,

$$N_z = 6 \begin{vmatrix} 1 & -1 \\ 2 & 1 \end{vmatrix} = 6(1 + 2) = 18$$

5. Finding x, y, and z,

$$x = \frac{N_x}{D} = \frac{12}{6} = 2$$

$$y = \frac{N_y}{D} = \frac{-6}{6} = -1$$

$$z = \frac{N_z}{D} = \frac{18}{6} = 3$$

The solution is $(2, -1, 3)$.

Check

$x + y + z = 4;$	$2 - 1 + 3 = 4$
$x - 2y - z = 1;$	$2 - 2(-1) - 3 = 1$
$2x - y - 2z = 1;$	$2(2) - (-1) - 2(3) = 4 + 1 - 6 = -1$

If $D = 0$ and $N_x \neq 0$ or $N_y \neq 0$ or $N_z \neq 0$, then the **solution set is empty. The equations are inconsistent.**

If $D = 0$ and $N_x = 0$ and $N_y = 0$ and $N_z = 0$, then the **solution set contains infinitely many solutions. The equations are dependent.**

EXERCISES 6.1

Use Cramer's Rule to solve each system. Check each solution.

1. $2x - 4y = 2$
$x + 2y = 9$

2. $3x + 4y = 2$
$2x + 3y = 0$

3. $3x + 2y - 1 = 0$
$5x + 3y - 3 = 0$

4. $5x + 2y - 70 = 0$
$2x + 5y - 70 = 0$

5. $4x = 3y + 7$
$3x = 24 - 4y$

6. $2x = y + 8$
$2y = x - 1$

7. $2x - 3y = 200$
$3x - 2y = 675$

8. $7x - 5y = 24$
$5x + 7y = 70$

9. $2x + y + 3z = 15$
$ 2x + 7z = 25$
$ 3x + 2y = 5$

10. $x + y - z = 7$
$x - y + z = 1$
$3x - y - z = 3$

11. $2x - y + 2z = 9$
$x + 2y - 2z = 1$
$3x - 2y - z = 5$

12. $x + 2y + z = 5$
$2x + 3y - z = 0$
$2x - y - 3z = 0$

13. $2x + 3y - 3 = 0$
$3x + 5z - 8 = 0$
$2y - 5z - 4 = 0$

14. $ x + y = z$
$2x + 2z = 5y + 1$
$2y + z = 5x - 1$

15. $x + 4z + 2 = 0$
$6y - z + 1 = 0$
$x + 3y + z = 0$

16. $2x - y + 3z = 0$
$x + 3y - 5z = 8$
$3x + 2y - 4z = 11$

17. $y + 10 = 3x + 2z$
$2y + 22 = -x + 3z$
$3y - 13 = 2x - 4z$

18. $y + z = 2x$
$2y - 3z = 3x - 4$
$7y + 2z = 2 - 5x$

19. $x - 2y + z = 0$
$2x + y - 5z = 5$
$x - 3y + 3z = 5$

20. $2x - y - z = 0$
$2x + y - 2z = 0$
$x + 3y - 2z = 10$

21. $x + y - z = 0.1$
$2x - 3y + 2z = 0.5$
$4x + 2y - 3z = 0.1$

22. $0.1x - 0.1y + 0.1z = 1$
$0.3x + 0.2y - 0.2z = 1$
$0.5x + 0.1y - 0.5z = 1$

23. $x + y + z = 1$
$x + 2y + 3z = 4$
$x + 4y + 9z = 16$

24. $x + 2y + 3z = 2$
$2x + 3y + z = 3$
$3x + y + 2z = 1$

6.2 MATRIX SOLUTION OF LINEAR SYSTEMS

In this section we learn another method for solving a linear system.
This method is called the **matrix method** or the **Gauss reduc-**

tion method in honor of the great German mathematician Carl Friedrick Gauss (1777–1855).

First let us review the elimination method by applying it to the system

$$x - y + z = 4$$
$$x + 2y - z = 1$$
$$2x - 3y + 3z = 10$$

Eliminating x by using the first two equations and then the first and third equations, we obtain

$$3y - 2z = -3$$
$$y - z = -2$$

Now eliminating y, we find that

$$z = 3$$

The original system can be replaced by the following equivalent system:

$$x - y + z = 4$$
$$y - z = -2$$
$$z = 3$$

This is a relatively simple system to solve.

Using $z = 3$ and $y - z = -2$, then $y = 1$.

Now for $z = 3$ and $y = 1$, the first equation becomes

$$x - 1 + 3 = 4 \quad \text{and} \quad x = 2$$

The solution is $(2, 1, 3)$.

Since the numerical calculations involve only the coefficients of x, y, and z and the constants, we can shorten the work by not writing the letters x, y, and z. We replace the system

$$
\begin{aligned}
x - y + z &= 4 \\
x + 2y - z &= 1 \\
2x - 3y + 3z &= 10
\end{aligned}
\quad \text{by the array} \quad
\begin{bmatrix}
1 & -1 & 1 & 4 \\
1 & 2 & -1 & 1 \\
2 & -3 & 3 & 10
\end{bmatrix}
$$

and we replace the system

$$
\begin{aligned}
x - y + z &= 4 \\
y - z &= -2 \\
z &= 3
\end{aligned}
\quad \text{by the array} \quad
\begin{bmatrix}
1 & -1 & 1 & 4 \\
0 & 1 & -1 & -2 \\
0 & 0 & 1 & 3
\end{bmatrix}
$$

By keeping in mind the meaning of the array, we can perform calculations on the numbers in the array that parallel the calculations used in the elimination solution method.

DEFINITION

Any array of numbers in n rows and k columns, such as those above, is called a **matrix** of dimensions n by k.

Each of the preceding arrays is a matrix of dimensions 3 by 4.

For the linear system
$$a_1x + b_1y + c_1z = d_1$$
$$a_2x + b_2y + c_2z = d_2$$
$$a_3x + b_3y + c_3z = d_3$$
the matrix
$$\begin{bmatrix} a_1 & b_1 & c_1 \\ a_2 & b_2 & c_2 \\ a_3 & b_3 & c_3 \end{bmatrix}$$
is called the **coefficient matrix** and the matrix
$$\begin{bmatrix} a_1 & b_1 & c_1 & d_1 \\ a_2 & b_2 & c_2 & d_2 \\ a_3 & b_3 & c_3 & d_3 \end{bmatrix}$$
is called the **augmented matrix.**

DEFINITION

Two augmented matrices of two linear systems are equivalent if and only if the linear systems have the same solution set.

The matrix method for solving a linear system involves performing operations on its augmented matrix to obtain an equivalent **matrix in triangular form;** that is, the form
$$\begin{bmatrix} a_1' & b_1' & c_1' & d_1' \\ 0 & b_2' & c_2' & d_2' \\ 0 & 0 & c_3' & d_3' \end{bmatrix}$$
The following theorem indicates operations that produce equivalent augmented matrices. The proof, while omitted here, is direct and merely involves showing that the associated linear systems have the same solution set.

THEOREM

Two augmented matrices of two linear systems are equivalent if one matrix is obtained from the other by

1. interchanging any two rows, or
2. multiplying each element of a row by a nonzero constant, or
3. adding a nonzero constant multiple of the elements of one row to the corresponding elements of another row.

EXAMPLE 1   Solve
$$x - y + z = 4$$
$$x + 2y - z = 1$$
$$2x - 3y + 3z = 10$$
by the matrix method.

Solution

1. Form the augmented matrix:
$$\begin{bmatrix} 1 & -1 & 1 & 4 \\ 1 & 2 & -1 & 1 \\ 2 & -3 & 3 & 10 \end{bmatrix}$$

2. Multiply row 1 by -1, and add to row 2; multiply row 1 by -2, and add to row 3:
$$\begin{bmatrix} 1 & -1 & 1 & 4 \\ 0 & 3 & -2 & -3 \\ 0 & -1 & 1 & 2 \end{bmatrix}$$

3. Interchange row 2 and row 3 (to obtain a 1 in the first nonzero position):
$$\begin{bmatrix} 1 & -1 & 1 & 4 \\ 0 & -1 & 1 & 2 \\ 0 & 3 & -2 & -3 \end{bmatrix}$$

4. Multiply row 2 by 3, and add to row 3:
$$\begin{bmatrix} 1 & -1 & 1 & 4 \\ 0 & -1 & 1 & 2 \\ 0 & 0 & 1 & 3 \end{bmatrix}$$

5. Multiply row 2 by -1:
$$\begin{bmatrix} 1 & -1 & 1 & 4 \\ 0 & 1 & -1 & -2 \\ 0 & 0 & 1 & 3 \end{bmatrix}$$

6. Write the associated linear system:
$$x - y + z = 4$$
$$y - z = -2$$
$$z = 3$$

7. Solve this system:
$(2, 1, 3)$

8. Check the solution in each equation of the original system (check left for student).

Note that the system
$x = a$
$y = b$
$z = c$
has the augmented matrix
$$\begin{bmatrix} 1 & 0 & 0 & a \\ 0 & 1 & 0 & b \\ 0 & 0 & 1 & c \end{bmatrix}$$
It is often possible to transform an augmented matrix into this form as the next example illustrates.

EXAMPLE 2 Solve
$3x - 2y - 3z = 1$
$2x - 3y - 2z = 4$
$x + 2y + z = 7$
by the matrix method.

Solution

1. Form the augmented matrix:
$$\begin{bmatrix} 3 & -2 & -3 & 1 \\ 2 & -3 & -2 & 4 \\ 1 & 2 & 1 & 7 \end{bmatrix}$$

2. Interchange rows 1 and 3 (to obtain a 1 in the upper left position):
$$\begin{bmatrix} 1 & 2 & 1 & 7 \\ 2 & -3 & -2 & 4 \\ 3 & -2 & -3 & 1 \end{bmatrix}$$

3. Multiply row 1 by -2 and add to row 2; multiply row 1 by -3 and add to row 3:
$$\begin{bmatrix} 1 & 2 & 1 & 7 \\ 0 & -7 & -4 & -10 \\ 0 & -8 & -6 & -20 \end{bmatrix}$$

4. Multiply rows 2 and 3 by -1:
$$\begin{bmatrix} 1 & 2 & 1 & 7 \\ 0 & 7 & 4 & 10 \\ 0 & 8 & 6 & 20 \end{bmatrix}$$

5. Multiply row 2 by -1 and add to row 3:
$$\begin{bmatrix} 1 & 2 & 1 & 7 \\ 0 & 7 & 4 & 10 \\ 0 & 1 & 2 & 10 \end{bmatrix}$$

6. Interchange rows 2 and 3:
$$\begin{bmatrix} 1 & 2 & 1 & 7 \\ 0 & 1 & 2 & 10 \\ 0 & 7 & 4 & 10 \end{bmatrix}$$

7. Multiply row 2 by -2 and add to row 1; multiply row 2 by -7 and add to row 3:
$$\begin{bmatrix} 1 & 0 & -3 & -13 \\ 0 & 1 & 2 & 10 \\ 0 & 0 & -10 & -60 \end{bmatrix}$$

8. Multiply row 3 by $\frac{-1}{10}$:
$$\begin{bmatrix} 1 & 0 & -3 & -13 \\ 0 & 1 & 2 & 10 \\ 0 & 0 & 1 & 6 \end{bmatrix}$$

9. Multiply row 3 by 3 and add to row 1; multiply row 3 by -2 and add to row 2:
$$\begin{bmatrix} 1 & 0 & 0 & 5 \\ 0 & 1 & 0 & -2 \\ 0 & 0 & 1 & 6 \end{bmatrix}$$

10. Form the associated system and solve:
$x = 5$
$y = -2$
$z = 6$
The solution is $(5, -2, 6)$.

11. **Check**
$3x - 2y - 3z = 15 + 4 - 18 = 1$ and $1 = 1$
$2x - 3y - 2z = 10 + 6 - 12 = 4$ and $4 = 4$
$x + 2y + z = 5 - 4 + 6 = 7$ and $7 = 7$

EXAMPLE 3 Solve
$x + y + 2z = 2$
$2x - 2y - z = 1$
$3x - y + z = 2$
by the matrix method.

Solution

1. Form augmented matrix:
$$\begin{bmatrix} 1 & 1 & 2 & 2 \\ 2 & -2 & -1 & 1 \\ 3 & -1 & 1 & 2 \end{bmatrix}$$

2. Add -2 times row 1 to row 2;
add -3 times row 1 to row 3:
$$\begin{bmatrix} 1 & 1 & 2 & 2 \\ 0 & -4 & -5 & -3 \\ 0 & -4 & -5 & -4 \end{bmatrix}$$

3. Multiply row 2 by -1; then add row 2 to row 3:
$$\begin{bmatrix} 1 & 1 & 2 & 2 \\ 0 & 4 & 5 & 3 \\ 0 & 0 & 0 & -1 \end{bmatrix}$$
Note the last equation of the associated linear system is $0 = -1$, a false statement. Therefore the system is inconsistent and the solution set is the empty set, $\emptyset$.

EXAMPLE 4 Solve
$x + 2y - 5z = 4$
$x + y - 3z = 1$
$2x + 3y - 8z = 5$
by the matrix method.

Solution

1. Form augmented matrix:
$$\begin{bmatrix} 1 & 2 & -5 & 4 \\ 1 & 1 & -3 & 1 \\ 2 & 3 & -8 & 5 \end{bmatrix}$$

2. Add -1 times row 1 to row 2;
   add -2 times row 1 to row 3:
   $$\begin{bmatrix} 1 & 2 & -5 & 4 \\ 0 & -1 & 2 & -3 \\ 0 & -1 & 2 & -3 \end{bmatrix}$$

3. Multiply row 2 by -1; then add row 2 to row 3:
   $$\begin{bmatrix} 1 & 2 & -5 & 4 \\ 0 & 1 & -2 & 3 \\ 0 & 0 & 0 & 0 \end{bmatrix}$$

4. Form the associated system and solve:
   $$x + 2y - 5z = 4$$
   $$y - 2z = 3$$
   $$0 = 0$$
   $$y = 2z + 3$$
   $$x + 2(2z + 3) - 5z = 4 \quad \text{and} \quad x = z - 2$$
   The solution set is $\{(z - 2, 2z + 3, z)\}$, where z is any real number and the system is dependent.

The matrix method is readily adaptable to the electronic computer. Also it can be extended to cover any number of equations in any number of variables.

EXERCISES 6.2

(1–10) Solve by the matrix method and check. See Examples 1 and 2.

1.
$$x + y - z = 1$$
$$2x + 5y - 3z = 4$$
$$3x + 8y - 4z = 9$$

2.
$$x - 3y + 2z = -6$$
$$3x - 2y - 3z = 14$$
$$5x - y + 8z = 2$$

3.
$$3x + y + 2z = 5$$
$$x + 2y + 2z = 4$$
$$4x - 2y + z = 0$$

4.
$$2x - 3y + 2z = 5$$
$$x - 2y + z = 6$$
$$x + y - 2z = 0$$

5.
$$2x - 2y - z = 10$$
$$2x + 4y + z = 20$$
$$4x - 5y - 2z = 25$$

6.
$$x + y + z = 1$$
$$2x - 3y + 4z = 6$$
$$2x + 5y - 2z = 8$$

7.
$$3x - 2y + 3z = 0$$
$$2x - 2y + 15z - 1 = 0$$
$$2x + y - 3z - 1 = 0$$

8.
$$2x - 3y + z = 1$$
$$3x + 2y + 4z = 5$$
$$3x + 4y - 2z = 1$$

9.
$$r - 2s + 3t = 1$$
$$r + 3s - t = 4$$
$$2r + s - 2t = 13$$

10.
$$r + s - t - 5 = 0$$
$$2r - 4s + t - 10 = 0$$
$$3r - 2s - t - 5 = 0$$

(11–18) Show that each system is inconsistent or dependent. Find the solution set. See Examples 3 and 4.

11.
$$x - 2y + z = 1$$
$$x + y - 5z = 4$$
$$2x - y - 4z = 5$$

12.
$$x + y + 2z = 7$$
$$4x - 2y + 3z = 1$$
$$9x - 3y + 8z = 4$$

13. $r + s - 2t = 2$
$3r - 2s + t = 3$
$2r - 3s + 3t = 4$

14. $r - s + t = 6$
$2r - s - 2t = 2$
$3r - 2s - t = 8$

15. $u + v - w = 5$
$2u - 4v + w = 10$
$3u + v - 2w = 15$

16. $2u - v + w = 4$
$u - v + 2w = 8$
$3u - 2v + 3w = 12$

17. $x - y + 2z - 8 = 0$
$2x - 3y + 6z - 16 = 0$
$3x - y + 2z - 24 = 0$

18. $x + y - z = 1$
$3x + 3y - 3z = 3$
$5x + 5y - 5z = 5$

(19–22) Solve and check.

EXAMPLE 5 Solve
$x + y - z + t = 4$
$x - y + z - 2t = 2$
$2x - y - z + 3t = 6$
$2x + 2y - 3z + 3t = 5$

Solution Form augmented matrix:

$$\begin{bmatrix} 1 & 1 & -1 & 1 & 4 \\ 1 & -1 & 1 & -2 & 2 \\ 2 & -1 & -1 & 3 & 6 \\ 2 & 2 & -3 & 3 & 5 \end{bmatrix}$$

Transform the matrix to the triangular form using the theorem in this section:

$$\begin{bmatrix} 1 & 1 & -1 & 1 & 4 \\ 0 & -2 & 2 & -3 & -2 \\ 0 & -3 & 1 & 1 & -2 \\ 0 & 0 & -1 & 1 & -3 \end{bmatrix} \rightarrow \begin{bmatrix} 1 & 1 & -1 & 1 & 4 \\ 0 & 2 & -2 & 3 & 2 \\ 0 & -1 & -1 & 4 & 0 \\ 0 & 0 & -1 & 1 & -3 \end{bmatrix} \rightarrow$$

$$\begin{bmatrix} 1 & 1 & -1 & 1 & 4 \\ 0 & -1 & -1 & 4 & 0 \\ 0 & 0 & -4 & 11 & 2 \\ 0 & 0 & -1 & 1 & -3 \end{bmatrix} \rightarrow \begin{bmatrix} 1 & 1 & -1 & 1 & 4 \\ 0 & 1 & 1 & -4 & 0 \\ 0 & 0 & 1 & -1 & 3 \\ 0 & 0 & 0 & 7 & 14 \end{bmatrix}$$

Write the equivalent system and solve:

$x + y - z + t = 4$ $t = 2$
$y + z - 4t = 0$ $z = t + 3 = 5$
$z - t = 3$ $y = -z + 4t = -5 + 8 = 3$
$7t = 14$ $x = -y + z - t + 4 = 4$

The solution is $(4, 3, 5, 2)$.

19. $x + y - z - t = 2$
$2x - y - z + t = 3$
$3x - 2y - 2z + t = -3$
$4x - y + z - t = 5$

20. $x + y + z + t = 2$
$2x - y + t = 9$
$2x - 2z - t = 1$
$2y + z - 2t = 3$

21. $a - b + 2c - d = 0$
$2b - 3c + 2d = 5$
$a - 2c - 3d = 2$
$b + 2c + 4d = 2$

22. $a + b + c = 14$
$a + c + d = 13$
$b + c + d = 10$
$a - b + c - d = 9$

6.3 DETERMINANTS OF ORDER *n*

In this section we generalize the ideas discussed in Section 6.1 concerning determinants or orders 2 and 3. First, in order to define a determinant of order *n*, we need to introduce the concept of an inversion.

DEFINITION

An **inversion** occurs in an arrangement of the natural numbers 1, 2, 3, $\cdots$, *n* whenever a larger number precedes a smaller number.

EXAMPLE 1 Find the number of inversions in the arrangement 25134.

Solution There are 4 inversions since 2 precedes 1, 5 precedes 1, 5 precedes 3, and 5 precedes 4.

DEFINITION

A **determinant of order *n*** is an expression having the form

$$\begin{vmatrix} a_{11} & a_{12} & a_{13} & \cdots & a_{1n} \\ a_{21} & a_{22} & a_{23} & \cdots & a_{2n} \\ a_{31} & a_{32} & a_{33} & \cdots & a_{3n} \\ & & \cdot & & \\ & & \cdot & & \\ & & \cdot & & \\ a_{n1} & a_{n2} & a_{n3} & \cdots & a_{nn} \end{vmatrix}$$

where each real number a_{ij} is called an **element.**

The determinant's value is the sum of all possible products of *n* factors such that:

1. each product contains exactly one element from each row and each column,
2. the factors of each product are written so that the row (first) subscripts are in increasing order, and
3. a plus or minus sign is written before each product depending on whether the number of inversions in the arrangement of the column (second) subscripts is even or odd.

The sum described in the definition of a determinant of order *n* is called an **expansion** of the determinant.

Note that for the determinant

$$\begin{vmatrix} a_{11} & a_{12} & a_{13} \\ a_{21} & a_{22} & a_{23} \\ a_{31} & a_{32} & a_{33} \end{vmatrix}$$

the expansion $a_{11}a_{22}a_{33} - a_{11}a_{23}a_{32} - a_{12}a_{21}a_{33} + a_{12}a_{23}a_{31} + a_{13}a_{21}a_{32} - a_{13}a_{22}a_{31}$ is formed in accordance with the definition.

DEFINITION

The **minor** M_{ij} of an element a_{ij} in a determinant of order n is the determinant of order $n - 1$ that remains after the row and column in which a_{ij} lies is deleted.

EXAMPLE 2 Find the minor M_{23} of a_{23} in the determinant

$$\begin{vmatrix} a_{11} & a_{12} & a_{13} & a_{14} \\ a_{21} & a_{22} & a_{23} & a_{24} \\ a_{31} & a_{32} & a_{33} & a_{34} \\ a_{41} & a_{42} & a_{43} & a_{44} \end{vmatrix}$$

Solution   Deleting the row and column in which a_{23} lies,

$$M_{23} = \begin{vmatrix} a_{11} & a_{12} & a_{14} \\ a_{31} & a_{32} & a_{34} \\ a_{41} & a_{42} & a_{44} \end{vmatrix}$$

DEFINITION

The **cofactor** A_{ij} of an element a_{ij} in a determinant of order n is given by
$$A_{ij} = (-1)^{i+j} M_{ij}$$

EXAMPLE 3 Find the cofactor A_{23} of a_{23} in the determinant of Example 2.
Solution

$$A_{23} = (-1)^{2+3} M_{23} = - \begin{vmatrix} a_{11} & a_{12} & a_{14} \\ a_{31} & a_{32} & a_{34} \\ a_{41} & a_{42} & a_{44} \end{vmatrix}$$

THEOREM 1   EXPANSION BY COFACTORS

The value of a determinant of order n is the sum of the n products formed by multiplying each element in a row (or column) by its cofactor.

EXAMPLE 4 Expand the determinant of Example 2 (a) about row 2; (b) about column 3.

Solution

(a) $D = a_{21}A_{21} + a_{22}A_{22} + a_{23}A_{23} + a_{24}A_{24}$

$$= -a_{21}\begin{vmatrix} a_{12} & a_{13} & a_{14} \\ a_{32} & a_{33} & a_{34} \\ a_{42} & a_{43} & a_{44} \end{vmatrix} + a_{22}\begin{vmatrix} a_{11} & a_{13} & a_{14} \\ a_{31} & a_{33} & a_{34} \\ a_{41} & a_{43} & a_{44} \end{vmatrix}$$

$$- a_{23}\begin{vmatrix} a_{11} & a_{12} & a_{14} \\ a_{31} & a_{32} & a_{34} \\ a_{41} & a_{42} & a_{44} \end{vmatrix} + a_{24}\begin{vmatrix} a_{11} & a_{12} & a_{13} \\ a_{31} & a_{32} & a_{33} \\ a_{41} & a_{42} & a_{43} \end{vmatrix}$$

(b) $D = a_{13}A_{13} + a_{23}A_{23} + a_{33}A_{33} + a_{43}A_{43}$
$$= a_{13}M_{13} - a_{23}M_{23} + a_{33}M_{33} - a_{43}M_{43}$$

The following theorems are useful for finding the value of a determinant of order n.

THEOREM 2

If each element of a row (or column) is 0, the value of the determinant is 0.

THEOREM 3

If the columns of one determinant are the same as the rows of another determinant, then the determinants are equal.

THEOREM 4

If two rows (or two columns) of a determinant are interchanged, the resulting determinant is equal to the negative of the original determinant.

THEOREM 5

If two rows (or two columns) of a determinant are identical, the determinant is equal to 0.

THEOREM 6

If each element of a row (or column) is multiplied by k, then the value of the determinant is multiplied by k.

THEOREM 7

If each element of a row (or column) is multiplied by k and the resulting product is added to the corresponding element of another row (or column), the resulting determinant is equal to the original determinant.

EXAMPLE 5 Evaluate

$$\begin{vmatrix} 2 & 4 & 3 & 2 \\ 1 & 3 & 2 & 5 \\ 3 & 5 & 4 & 2 \\ 4 & 2 & 2 & 3 \end{vmatrix}$$

Solution

1. Multiply row 2 by -2 and add to row 1.
2. Multiply row 2 by -3 and add to row 3.
3. Multiply row 2 by -4 and add to row 4.

$$\begin{vmatrix} 0 & -2 & -1 & -8 \\ 1 & 3 & 2 & 5 \\ 0 & -4 & -2 & -13 \\ 0 & -10 & -6 & -17 \end{vmatrix}$$

4. Multiply each of rows 1, 3, and 4 by -1.
   (Since three rows are being multiplied by -1 the determinant's value is multiplied by $(-1)(-1)(-1) = -1$, by Theorem 6.)

$$(-1)\begin{vmatrix} 0 & 2 & 1 & 8 \\ 1 & 3 & 2 & 5 \\ 0 & 4 & 2 & 13 \\ 0 & 10 & 6 & 17 \end{vmatrix}$$

5. Expand, using column 1.

$$(-1)\left[0A_{11} + (-1)\begin{vmatrix} 2 & 1 & 8 \\ 4 & 2 & 13 \\ 10 & 6 & 17 \end{vmatrix} + 0A_{31} + 0A_{41} \right] = \begin{vmatrix} 2 & 1 & 8 \\ 4 & 2 & 13 \\ 10 & 6 & 17 \end{vmatrix}$$

6. Multiply column 2 by -2 and add to column 1.
   Multiply column 2 by -8 and add to column 3.

$$\begin{vmatrix} 0 & 1 & 0 \\ 0 & 2 & -3 \\ -2 & 6 & -31 \end{vmatrix}$$

7. Expand, using row 1.

$$(-1)\begin{vmatrix} 0 & -3 \\ -2 & -31 \end{vmatrix} = -(-6) = 6.$$

There is also a Cramer's Rule for solving a system of n linear equations in n variables.

THEOREM 8   CRAMER'S RULE

Given the linear system in the n variables, $x_1, x_2, \cdots, x_n$,

$$a_{11}x_1 + a_{12}x_2 + \cdots + a_{1n}x_n = c_1$$
$$a_{21}x_1 + a_{22}x_2 + \cdots + a_{2n}x_n = c_2$$
$$\cdots\cdots\cdots\cdots\cdots\cdots\cdots\cdots\cdots\cdots$$
$$a_{n1}x_1 + a_{n2}x_2 + \cdots + a_{nn}x_n = c_n$$

where the determinant of the coefficients D is given by

$$D = \begin{vmatrix} a_{11} & a_{12} & \cdots & a_{1n} \\ a_{21} & a_{22} & \cdots & a_{2n} \\ \cdots & \cdots & \cdots & \cdots \\ a_{n1} & a_{n2} & \cdots & a_{nn} \end{vmatrix}, \qquad \text{where } D \neq 0$$

The solutions of this system are given by

$$x_1 = \frac{N_1}{D}, \; x_2 = \frac{N_2}{D}, \; \cdots, \; x_n = \frac{N_n}{D}$$

where N_i is the determinant obtained from D by replacing the column of the coefficients of x_i by the column of constants, $c_1, c_2, \cdots, c_n$.

EXAMPLE 6   Solve, using Cramer's Rule:

$$x - y + z - 2t = 2$$
$$2x - y - z + 3t = 6$$
$$x + y - z + t = 4$$
$$2x + 2y - 3z + 3t = 5$$

Solution

1.
$$D = \begin{vmatrix} 1 & -1 & 1 & -2 \\ 2 & -1 & -1 & 3 \\ 1 & 1 & -1 & 1 \\ 2 & 2 & -3 & 3 \end{vmatrix} = \begin{vmatrix} 1 & -1 & 1 & -2 \\ 0 & 1 & -3 & 7 \\ 0 & 2 & -2 & 3 \\ 0 & 4 & -5 & 7 \end{vmatrix} = -7$$

2.
$$N_1 = \begin{vmatrix} 2 & -1 & 1 & -2 \\ 6 & -1 & -1 & 3 \\ 4 & 1 & -1 & 1 \\ 5 & 2 & -3 & 3 \end{vmatrix} = -28 \quad \text{and} \quad x = \frac{-28}{-7} = 4$$

3.
$$N_2 = \begin{vmatrix} 1 & 2 & 1 & -2 \\ 2 & 6 & -1 & 3 \\ 1 & 4 & -1 & 1 \\ 2 & 5 & -3 & 3 \end{vmatrix} = -21 \quad \text{and} \quad y = \frac{-21}{-7} = 3$$

4.
$$N_3 = \begin{vmatrix} 1 & -1 & 2 & -2 \\ 2 & -1 & 6 & 3 \\ 1 & 1 & 4 & 1 \\ 2 & 2 & 5 & 3 \end{vmatrix} = -35 \quad \text{and} \quad z = \frac{-35}{-7} = 5$$

5.
$$N_4 = \begin{vmatrix} 1 & -1 & 1 & 2 \\ 2 & -1 & -1 & 6 \\ 1 & 1 & -1 & 4 \\ 2 & 2 & -3 & 5 \end{vmatrix} = -14 \quad \text{and} \quad t = \frac{-14}{-7} = 2$$

The solution is $(4, 3, 5, 2)$.

HOMOGENEOUS LINEAR SYSTEMS

A **homogeneous linear equation** in two or more variables is a linear equation whose constant term is 0.

Consider the following system of homogeneous linear equations:

$$a_1x + b_1y + c_1z = 0$$
$$a_2x + b_2y + c_2z = 0$$
$$a_3x + b_3y + c_3z = 0$$

Obviously, $(0, 0, 0)$ is a solution of the system. It is called the trivial solution. In general, the **trivial solution** of a system of n homogeneous linear equations in n variables is the solution for which each variable has the value 0. If D is the determinant of the coefficients and if $D \neq 0$, then by using Cramer's Rule, the trivial solution is the only solution of the system. However, if $D = 0$, then there are infinitely many nontrivial solutions. Theorem 9, which follows, establishes this fact for the case when D has a nonzero minor, and the proof of the theorem provides the method for the solution.

THEOREM 9

A system of n homogeneous linear equations in n variables has infinitely many solutions if $D = 0$ and if the minor of at least one element of D is not zero, where D is the determinant of the coefficients.

Proof For
$$a_{11}x_1 + a_{12}x_2 + \cdots + a_{1n}x_n = 0$$
$$a_{21}x_1 + a_{22}x_2 + \cdots + a_{2n}x_n = 0$$
$$\cdots \cdots \cdots \cdots \cdots \cdots \cdots \cdots$$
$$a_{n1}x_1 + a_{n2}x_2 + \cdots + a_{nn}x_n = 0$$
given $D = 0$ and M_{11}, the minor of a_{11}, is not zero. Then the system
$$a_{22}x_2 + \cdots + a_{2n}x_n = -a_{21}x_1$$
$$a_{32}x_2 + \cdots + a_{3n}x_n = -a_{31}x_1$$
$$\cdots \cdots \cdots \cdots \cdots \cdots \cdots \cdots$$
$$a_{n2}x_2 + \cdots + a_{nn}x_n = -a_{n1}x_1$$
has a solution in terms of the variable x_1, by using Cramer's Rule. Substituting the values of this solution into the first equation of the given system,
$$a_{11}x_1 + a_{12}\frac{-M_{12}}{M_{11}}x_1 + a_{13}\frac{M_{13}}{M_{11}}x_1 + \cdots + a_{1n}\frac{(-1)^{n+1}M_{1n}}{M_{11}}x_1 = 0$$
and
$$x_1(a_{11}M_{11} - a_{12}M_{12} + a_{13}M_{13} + \cdots + a_{1n}(-1)^{n+1}M_{1n}) = 0$$
Thus $x_1 D = 0$ and since $D = 0$, this is a true statement, and the solution in terms of x_1 is a solution of all the equations of the system. Since the value of x_1 can be assigned in infinitely many ways, there are infinitely many solutions.

EXAMPLE 7 Find the solution set of
$$2x - 3y - 2z = 0$$
$$x + 2y + 3z = 0$$
$$3x - y + z = 0$$

Solution

1. Find D, the determinant of the coefficients.
$$D = \begin{vmatrix} 2 & -3 & -2 \\ 1 & 2 & 3 \\ 3 & -1 & 1 \end{vmatrix} = \begin{vmatrix} 0 & -7 & -8 \\ 1 & 2 & 3 \\ 0 & -7 & -8 \end{vmatrix} = 0$$

2. Find a minor that is not zero. Using M_{11}, and rewriting the last two equations,
$$2y + 3z = -x$$
$$-y + z = -3x$$

3. Solving for y and z in terms of x,
$$y = \frac{8x}{5} \quad \text{and} \quad z = \frac{-7x}{5}$$

4. The solution set is $\left\{ \left(x, \frac{8x}{5}, \frac{-7x}{5} \right) \right\}$.

5. Checking in the first equation,
$$2x - 3\left(\frac{8x}{5}\right) - 2\left(\frac{-7x}{5}\right) = \frac{1}{5}(10x - 24x + 14x) = 0$$

EXAMPLE 8 Find the solution set of
$$x + 2y - z = 0$$
$$2x - 3y + 2z = 0$$
$$3x + y - 2z = 0$$

Solution

$$D = \begin{vmatrix} 1 & 2 & -1 \\ 2 & -3 & 2 \\ 3 & 1 & -2 \end{vmatrix} = \begin{vmatrix} 1 & 0 & 0 \\ 2 & -7 & 4 \\ 3 & -5 & 1 \end{vmatrix} = 13$$

Since $D \neq 0$, the trivial solution is the only solution. The solution set is $\{(0, 0, 0)\}$.

EXERCISES 6.3

(1–8) Evaluate.

1. $\begin{vmatrix} 3 & 2 & 3 & 4 \\ -2 & 4 & -2 & 3 \\ 1 & 2 & 1 & 5 \\ -3 & 6 & -3 & 7 \end{vmatrix}$

2. $\begin{vmatrix} 3 & -4 & 7 & 6 \\ 2 & -3 & 4 & 5 \\ 2 & -5 & 1 & 5 \\ 4 & -6 & 8 & 10 \end{vmatrix}$

3. $\begin{vmatrix} 1 & 2 & -3 & -1 \\ 3 & 5 & 2 & 4 \\ 2 & -1 & 1 & 5 \\ -2 & 1 & 6 & 2 \end{vmatrix}$

4. $\begin{vmatrix} 1 & 1 & 1 & 1 \\ 1 & 2 & 3 & 4 \\ 1 & 4 & 9 & 16 \\ 1 & 8 & 27 & 64 \end{vmatrix}$

5. $\begin{vmatrix} 1 & 2 & 3 & 4 \\ 1 & 4 & 9 & 16 \\ 1 & 8 & 27 & 64 \\ 1 & 16 & 81 & 256 \end{vmatrix}$

6. $\begin{vmatrix} 1 & 1 & 1 & 3x - 3y \\ 0 & 0 & x & x^2 - xy \\ x & y & 0 & x^2 - y^2 \\ -y & 0 & 0 & y^2 - xy \end{vmatrix}$

7. $\begin{vmatrix} 1 & 5 & 4 & 3 & 2 \\ 0 & 2 & 3 & 6 & 1 \\ 0 & 0 & 4 & 5 & 3 \\ 0 & 0 & 0 & 5 & 7 \\ 0 & 0 & 0 & 0 & 6 \end{vmatrix}$

8. $\begin{vmatrix} 0 & -2 & -1 & 0 & 4 \\ 5 & 3 & 2 & 4 & 1 \\ 0 & 1 & 3 & 0 & 2 \\ 0 & 4 & 0 & 3 & -2 \\ 0 & 2 & 4 & 1 & 1 \end{vmatrix}$

(9–12) Solve, using Cramer's Rule.

9. $x + y + z + t = 0$
$x - 2z = 4$
$x + 3t = 6$
$y + z + t = 6$

10. $2x + y - z + t = 1$
$x - y + 2z - t = 2$
$x + y - z + t = 1$
$3x + y - z + 2t = 0$

11. $x_1 + x_2 + x_3 + x_4 = 2$
$2x_1 + 2x_2 - x_3 = -3$
$x_2 - 2x_3 - 2x_4 = 1$
$x_1 + x_2 - x_4 = 5$

12. $x_1 + x_2 + x_3 + x_4 + x_5 = 15$
$x_1 + x_3 + x_5 = 9$
$x_2 + x_4 + x_5 = 10$
$x_1 + x_2 + x_4 = 8$
$x_2 + x_3 + x_4 = 9$

(13–18) Find the solution set of each system.

13. $x - 2y + 3z = 0$
$3x + y - 2z = 0$
$4x - y + z = 0$

14. $x + 3y - 4z = 0$
$2x - 4y + 3z = 0$
$4x + 2y - 5z = 0$

15. $2x + y + z = 0$
$x - 2y + 2z = 0$
$x + 3y - 3z = 0$

16. $x + y + z = 0$
$x - 2y - z = 0$
$2x - y - 2z = 0$

17. $x + y + z + t = 0$
$2x - y + z - t = 0$
$x - 2y - z + 2t = 0$
$4x - 2y + z + 2t = 0$

18. $x + 2y + z - 3t = 0$
$x - y - 2z + 2t = 0$
$2x - y + 2z - 2t = 0$
$2x + y - z - 2t = 0$

6.4 MATRIX ALGEBRA

In Section 6.2 we used some special types of matrices to solve linear systems. This section is concerned with matrices and their properties.

DEFINITION

A **matrix** is a rectangular (or square) array of numbers (or elements) arranged in rows and columns enclosed by brackets, double vertical lines, or by large parentheses.

Some examples of matrices are

$$\begin{bmatrix} 1 & 0 & 2 \\ 3 & -1 & 5 \\ 4 & 2 & 6 \end{bmatrix}, \quad \begin{Vmatrix} 5 & 1 & 2 \\ 3 & 2 & -7 \end{Vmatrix}, \quad \begin{pmatrix} 2 & 0 \\ 1 & 3 \end{pmatrix}, \quad \begin{bmatrix} 2 \\ 5 \end{bmatrix}, \quad [3 \quad 1]$$

DEFINITION

The **dimensions** of a matrix are $n \times k$ (read "*n* by *k*") if the matrix is an array of *n* rows and *k* columns.

If the number of rows is equal to the number of columns, then the matrix is called a **square** matrix.

If a matrix has only one row or one column, then it is called a **row matrix** or a **column matrix** (or a **row vector** or a **column vector**).

In the examples above, the first matrix is a 3×3 square matrix, the second is a 2×3 matrix, the third is a 2×2 square matrix, the fourth is a column matrix or column vector, and the fifth is a row matrix or a row vector.

DEFINITION EQUAL MATRICES

Two matrices are equal if and only if they have the same dimensions and their corresponding elements are equal.

For example,

$$\begin{bmatrix} 2 & 1 \\ 3 & 0 \end{bmatrix} = \begin{bmatrix} \frac{6}{3} & \frac{3}{3} \\ \frac{9}{3} & 0 \end{bmatrix} \quad \text{and} \quad \begin{bmatrix} 4 & -1.5 & 2.5 \\ 1.5 & 3 & 5 \end{bmatrix} = \begin{bmatrix} 4 & -\frac{3}{2} & \frac{5}{2} \\ \frac{3}{2} & 3 & 5 \end{bmatrix}$$

but

$$\begin{bmatrix} 1 & 4 \\ 3 & 2 \end{bmatrix} \neq \begin{bmatrix} 1 & 3 \\ 4 & 2 \end{bmatrix} \quad \text{and} \quad \begin{bmatrix} 0 & 0 & 0 \\ 0 & 0 & 0 \end{bmatrix} \neq \begin{bmatrix} 0 & 0 \\ 0 & 0 \end{bmatrix}$$

DEFINITION SUM OF TWO MATRICES

The sum of two matrices having the same dimensions is the matrix whose elements are the sums of the corresponding elements of the matrices being added.

EXAMPLE 1 Find the indicated sum.

$$\begin{bmatrix} a & b \\ c & d \end{bmatrix} + \begin{bmatrix} 1 & 2 \\ 3 & 4 \end{bmatrix}$$

Solution

$$\begin{bmatrix} a & b \\ c & d \end{bmatrix} + \begin{bmatrix} 1 & 2 \\ 3 & 4 \end{bmatrix} = \begin{bmatrix} a+1 & b+2 \\ c+3 & d+4 \end{bmatrix}$$

EXAMPLE 2 Find the indicated sum.

$$\begin{bmatrix} 1 & 0 & 2 \\ 2 & 1 & -3 \end{bmatrix} + \begin{bmatrix} 2 & 4 & -1 \\ 0 & 2 & 1 \end{bmatrix}$$

Solution

$$\begin{bmatrix} 1 & 0 & 2 \\ 2 & 1 & -3 \end{bmatrix} + \begin{bmatrix} 2 & 4 & -1 \\ 0 & 2 & 1 \end{bmatrix} = \begin{bmatrix} (1+2) & (0+4) & (2-1) \\ (2+0) & (1+2) & (-3+1) \end{bmatrix}$$
$$= \begin{bmatrix} 3 & 4 & 1 \\ 2 & 3 & -2 \end{bmatrix}$$

It should be noted that addition is defined only for matrices having the same dimensions. It is left for the student to verify that matrix addition is commutative and associative.

An element of a matrix is also called a **scalar.**

If the elements of a matrix are members of the set of real numbers, then a **scalar** is any real number.

DEFINITION   SCALAR MULTIPLICATION

The product of a scalar and a matrix is the matrix each of whose elements is the product of the scalar and the corresponding element of the matrix being multiplied.

For example,

$$k\begin{bmatrix} 2 & 1 \\ 3 & 0 \end{bmatrix} = \begin{bmatrix} 2k & k \\ 3k & 0 \end{bmatrix} \text{ and } 2\begin{bmatrix} 1 & -1 & 2 \\ 3 & 0 & 4 \end{bmatrix} = \begin{bmatrix} 2 & -2 & 4 \\ 6 & 0 & 8 \end{bmatrix}$$

Scalar multiplication may be shown to be commutative and associative.

DEFINITION   MATRIX MULTIPLICATION

The product of an $n \times k$ matrix, A, and a $k \times m$ matrix, B, is an $n \times m$ matrix, AB, in which the element in the ath row and bth column is the sum of the products of the elements of the ath row of A multiplied by the corresponding elements of the bth row of B.

EXAMPLE 3   If $A = \begin{bmatrix} 2 & 3 \\ 4 & 5 \end{bmatrix}$ and $B = \begin{bmatrix} a & b & c \\ x & y & z \end{bmatrix}$,

find the matrix product AB.

Solution

$$AB = \begin{bmatrix} 2a + 3x & 2b + 3y & 2c + 3z \\ 4a + 5x & 4b + 5y & 4c + 5z \end{bmatrix}$$

EXAMPLE 4   Multiply

$$\begin{bmatrix} 2 & -1 & 3 \\ 3 & 4 & 1 \end{bmatrix} \text{ by } \begin{bmatrix} x \\ y \\ z \end{bmatrix}$$

Solution

$$\begin{bmatrix} 2 & -1 & 3 \\ 3 & 4 & 1 \end{bmatrix} \begin{bmatrix} x \\ y \\ z \end{bmatrix} = \begin{bmatrix} 2x - y + 3z \\ 3x + 4y + z \end{bmatrix}$$

EXAMPLE 5   If $A = \begin{bmatrix} 2 & 0 \\ 1 & 3 \end{bmatrix}$ and $B = \begin{bmatrix} 1 & 5 \\ 3 & 2 \end{bmatrix}$, find AB and BA.

Solution

$$AB = \begin{bmatrix} 2 & 0 \\ 1 & 3 \end{bmatrix} \begin{bmatrix} 1 & 5 \\ 3 & 2 \end{bmatrix} = \begin{bmatrix} 2 \cdot 1 + 0 \cdot 3 & 2 \cdot 5 + 0 \cdot 2 \\ 1 \cdot 1 + 3 \cdot 3 & 1 \cdot 5 + 3 \cdot 2 \end{bmatrix} = \begin{bmatrix} 2 & 10 \\ 10 & 11 \end{bmatrix}$$

$$BA = \begin{bmatrix} 1 & 5 \\ 3 & 2 \end{bmatrix} \begin{bmatrix} 2 & 0 \\ 1 & 3 \end{bmatrix} = \begin{bmatrix} 2 + 5 & 0 + 15 \\ 6 + 2 & 0 + 6 \end{bmatrix} = \begin{bmatrix} 7 & 15 \\ 8 & 6 \end{bmatrix}$$

Example 5 illustrates that matrix multiplication is *not commutative*. Note that it is not necessary to compute all entries in BA to show that $AB \neq BA$. Only two corresponding entries need to be shown unequal; for example, the entries in row 1, column 1 are not equal, $2 \neq 7$ and thus $AB \neq BA$.

On the other hand, matrix multiplication *is* associative and distributive from both the right and the left; that is,

$$(AB)C = A(BC)$$
$$A(B + C) = AB + AC$$
$$(B + C)A = BA + CA$$

It should be noted that the product of two matrices is defined *only* for the case where the number of columns of the left matrix is equal to the number of rows of the right matrix.

It may seem strange that matrix multiplication is defined as it is and not some other way; for example, as the matrix whose elements are the products of the corresponding elements. There are very few applications of such a definition whereas there is a wide range of applications for the definition as stated, particularly in the solution of systems of equations.

DEFINITION

The **main diagonal** (or **principal diagonal**) of a square matrix is the diagonal from the upper left element to the lower right element.

For example, the main diagonal of the matrix $\begin{bmatrix} 1 & 2 & 3 \\ 4 & 5 & 6 \\ 7 & 8 & 9 \end{bmatrix}$ contains the elements 1, 5, and 9.

DEFINITION

An **identity matrix, *I*,** is a square matrix whose elements along the main diagonal are ones and the rest of whose elements are zeros.
For example,

$$\begin{bmatrix} 1 & 0 \\ 0 & 1 \end{bmatrix}, \quad \begin{bmatrix} 1 & 0 & 0 \\ 0 & 1 & 0 \\ 0 & 0 & 1 \end{bmatrix}, \quad \text{and} \quad \begin{bmatrix} 1 & 0 & 0 & 0 \\ 0 & 1 & 0 & 0 \\ 0 & 0 & 1 & 0 \\ 0 & 0 & 0 & 1 \end{bmatrix}$$

are identity matrices.

THEOREM

If *I* is an identity matrix and if *A* is any matrix having the same dimensions as *I*, then $IA = AI = A$.
For example,

$$\begin{bmatrix} 1 & 0 \\ 0 & 1 \end{bmatrix} \begin{bmatrix} a & b \\ c & d \end{bmatrix} = \begin{bmatrix} a & b \\ c & d \end{bmatrix} = \begin{bmatrix} a & b \\ c & d \end{bmatrix} \begin{bmatrix} 1 & 0 \\ 0 & 1 \end{bmatrix}$$

DEFINITION

An **inverse matrix, A^{-1},** of a square matrix *A* is a square matrix having the same dimensions as *A*, such that $AA^{-1} = I$ and $A^{-1}A = I$.

EXAMPLE 6 Find A^{-1}, if

$$A = \begin{bmatrix} 4 & 2 \\ 1 & 1 \end{bmatrix}$$

Solution   Let

$$A^{-1} = \begin{bmatrix} a & b \\ c & d \end{bmatrix}$$

Then

$$AA^{-1} = \begin{bmatrix} 4 & 2 \\ 1 & 1 \end{bmatrix} \cdot \begin{bmatrix} a & b \\ c & d \end{bmatrix}$$

$$= \begin{bmatrix} 4a + 2c & 4b + 2d \\ a + c & b + d \end{bmatrix} = \begin{bmatrix} 1 & 0 \\ 0 & 1 \end{bmatrix} = I$$

Equating corresponding elements,

$$4a + 2c = 1 \qquad 4b + 2d = 0$$
$$a + c = 0 \qquad b + d = 1$$

Solving,

$$a = \tfrac{1}{2}, \qquad b = -1, \qquad c = -\tfrac{1}{2}, \qquad d = 2$$

$$A^{-1} = \begin{bmatrix} \tfrac{1}{2} & -1 \\ -\tfrac{1}{2} & 2 \end{bmatrix}$$

Note that

$$A^{-1}A = \begin{bmatrix} \tfrac{1}{2} & -1 \\ -\tfrac{1}{2} & 2 \end{bmatrix} \cdot \begin{bmatrix} 4 & 2 \\ 1 & 1 \end{bmatrix} = \begin{bmatrix} 1 & 0 \\ 0 & 1 \end{bmatrix}$$

EXERCISES 6.4

(1–23) Perform the indicated operations, if possible. If not possible, state why not.

1. $\begin{bmatrix} 3 & 5 \\ -2 & 4 \end{bmatrix} + \begin{bmatrix} 1 & 0 \\ -5 & -5 \end{bmatrix}$

2. $\begin{bmatrix} 2 & 4 & -3 \\ -1 & 0 & 1 \end{bmatrix} + \begin{bmatrix} 3 & 4 & 5 \\ 6 & 7 & 8 \end{bmatrix}$

3. $\begin{bmatrix} 4 & 2 & 9 \\ 2 & -1 & 0 \end{bmatrix} + \begin{bmatrix} 3 & 5 \\ 1 & -7 \end{bmatrix}$

4. $[5 \quad -2 \quad 7] + \begin{bmatrix} 3 \\ 8 \\ 2 \end{bmatrix}$

5. $\begin{bmatrix} 3 & -2 & 4 \\ 1 & 0 & 2 \\ 1 & 0 & 1 \end{bmatrix} + \begin{bmatrix} 2 & 5 & -8 \\ 1 & -3 & -2 \\ 3 & -2 & -2 \end{bmatrix}$

6. $\begin{bmatrix} 4 & 6 & -3 & 5 \\ 1 & -4 & 2 & 1 \\ 3 & 0 & 0 & 1 \end{bmatrix} + \begin{bmatrix} 7 & 0 & -5 \\ 2 & -1 & 0 \end{bmatrix}$

7. $[4 \quad 7 \quad 9] + [2 \quad 8 \quad 5]$

8. $-2\begin{bmatrix} \frac{1}{2} & 3 & -\frac{5}{2} \\ 1 & 0 & -1 \\ 3 & -2 & 1 \end{bmatrix}$

9. $3[2 \quad -3 \quad 4]$

10. $5\begin{bmatrix} 2 & -3 & 5 \\ 1 & 4 & -2 \end{bmatrix}$

11. $\begin{bmatrix} 3 & 1 \\ 2 & 5 \end{bmatrix} \begin{bmatrix} -2 & 3 \\ 4 & 1 \end{bmatrix}$

12. $\begin{bmatrix} 2 & 3 \\ 1 & -2 \\ 4 & 3 \end{bmatrix} \begin{bmatrix} a & b & c \\ 4 & 2 & 3 \end{bmatrix}$

13. $\begin{bmatrix} 2 & 1 & 4 \\ 3 & -2 & 3 \end{bmatrix} \begin{bmatrix} a & b & c \\ 4 & 2 & 3 \end{bmatrix}$

14. $\begin{bmatrix} 1 & -1 & 1 \\ 2 & 1 & -3 \\ 3 & 2 & -1 \end{bmatrix} \begin{bmatrix} x \\ y \\ z \end{bmatrix}$

15. $\begin{bmatrix} 1 & -2 & 1 & 3 \\ 2 & 0 & 5 & 4 \end{bmatrix} \begin{bmatrix} a & 2 \\ b & -3 \\ c & -1 \\ d & 0 \end{bmatrix}$

16. $\begin{bmatrix} 1 & 3 & -4 \\ 5 & -1 & 2 \end{bmatrix} \begin{bmatrix} x \\ y \\ z \end{bmatrix}$

17. $\begin{bmatrix} 1 & 0 & 0 \\ 0 & 1 & 0 \\ 0 & 0 & 1 \end{bmatrix} \begin{bmatrix} a & b & c \\ d & e & f \\ g & h & i \end{bmatrix}$

18. $\begin{bmatrix} 1 & 2 & 1 & 3 \\ 1 & 1 & -2 & 1 \end{bmatrix} \begin{bmatrix} a & 0 \\ b & 1 \\ c & -1 \\ d & 0 \end{bmatrix}$

19. $\begin{bmatrix} 1 & 2 & 1 & 3 \\ 0 & 1 & -2 & 1 \end{bmatrix} + \begin{bmatrix} a & 0 \\ b & 1 \\ c & -1 \\ d & 0 \end{bmatrix}$

20. $\begin{bmatrix} 4 & 2 & 4 & -1 \\ 3 & -1 & 0 & 2 \end{bmatrix} \begin{bmatrix} 2 & x & 0 & 1 \\ y & 1 & -1 & 0 \end{bmatrix}$

21. $\begin{bmatrix} 4 & 2 & 4 & -1 \\ 3 & -1 & 0 & 2 \end{bmatrix} + \begin{bmatrix} 2 & x & 0 & 1 \\ y & 1 & -1 & 0 \end{bmatrix}$

22. $\begin{bmatrix} 1 & 2 \\ 2 & 4 \end{bmatrix} \begin{bmatrix} 6 & -2 \\ -3 & 1 \end{bmatrix}$

23. $\begin{bmatrix} 1 & 0 & 0 & 0 \\ 0 & 1 & 0 & 0 \\ 0 & 0 & 1 & 0 \\ 0 & 0 & 0 & 1 \end{bmatrix} \begin{bmatrix} w \\ x \\ y \\ z \end{bmatrix} = \begin{bmatrix} -2 \\ 3 \\ 1 \\ 5 \end{bmatrix}$

(24–28) Solve for the variables.

24. $\begin{bmatrix} 1 & x & 3 \\ y & 2 & z \end{bmatrix} = \begin{bmatrix} 1 & 0 & 3 \\ 6 & 2 & 9 \end{bmatrix}$

25. $5\begin{bmatrix} x & y \\ 3 & -2 \end{bmatrix} + \begin{bmatrix} 3 & -2 \\ -1 & 4 \end{bmatrix} = 2\begin{bmatrix} -6 & 5 \\ 7 & -3 \end{bmatrix}$

26. $\begin{bmatrix} 1 & 0 & 0 \\ 0 & -1 & 0 \\ 0 & 0 & 2 \end{bmatrix} \begin{bmatrix} x \\ y \\ z \end{bmatrix} = \begin{bmatrix} 2 \\ 3 \\ 8 \end{bmatrix}$

27. $\begin{bmatrix} 2 & 0 & 0 \\ 0 & 3 & 0 \\ 0 & 0 & -4 \end{bmatrix} \begin{bmatrix} x & -1 \\ y & 2 \\ z & -3 \end{bmatrix} = 3\begin{bmatrix} 3 & -\frac{2}{3} \\ -2 & 2 \\ 5 & 4 \end{bmatrix}$

28. $\begin{bmatrix} 1 & 2 & -1 \\ 0 & 1 & 1 \\ 0 & 0 & 1 \end{bmatrix} \begin{bmatrix} x \\ y \\ z \end{bmatrix} = \begin{bmatrix} -2 \\ 5 \\ 3 \end{bmatrix}$

(29–37) Let $A = \begin{bmatrix} 1 & 2 \\ -1 & 1 \end{bmatrix}$ and $B = \begin{bmatrix} 1 & -1 \\ 2 & 1 \end{bmatrix}$. Find each of the following.

29. $-B$
31. AB
33. $A(-B)$
35. $(AB)^{-1}$
37. $-A(B)^{-1}$

30. $-BA$
32. BA
34. B^{-1}
36. $B^{-1}A^{-1}$

(38–40) Determine the relationship between each of the following.

38. AB and BA
39. $A(-B)$ and $(-B)A$
40. $(AB)^{-1}$ and $B^{-1}A^{-1}$

41. Let $A = \begin{bmatrix} x & y \\ 2x & 2y \end{bmatrix}$ and $B = \begin{bmatrix} y & 3y \\ -x & -3x \end{bmatrix}$. Find AB. If $AB = 0$, that is,

$AB = \begin{bmatrix} 0 & 0 \\ 0 & 0 \end{bmatrix}$, does it follow that $A = 0$ or $B = 0$?

(42–52) Let $A = \begin{bmatrix} a & b \\ c & d \end{bmatrix}$, $B = \begin{bmatrix} 1 & 1 \\ -1 & 1 \end{bmatrix}$, $C = \begin{bmatrix} 1 & 1 \\ 1 & -1 \end{bmatrix}$, and let $S = $ the set of 2×2 square matrices whose elements are real numbers.

42. Show that $A + B \in S$ (closure, addition), and that $AB \in S$ (closure, multiplication).

43. Show that $A + B = B + A$ (commutativity, addition).

44. Show that $BC \neq CB$ (multiplication is not commutative).

45. Show that $(A + B) + C = A + (B + C)$ and $(AB)C = A(BC)$ (associativity, addition and multiplication).

46. Find 0 so that $0 + A = A = A + 0$ (addition identity).

47. Find I so that $IA = A = AI$ (multiplication identity).

48. Find $-A$ so that $A + (-A) = 0$ (addition inverse).

49. Assuming that $ad - bc \neq 0$, find A^{-1} so that $AA^{-1} = I$ (multiplication inverse). Does $A^{-1}A = AA^{-1}$? Justify your answer.

50. Find X so that $A = B + X$. Define subtraction.

51. Find Y so that $AY = B$. Define division.

52. Assuming $ad - bc \neq 0$, if $AX = B$, then which of the following is correct: $X = BA^{-1}$ or $X = A^{-1}B$? Justify your answer.

6.5 INVERSES OF MATRICES, LINEAR SYSTEMS

In the previous section we defined an inverse matrix A^{-1} of a square matrix A as a matrix having the property $AA^{-1} = A^{-1}A = I$, where I is the identity matrix.

Now let us find A^{-1} for

$$A = \begin{bmatrix} a & b \\ c & d \end{bmatrix}$$

Assuming A^{-1} exists, let

$$A^{-1} = \begin{bmatrix} x & z \\ y & t \end{bmatrix}$$

with $x, y, z,$ and t to be determined. Then we want

$$\begin{bmatrix} a & b \\ c & d \end{bmatrix} \cdot \begin{bmatrix} x & z \\ y & t \end{bmatrix} = \begin{bmatrix} 1 & 0 \\ 0 & 1 \end{bmatrix}$$

and thus

$$\begin{bmatrix} ax + by & az + bt \\ cx + dy & cz + dt \end{bmatrix} = \begin{bmatrix} 1 & 0 \\ 0 & 1 \end{bmatrix}$$

This means that

$$ax + by = 1 \quad \text{and} \quad az + bt = 0$$
$$cx + dy = 0 \qquad\qquad cz + dt = 1$$

Now, if $ad - bc \neq 0$, then we can solve for $x, y, z,$ and t.

$$x = \frac{d}{ad - bc}, \quad y = \frac{-c}{ad - bc}, \quad z = \frac{-b}{ad - bc}, \quad t = \frac{a}{ad - bc}$$

Note that $ad - bc$ is the determinant $\begin{vmatrix} a & b \\ c & d \end{vmatrix}$ which we call $D(A)$,

the **determinant of the matrix A.**

Then

$$A^{-1} = \begin{bmatrix} \dfrac{d}{ad - bc} & \dfrac{-b}{ad - bc} \\ \dfrac{-c}{ad - bc} & \dfrac{a}{ad - bc} \end{bmatrix}$$

and

$$A^{-1} = \frac{1}{D(A)} \begin{bmatrix} d & -b \\ -c & a \end{bmatrix}$$

THEOREM 1

Inverse of a 2×2 matrix A.

1. If $D(A) = 0$, then the inverse of matrix A does not exist.
2. If $D(A) \neq 0$, then the inverse of a 2×2 matrix A is the product of $\frac{1}{D(A)}$ and the matrix obtained from A by interchanging the elements on the main diagonal and multiplying each of the other two elements by -1.

EXAMPLE 1 Find A^{-1}, if possible, for

(a) $A = \begin{bmatrix} 8 & 3 \\ 4 & 2 \end{bmatrix}$;   (b) $A = \begin{bmatrix} 8 & 6 \\ 4 & 3 \end{bmatrix}$

Solution

(a) $D(A) = \begin{vmatrix} 8 & 3 \\ 4 & 2 \end{vmatrix} = 16 - 12 = 4.$

Since $D(A) \neq 0$, the inverse exists.

$$A^{-1} = \frac{1}{4} \begin{bmatrix} 2 & -3 \\ -4 & 8 \end{bmatrix}$$

Check

$$A^{-1}A = \frac{1}{4} \begin{bmatrix} 2 & -3 \\ -4 & 8 \end{bmatrix} \cdot \begin{bmatrix} 8 & 3 \\ 4 & 2 \end{bmatrix} = \frac{1}{4} \begin{bmatrix} 4 & 0 \\ 0 & 4 \end{bmatrix} = \begin{bmatrix} 1 & 0 \\ 0 & 1 \end{bmatrix} = I$$

(b) $D(A) = \begin{vmatrix} 8 & 6 \\ 4 & 3 \end{vmatrix} = 24 - 24 = 0$

Therefore A^{-1} does not exist.

THEOREM 2

Inverse of a 3×3 matrix A.
Let

$$A = \begin{bmatrix} a_1 & b_1 & c_1 \\ a_2 & b_2 & c_2 \\ a_3 & b_3 & c_3 \end{bmatrix}$$

and let

$$D(A) = \begin{vmatrix} a_1 & b_1 & c_1 \\ a_2 & b_2 & c_2 \\ a_3 & b_3 & c_3 \end{vmatrix}$$

with $D(A)$ called the determinant of A.

1. If $D(A) = 0$, then A^{-1} does not exist.
2. If $D(A) \neq 0$, then

$$A^{-1} = \frac{1}{D(A)} \begin{bmatrix} A_1 & A_2 & A_3 \\ B_1 & B_2 & B_3 \\ C_1 & C_2 & C_3 \end{bmatrix}$$

where A_i is the cofactor of a_i, B_i is the cofactor of b_i, and C_i is the cofactor of c_i, for $i = 1, 2,$ and 3.

Theorem 2 can be proved by forming $A^{-1}A$ and showing that $A^{-1}A = I$. Note that each **row in A^{-1}** consists of the cofactors of the corresponding **column of A.**

EXAMPLE 2   Find A^{-1} if

$$A = \begin{bmatrix} 1 & -1 & 3 \\ 2 & 1 & 0 \\ 1 & 0 & 2 \end{bmatrix}$$

Solution

1. $D(A) = \begin{vmatrix} 1 & -1 & 3 \\ 2 & 1 & 0 \\ 1 & 0 & 2 \end{vmatrix} = \begin{vmatrix} 1 & 0 & 0 \\ 2 & 3 & -6 \\ 1 & 1 & -1 \end{vmatrix} = 3$

2. Form the matrix in which each entry is the cofactor of the corresponding element of A:

$$\begin{bmatrix} 2 & -4 & -1 \\ 2 & -1 & -1 \\ -3 & 6 & 3 \end{bmatrix}$$

3. Interchange the rows and columns of the matrix in Step 2 and multiply by $\dfrac{1}{D(A)}$:

$$A^{-1} = \frac{1}{3}\begin{bmatrix} 2 & 2 & -3 \\ -4 & -1 & 6 \\ -1 & -1 & 3 \end{bmatrix}$$

4. Check that $A^{-1}A = I$:

$$\frac{1}{3}\begin{bmatrix} 2 & 2 & -3 \\ -4 & -1 & 6 \\ -1 & -1 & 3 \end{bmatrix}\begin{bmatrix} 1 & -1 & 3 \\ 2 & 1 & 0 \\ 1 & 0 & 2 \end{bmatrix} = \frac{1}{3}\begin{bmatrix} 3 & 0 & 0 \\ 0 & 3 & 0 \\ 0 & 0 & 3 \end{bmatrix} = \begin{bmatrix} 1 & 0 & 0 \\ 0 & 1 & 0 \\ 0 & 0 & 1 \end{bmatrix}$$

The inverse of a matrix can be used to solve a linear system. We first note that the system

$$a_1 x + a_2 y = c_1$$
$$b_1 x + b_2 y = c_2$$

can be written in matrix notation as

$$\begin{bmatrix} a_1 & a_2 \\ b_1 & b_2 \end{bmatrix}\begin{bmatrix} x \\ y \end{bmatrix} = \begin{bmatrix} c_1 \\ c_2 \end{bmatrix}$$

Also the linear system

$$a_{11} x + a_{12} y + a_{13} z = c_1$$
$$a_{21} x + a_{22} y + a_{23} z = c_2$$
$$a_{31} x + a_{32} y + a_{33} z = c_3$$

can be written in matrix notation as

$$\begin{bmatrix} a_{11} & a_{12} & a_{13} \\ a_{21} & a_{22} & a_{23} \\ a_{31} & a_{32} & a_{33} \end{bmatrix}\begin{bmatrix} x \\ y \\ z \end{bmatrix} = \begin{bmatrix} c_1 \\ c_2 \\ c_3 \end{bmatrix}$$

Now if we let A = the matrix of the coefficients of the variables, X = the column matrix of the variables, and C = the column matrix of the constants, then we can write these systems more simply as

$$AX = C$$

Multiplying each side by A^{-1}, assuming that it exists,

$$A^{-1}AX = A^{-1}C$$
$$IX = A^{-1}C$$
$$X = A^{-1}C$$

This result provides a method for solving a linear system when A^{-1} exists.

EXAMPLE 3   Solve by using an inverse matrix.

$$x - y + 2z = 7$$
$$x + y - 3z = 3$$
$$2x - y - 2z = 5$$

Solution

1. Write the system in matrix notation as $AX = C$:

$$\begin{bmatrix} 1 & -1 & 2 \\ 1 & 1 & -3 \\ 2 & -1 & -2 \end{bmatrix} \begin{bmatrix} x \\ y \\ z \end{bmatrix} = \begin{bmatrix} 7 \\ 3 \\ 5 \end{bmatrix}$$

2. Find $D(A)$:

$$D(A) = \begin{vmatrix} 1 & -1 & 2 \\ 1 & 1 & -3 \\ 2 & -1 & -2 \end{vmatrix} = \begin{vmatrix} 2 & 0 & -1 \\ 1 & 1 & -3 \\ 3 & 0 & -5 \end{vmatrix} = -7$$

3. Find A^{-1} as shown in Example 2:

$$A^{-1} = \frac{-1}{7} \begin{bmatrix} -5 & -4 & 1 \\ -4 & -6 & 5 \\ -3 & -1 & 2 \end{bmatrix}$$

4. Now find $X = A^{-1}C$:

$$\begin{bmatrix} x \\ y \\ z \end{bmatrix} = \frac{-1}{7} \begin{bmatrix} -5 & -4 & 1 \\ -4 & -6 & 5 \\ -3 & -1 & 2 \end{bmatrix} \begin{bmatrix} 7 \\ 3 \\ 5 \end{bmatrix} = \frac{-1}{7} \begin{bmatrix} -42 \\ -21 \\ -14 \end{bmatrix} = \begin{bmatrix} 6 \\ 3 \\ 2 \end{bmatrix}$$

5. State the solution:

$$x = 6, \qquad y = 3, \qquad z = 2$$

The solution set is $\{(6, 3, 2)\}$.

EXERCISES 6.5

Solve by using an inverse matrix.

1. $2x - y = 2$
    $3x - 2y = -1$

2. $3x + 5y = 6$
    $2x + 3y = 5$

3. $3x + 4y = 2$
    $4x - 3y = 36$

4. $3x - 2y = 15$
    $x + 3y = 27$

5. $2x + 5y = 2$
    $30x - 40y = 7$

6. $3x - 3y = 2$
    $6x + 12y = 7$

7. $x - y + z = 1$
    $3x - y - z = 3$
    $x + y - z = 7$

8. $x + 2y - 2z = 3$
    $3x - 2y - z = 5$
    $2x - y + 2z = 9$

9. $r + s + t = 1$
$2r - 3s + 4t = 6$
$2r + 5s - 2t = 8$

10. $r - s + t = 1$
$2r - 5s + 3t = 4$
$3r - 8s + 4t = 9$

11. $2u - 3v + w = 1$
$3u + 4v - 2w = 1$
$3u + 2v + 4w = 5$

12. $2a + b - 3c = 1$
$2a - 2b + 15c = 1$
$3a - 2b + 3c = 0$

6.6   CHAPTER REVIEW

(1–2)   Solve by using Cramer's Rule and check. (6.1)

1. $x + y - 2z = 1$
$2x + y - 2z = 6$
$4x + y - 4z = 4$

2. $2x + y + z = 0$
$2x - 3y - 3z = 2$
$4x + 3y - z = 3$

(3–6)   Solve by the matrix method and check. (6.2)

3. $x + 2y - z = 5$
$2x - 5y + 2z = 1$
$5x - 2y - z = 1$

4. $a + b - c = 10$
$3a - 2b + 2c = 10$
$5a - 5b + c = -10$

5. $2r - s - t = 4$
$3r + 2s - 4t = 3$
$4r + 5s - 7t = 5$

6. $x - 2y + 3z = 8$
$2x - 5y + 5z = 10$
$x - 3y + 2z = 2$

(7–9)   Find the solution set of each system. (6.3)

7. $x + y - z - t = 2$
$2x - y + z - t = 3$
$2x + y - 2z + t = 5$
$3x - 2y - z + 3t = 1$

8. $2x - 3y + 4z = 0$
$3x + 2y - 3z = 0$
$5x - y + z = 0$

9. $x + 2y - 2z = 0$
$2x - y + 3z = 0$
$4x + 3y - 4z = 0$

(10–12)   Do the indicated operations. (6.4)

10. $\dfrac{1}{6}\begin{bmatrix} 18 & -12 & 24 \\ 6 & 30 & -18 \\ 36 & 0 & -6 \end{bmatrix}$

11. $\begin{bmatrix} 3 & 5 & -2 \\ 1 & 4 & 2 \end{bmatrix} + \begin{bmatrix} 1 & -2 & 6 \\ 4 & 4 & -2 \end{bmatrix}$

12. $\begin{bmatrix} 2 & 1 & 3 \\ 1 & 4 & -2 \end{bmatrix} \cdot \begin{bmatrix} 1 & 2 \\ -2 & 5 \\ 1 & -3 \end{bmatrix}$

(13–15) Let $A = \begin{bmatrix} 1 & 3 \\ 2 & 4 \end{bmatrix}$. Find each of the following. (6.4)

13. $-A$ **14.** A^{-1} **15.** A^2

(16–17) Solve by using an inverse matrix. (6.5)

16. $3x + 5y = 17$
$5x + 3y = 15$

17. $x + y + z = 8$
$2x - 3y - 2z = 2$
$3x - 4y - z = 6$

7

EXPONENTS AND RADICALS

The power b^n was defined in Chapter 1 as n factors of b when n is a positive integer. In this chapter b^r will be defined for any rational number r, and properties of exponents and radicals will be explained. Finally, we will discuss the solution of equations containing radical expressions.

7.1 EXPONENT THEOREMS

Before stating the theorems that describe the properties of exponents in general, let us examine some special cases.

$$3^4 \cdot 3^2 = (3 \cdot 3 \cdot 3 \cdot 3)(3 \cdot 3) = 3^6 = 3^{4+2}$$

$$x^4 \cdot x^2 = (xxxx)(xx) = x^6 = x^{4+2}$$

$$(3^4)^2 = (3 \cdot 3 \cdot 3 \cdot 3)(3 \cdot 3 \cdot 3 \cdot 3) = 3^8 = 3^{4(2)}$$

$$(x^4)^2 = (xxxx)(xxxx) = x^8 = x^{4(2)}$$

$$(3x)^4 = 3x \cdot 3x \cdot 3x \cdot 3x = 3^4 x^4$$

$$\left(\frac{3}{x}\right)^4 = \frac{3}{x} \cdot \frac{3}{x} \cdot \frac{3}{x} \cdot \frac{3}{x} = \frac{3^4}{x^4}$$

$$\frac{3^6}{3^4} = \frac{3 \cdot 3 \cdot 3 \cdot 3 \cdot 3 \cdot 3}{3 \cdot 3 \cdot 3 \cdot 3} = 3^2 = 3^{6-4}$$

$$\frac{x^6}{x^4} = \frac{xxxxxx}{xxxx} = x^2 = x^{6-4}$$

$$\frac{x^4}{x^6} = \frac{xxxx}{xxxxxx} = \frac{1}{x^2} = \frac{1}{x^{6-4}}$$

Generalizing the results obtained for these special cases, we have the following theorems describing the properties of exponents.

EXPONENT THEOREMS

For a and b real numbers and m and n positive integers,

1. $b^m b^n = b^{m+n}$

2. $(b^m)^n = b^{mn}$

3. $(ab)^n = a^n b^n$

4. $\left(\dfrac{a}{b}\right)^n = \dfrac{a^n}{b^n}$,   for $b \neq 0$

5. $\dfrac{b^m}{b^n} = \begin{cases} b^{m-n}, & \text{if } m > n \quad \text{for } b \neq 0 \\ \dfrac{1}{b^{n-m}}, & \text{if } n > m \\ 1, & \text{if } m = n \end{cases}$

A formal proof of the exponent theorems requires the use of mathematical induction which is discussed in a later chapter. However, we can give intuitive arguments as follows:

$$b^m b^n = \underbrace{(bb \cdots b)(bb \cdots b)}_{m+n \text{ factors}} = b^{m+n}$$

$$(b^m)^n = \underbrace{b^m b^m \cdots b^m}_{n \text{ times } (m \text{ factors of } b)} = b^{mn}$$

$$(ab)^n = \underbrace{ab \cdot ab \cdots ab}_{n \text{ factors of } ab} = a^n b^n$$

$$\left(\frac{a}{b}\right)^n = \underbrace{\frac{a}{b} \cdot \frac{a}{b} \cdots \frac{a}{b}}_{n \text{ factors of } \frac{a}{b}} = \frac{a^n}{b^n}$$

For $m > n$,

$$\frac{b^m}{b^n} = \frac{\overbrace{(bb \cdots b)}^{m \text{ factors}}\overbrace{(bb \cdots b)}^{m-n \text{ factors}}}{\underbrace{(bb \cdots b)}_{n \text{ factors}}} = b^{m-n}$$

Many expressions involving exponents can be simplified by using one or more of the exponent theorems.

EXAMPLE 1 Simplify $\dfrac{(x^2)^3}{xx^2}$.

Solution   Using Theorem 2, $(b^m)^n = b^{mn}$,

$$\frac{x^6}{xx^2}$$

Using Theorem 1, $b^m b^n = b^{m+n}$
and the definition $x = x^1$,

$$\frac{x^6}{x^3}$$

Using Theorem 5, $\dfrac{b^m}{b^n} = b^{m-n}$,

$$x^{6-3}$$

The simplified result is
$$x^3$$

EXAMPLE 2 Simplify $\dfrac{(x^2 y)(xy^3)}{(xy^2)^3}$.

Solution   Rearranging factors,

$$\frac{(x^2 x)(yy^3)}{(xy^2)^3}$$

Using Theorem 1,

$$\frac{x^3 y^4}{(xy^2)^3}$$

Using Theorem 3,

$$\frac{x^3 y^4}{x^3 (y^2)^3}$$

Using Theorem 2,

$$\frac{x^3 y^4}{x^3 y^6}$$

Using Theorem 5,

$$\left(\frac{x^3}{x^3}\right)\left(\frac{y^4}{y^6}\right) = 1\left(\frac{1}{y^{6-4}}\right)$$

$$= \frac{1}{y^2}$$

EXAMPLE 3 Simplify $\dfrac{(12x^3)^4}{(30x)^4}$.

Solution   Using Theorem 4,

$$\left(\frac{12x^3}{30x}\right)^4$$

Using Theorem 5 and reducing $\frac{12}{30}$,

$$\left(\frac{2x^{3-1}}{5}\right)^4 = \left(\frac{2x^2}{5}\right)^4$$

Using Theorem 4,

$$\frac{(2x^2)^4}{5^4}$$

Using Theorem 3,

$$\frac{2^4 (x^2)^4}{5^4}$$

Using Theorem 2 and evaluating the powers,

$$\frac{16x^8}{625}$$

EXERCISES 7.1

(1–39) Simplify each of the following by using one of the exponent theorems. Assume $x \neq 0$, $y \neq 0$, and $t \neq 0$.

1. $2^4 2^3$
2. $x^5 x^4$
3. $x^k x^n$

4. xx^5
5. xx^n
6. $3 \cdot 3^4$

7. $(5^2)^3$
8. $(x^2)^3$
9. $(x^3)^2$

10. $(x^2)^k$
11. $(x^n)^3$
12. $(y^k)^n$

13. $(2x)^3$
14. $(3y)^4$
15. $(xy)^5$

16. $(5y)^n$
17. $(xy)^n$
18. $(10t)^6$

19. $\left(\dfrac{6}{x}\right)^3$
20. $\left(\dfrac{y}{2}\right)^4$
21. $\left(\dfrac{x}{y}\right)^5$

22. $\left(\dfrac{3}{y}\right)^n$
23. $\left(\dfrac{x}{5}\right)^k$
24. $\left(\dfrac{r}{t}\right)^n$

25. $\dfrac{7^5}{7^3}$
26. $\dfrac{8^6}{8^3}$
27. $\dfrac{10^8}{10^8}$

28. $\dfrac{x^3}{x^3}$
29. $\dfrac{x^{12}}{x^4}$
30. $\dfrac{y^6}{y^2}$

31. $\dfrac{9^3}{9^7}$
32. $\dfrac{2^5}{2^{10}}$
33. $\dfrac{x^5}{x^{10}}$

34. $\dfrac{y^7}{y^{10}}$
35. $\dfrac{y^n}{y^n}$
36. $\dfrac{x^2}{x^n}$ for $n > 2$

37. $\dfrac{x^{k-1}}{x^{k+1}}$ for $k > 1$
38. $\dfrac{y}{y^n}$ for $n > 1$
39. $\dfrac{x^n}{x^k}$ for $k > n$

(40–73) Simplify each of the following by using one or more of the exponent theorems. Assume $x \neq 0$ and $y \neq 0$.

40. $(2x^3)^4$
41. $x(x^2)^3$
42. $-2y(y^2)^2$

43. $(-2y^3)^2$
44. $-(-5x^2)^3$
45. $(3xy^3)^4$

46. $(5x^2)^4(2x^3)^2$
47. $x^3y^2(xy^3)^2$

48. $(2x^2y)^3(5xy^2)^4$
49. $\left(\dfrac{x^3}{10}\right)^4$

50. $\left(\dfrac{x^4}{y^2}\right)^3$
51. $\left(\dfrac{3}{x^5}\right)^2$

52. $\dfrac{(xy^3)^2}{(x^2y^2)^3}$
53. $\left(\dfrac{5^3}{x^2}\right)^3\left(\dfrac{x}{5^2}\right)^4$

54. $\left(\dfrac{x^3}{x^{12}}\right)^2$
55. $\dfrac{x(-2xy^2)^3}{(-x^2y)^5}$

56. $(xy^5)(x^3y)$
57. $\dfrac{(125x^2y)^4}{(25xy^2)^4}$

58. $\dfrac{(41x^2y^3)^3}{(123x^4y^2)^3}$

59. $\left(\dfrac{x^{n+1}}{x^n}\right)^3$

60. $(y^{1+n}y^{1-n})^2$

61. $\dfrac{5^{6x}}{5^{2x}}$

62. $\left(\dfrac{y^k}{y^{k+1}}\right)^2$

63. $\dfrac{(x+6)^8}{(x+6)^4}$, where $x \neq -6$

64. $\dfrac{12x^{3n}}{3x^n}$

65. $\dfrac{(4x^n)^2}{2^4x^n}$

66. $\dfrac{(y^{n+1})^5}{(y^n)^5}$

67. $\dfrac{(x^{b-1})^4}{(x^{b+1})^4}$

68. $\dfrac{(x^{k+3})^2}{(x^k)^3(x^2)^3}$

69. $\dfrac{(y^2)^{n+1}}{(y^n)^2}$

70. $\dfrac{xy(x^ny^{n+1})}{(xy)^n}$

71. $\dfrac{(2^x3^x)^3}{(6^x)^2}$

72. $\dfrac{(10^x)^4}{(2^4)^x5^{4x+1}}$

73. $\dfrac{10^{x-1}10^{x+6}}{10^{x+2}}$

7.2 INTEGRAL EXPONENTS

Now we want to give a meaning to expressions such as 3^0, 4^{-1}, 2^{-3}, and in general b^0 and b^{-n} where the exponent is either 0 or a negative integer. However, we want the definitions to be such that the exponent theorems remain valid; otherwise the definitions would not be very useful.

If n is a positive integer and if Theorem 1, $b^mb^n = b^{m+n}$, remains valid when $m = 0$, then

$$b^0b^n = b^{0+n} = b^n$$

For $b \neq 0$,

$$b^0 = \frac{b^n}{b^n} = 1$$

If $b = 0$, then $b^n = 0^n = 0$ and the division by b^n is not possible. Therefore we define b^0 as 1 for $b \neq 0$ and state that 0^0 is undefined.

DEFINITION OF b^0

If b is any real number and $b \neq 0$,
$$b^0 = 1$$

As examples, $3^0 = 1$, $\left(\frac{1}{2}\right)^0 = 1$, and $(x + 3)^0 = 1$ for $x \neq -3$.

Now if Theorem 1 remains valid when the exponent is a negative integer, $-n$, then

$$b^nb^{-n} = b^{n-n} = b^0$$

If $b \neq 0$, then
$$b^n b^{-n} = 1$$
and dividing each side by b^n,
$$b^{-n} = \frac{1}{b^n}$$

Therefore we define b^{-n} as the reciprocal of b^n when $b \neq 0$.

DEFINITION OF b^{-n}

If n is an integer and b is any nonzero real number, then
$$b^{-n} = \frac{1}{b^n}$$

As examples,
$$4^{-1} = \frac{1}{4}, \quad 2^{-3} = \frac{1}{2^3} = \frac{1}{8}, \quad \text{and} \quad \left(\frac{3}{5}\right)^{-2} = \left(\frac{5}{3}\right)^2 = \frac{25}{9}$$

Note that
$$\frac{1}{b^{-n}} = \frac{1}{\frac{1}{b^n}} = b^n$$

Thus
$$\frac{1}{2^{-5}} = 2^5 \quad \text{and} \quad \left(\frac{1}{3}\right)^{-2} = 3^2$$

The five exponent theorems remain valid when the exponents are any integers provided we use the definitions stated.

Let m and n be positive integers and b any nonzero real number. Then for Theorem 1,
$$b^m b^{-n} = b^m \cdot \frac{1}{b^n} \begin{cases} = b^{m-n}, & \text{for } m > n \\ = 1 = b^{m-n}, & \text{for } m = n \\ = \frac{1}{b^{n-m}}, & \text{for } m < n \end{cases}$$

But
$$\frac{1}{b^{n-m}} = b^{-(n-m)} = b^{m-n}$$

As a result, Theorem 5 can be condensed to one equation,
$$\frac{b^m}{b^n} = b^{m-n}$$
since a meaning has been established when $m - n$ is zero or negative.

Now
$$b^{-m} b^{-n} = \frac{1}{b^m} \cdot \frac{1}{b^n}$$
$$= \frac{1}{b^{m+n}}$$
$$= b^{-m-n}$$

Thus Theorem 1 has been shown to be valid for all integral exponents.

The other theorems can be verified in a similar way.

The following examples show how the theorems are used to simplify expressions containing negative exponents.

EXAMPLE 1   Simplify $(3^{-1}6^{-1})^{-1}$.

Solution   Using $(ab)^n = a^n b^n$, $(3^{-1})^{-1}(6^{-1})^{-1}$.
Using $(b^m)^n = b^{mn}$, $3^1 6^1 = 18$.

EXAMPLE 2   Simplify $(3^{-1} + 6^{-1})^{-1}$.

Solution   There is no exponent theorem for a power of a sum, so the definition is used.

$$(3^{-1} + 6^{-1})^{-1} = \frac{1}{3^{-1} + 6^{-1}} = \frac{1}{\frac{1}{3} + \frac{1}{6}} = \frac{1}{\frac{3}{6}}$$

$$= \frac{6}{3} = 2$$

EXAMPLE 3   Simplify $5^{-2}(5^2 \cdot 5^3)$.

Solution   Using Theorem 1,
$$5^{-2}(5^{2+3}) = 5^{-2}5^5$$
$$= 5^{-2+5}$$
$$= 5^3 = 125$$

EXAMPLE 4   Simplify $5^{-2}(5^2 + 5^3)$.

Solution   No exponent theorem can be used directly but the distributive axiom applies:
$$5^{-2}(5^2 + 5^3) = 5^{-2}5^2 + 5^{-2}5^3$$
Now Theorem 1 can be applied to each product.
$$= 5^0 + 5^1$$
$$= 1 + 5 = 6$$

EXAMPLE 5   Simplify $\dfrac{x^{-2}y^3}{x^{-3}y^{-2}}$. Assume $x \neq 0$ and $y \neq 0$.

Solution   Using Theorem 5, $\dfrac{b^m}{b^n} = b^{m-n}$

$$x^{-2-(-3)}y^{3-(-2)} = xy^5$$

EXAMPLE 6   Simplify $\dfrac{(a^{-1}b^{-2})^{-2}}{(ab^{-2})^3}$. Assume $a \neq 0$ and $b \neq 0$.

Solution

Using Theorem 3, $\quad \dfrac{(a^{-1})^{-2}(b^{-2})^{-2}}{a^3(b^{-2})^3}$
$(ab)^n = a^n b^n$

Using Theorem 2, $\quad = \dfrac{a^2 b^4}{a^3 b^{-6}}$
$(b^m)^n = b^{mn}$

Using Theorem 5, $\quad = a^{2-3}b^{4-(-6)}$

$\dfrac{b^m}{b^n} = b^{m-n} \quad\quad = \dfrac{b^{4-(-6)}}{a} = \dfrac{b^{10}}{a}$

EXERCISES 7.2

Use the exponent definitions and theorems to express each of the following as simply as possible without using zero or negative exponents.
Assume all bases and all denominators to be nonzero.

1. 6^0 **2.** 5^0 **3.** $\left(\dfrac{1}{5}\right)^0$ **4.** $\left(\dfrac{3}{8}\right)^0$

5. $(x - 2)^0$ **6.** $(x + y)^0$ **7.** 2^{-1} **8.** 5^{-1}

9. $\left(\dfrac{1}{4}\right)^{-1}$ **10.** $\left(\dfrac{-3}{5}\right)^{-1}$ **11.** $\left(\dfrac{7}{12}\right)^{-1}$ **12.** $\dfrac{1}{3^{-1}}$

13. $\dfrac{1}{10^{-1}}$ **14.** $\dfrac{1}{x^{-1}}$ **15.** 2^{-5} **16.** 3^{-4}

17. $\dfrac{1}{5^{-2}}$ **18.** $\dfrac{1}{4^{-3}}$ **19.** $\left(\dfrac{2}{5}\right)^{-2}$ **20.** $\left(\dfrac{3}{2}\right)^{-3}$

21. $\left(\dfrac{-10}{7}\right)^{-2}$ **22.** $\left(\dfrac{-x}{5}\right)^{-4}$ **23.** $\left(\dfrac{x}{y}\right)^{-1}$ **24.** $\left(\dfrac{-x}{y}\right)^{-3}$

25. $2^6 2^{-4}$ **26.** $5^3 5^{-6}$ **27.** $6^{-5} 6^5$ **28.** $10^{-2} 10^{-1}$

29. $2^{-3} 2^{-2}$ **30.** $5^4 (5^{-3} 5^{-4})$ **31.** $10^{-3} (10^4 10^{-1})$

32. $5^4 (5^{-3} + 5^{-4})$ **33.** $10^{-3} (10^4 + 10^3)$

34. $10^{-4} (10^3 + 10^4)$ **35.** $(10^{-1} 15^{-1})^{-1}$

36. $(10^{-1} + 15^{-1})^{-1}$ **37.** $(5^{-1} + 20^{-1})^{-2}$

38. $(5^{-1} 20^{-1})^{-2}$ **39.** $\dfrac{10^{-3}}{10^2}$

40. $\dfrac{10^2}{10^{-1}}$ **41.** $\dfrac{10^{-4}}{10^{-6}}$ **42.** $\dfrac{10^{-5}}{10^{-5}}$ **43.** $\dfrac{5^{-5}}{5^{-2}}$

44. $\dfrac{5^{-7}}{5^{-4}}$ **45.** $2x^{-3}$ **46.** $(2x)^{-3}$ **47.** $(3x^{-2})^{-4}$

48. $3(x^{-2})^{-4}$ **49.** $\left(\dfrac{x}{4}\right)^{-2}$ **50.** $\left(\dfrac{5^{-1}}{x^{-2}}\right)^{-2}$

51. $\left(\dfrac{2^{-2}}{5^{-2}}\right)^{-1}$ **52.** $\dfrac{(-2)^{-2}}{(-2)^{-3}}$ **53.** $x^{-2} y^{-3} (xy^{-1})^{-3}$

54. $\dfrac{x^2 y^{-4}}{x^{-2} y^{-1}}$ **55.** $\dfrac{x^{-1} y^{-2}}{5^{-3}}$ **56.** $\dfrac{4^{-2} x^{-3}}{y^{-4}}$

57. $(x + y)^{-1}$ **58.** $(x^{-1} + y^{-1})^{-1}$

59. $\dfrac{x^{-1} + y^{-1}}{y^{-1}}$ **60.** $xy^{-1} + x^{-1} y$

61. $(2x^3 y^{-2})^{-3}$ **62.** $(3x^{-4} y^0)^{-2}$

63. $\dfrac{a^{-1} + b^{-1}}{(a + b)^{-1}}$ **64.** $\dfrac{a^2 - b^2}{a^{-1} + b^{-1}}$

65. $\left(\dfrac{2a^2 b^{-1}}{5a^{-2} b^{-2}}\right)^2$ **66.** $\left(\dfrac{3a^{-3} b^2}{4a^{-4} b^{-1}}\right)^{-2}$

67. $\dfrac{(250 x^4 y^2)^{-4}}{(125 x^4 y^{-1})^{-4}}$ **68.** $\dfrac{(16^{-2} a^{-2} b^{-1})^3}{(8^{-2} a^{-2} b^{-2})^3}$

69. $\dfrac{a^{-n} b}{ab^{-n}}$ **70.** $\dfrac{b^{-n-2}}{b^{-n-1}}$

71. $\left(\dfrac{x^n}{y^{2n}}\right)^{-2}$ **72.** $\left(\dfrac{x^n y^{n-1}}{x^{n-1} y^n}\right)^{-n}$

73. $(a^{-n} b^{1-n})^{-2}$ **74.** $(ab^{-2})^{-n} (ab)^{-2n}$

75. $(a^{-n} + b^{-n})(a^{-n} - b^{-n})$ **76.** $a^{1-n} b^{n-1} (a^{n-2} b^{1-n})$

7.3 RATIONAL EXPONENTS

In this section we are concerned with powers of the form b^x where x is a rational number; that is, such expressions as $5^{1/2}$, $8^{1/3}$, $4^{-3/2}$.

First we seek a definition of $b^{1/n}$, where n is a positive integer, so that the five exponent theorems remain valid.

For Theorem 2 to be valid, $(b^m)^n = b^{mn}$, for $m = \frac{1}{n}$,

$$(b^{1/n})^n = b^{(1/n)(n)} = b^1 = b$$

As special cases,

$(b^{1/2})^2 = b$

$(5^{1/2})^2 = 5$ and $5^{1/2}$ is a square root of 5

$(b^{1/3})^3 = b$

$(8^{1/3})^3 = 8$ and $8^{1/3} = 2$, a cube root of 8

In general, we want a number x such that $x^n = b$. If b is positive, we know that b has a real positive nth root, $\sqrt[n]{b}$, and $(\sqrt[n]{b})^n = b$.

We also know that a negative real number does not have a square root that is a real number; for example, if $x^2 = -4$, then $x = 2i$ or $x = -2i$. In general, if b is negative and if n is an even positive integer such as 2, 4, 6, etc., the nth roots of b are imaginary. Because of restrictions on the domain when the base is negative, we will consider only the cases where the base is positive.

DEFINITION

If n is a positive integer and b is a positive real number, then
$$b^{1/n} = \sqrt[n]{b} \quad \text{and} \quad 0^{1/n} = \sqrt[n]{0} = 0$$

With the above definition it can be shown that the five exponent theorems remain valid.

We also want to assign meanings to $8^{2/3}$, $9^{-1/2}$, $25^{-3/2}$, and in general to $b^{m/n}$ where m and n are any nonzero integers. For Theorem 2 to remain valid, we must have
$$b^{m/n} = (b^m)^{1/n} = (b^{1/n})^m$$
Accordingly, we select this result for our definition. Again it can be shown that the five exponent theorems remain valid for this definition.

DEFINITION

If m is any integer, n is any positive integer, and b is any positive real number, then
$$b^{m/n} = (b^{1/n})^m$$

In Examples 1–10, use the exponent definitions and theorems to simplify. Assume x and y to be positive real numbers.

EXAMPLE 1
$8^{2/3}$

Solution

$8^{2/3} = (8^{1/3})^2 = (\sqrt[3]{8})^2 = 4$

EXAMPLE 2
$9^{-1/2}$

Solution

$9^{-1/2} = (9^{1/2})^{-1} = (\sqrt{9})^{-1} = \frac{1}{3}$

EXAMPLE 3
$25^{-3/2}$

Solution

$25^{-3/2} = (25^{1/2})^{-3} = (\sqrt{25})^{-3} = \frac{1}{5^3} = \frac{1}{125}$

EXAMPLE 4
$5^{1/3}5^{1/6}$

Solution

$5^{1/3+1/6} = 5^{3/6} = 5^{1/2} = \sqrt{5}$

EXAMPLE 5
$(5^{1/2})^{1/3}$

Solution

$(5^{1/2})^{1/3} = 5^{1/2 \cdot 1/3} = 5^{1/6} = \sqrt[6]{5}$

EXAMPLE 6
$10^{-2.4}10^{4.9}$

Solution

$10^{-2.4+4.9} = 10^{2.5}$

$10^{2.5} = 10^2 10^{0.5} = 10^2 10^{1/2} = 100\sqrt{10}$

EXAMPLE 7
$(10^{-3}10^{0.6})^{1/3}$

Solution

$(10^{-3})^{1/3}(10^{0.6})^{1/3} = 10^{-1}10^{0.2}$

$= \frac{10^{1/5}}{10} = \frac{\sqrt[5]{10}}{10}$

EXAMPLE 8
$(x^{-12}y^6)^{-2/3}$

Solution

$(x^{-12})^{-2/3}(y^6)^{-2/3} = x^8 y^{-4} = \frac{x^8}{y^4}$

EXAMPLE 9

$\left(\dfrac{x^{-3/2}}{y^{-1/2}}\right)^2$

Solution

$$\dfrac{(x^{-3/2})^2}{(y^{-1/2})^2} = \dfrac{x^{-3}}{y^{-1}} = \dfrac{y}{x^3}$$

EXAMPLE 10

$(x^{-3/4}y^{5/4})(x^{5/4}y^{-1/4})$

Solution

$$x^{-3/4+5/4}y^{5/4-1/4} = x^{2/4}y^{4/4} = x^{1/2}y^1 = y\sqrt{x}$$

EXERCISES 7.3

(1–40) Find the value of each of the following.

1. $4^{1/2}$
2. $9^{1/2}$
3. $25^{1/2}$
4. $8^{1/3}$
5. $125^{1/3}$
6. $81^{1/4}$
7. $32^{1/5}$
8. $(64)^{-1/3}$
9. $(625)^{-1/4}$
10. $\left(\dfrac{64}{25}\right)^{1/2}$
11. $\left(\dfrac{27}{125}\right)^{1/3}$
12. $(36)^{-1/2}$
13. $(27)^{-1/3}$
14. $\left(\dfrac{1}{1000}\right)^{-1/3}$
15. $\left(\dfrac{1}{49}\right)^{-1/2}$
16. $(32)^{3/5}$
17. $(64)^{2/3}$
18. $(100)^{-3/2}$
19. $(81)^{-3/4}$
20. $8^{-4/3}$
21. $16^{-5/4}$
22. $\left(\dfrac{16}{81}\right)^{3/4}$
23. $\left(\dfrac{216}{343}\right)^{2/3}$
24. $\left(\dfrac{125}{1000}\right)^{-4/3}$
25. $\left(\dfrac{81}{100}\right)^{-3/2}$
26. $(5^{1/6})^3$
27. $(3^4)^{-3/4}$
28. $\dfrac{5^{5/4}}{5^{3/4}}$
29. $\dfrac{10^{3.75}}{10^{3.25}}$
30. $\dfrac{10^{1.6}10^{4.2}}{10^{3.3}}$
31. $10^{-2.4}10^{3.9}$
32. $(10^{-3}10^{0.75})^{-1/3}$
33. $(10^{-2.4})^{-1/4}$
34. $\dfrac{10^{-3.6}10^{1.2}}{10^{-4.9}}$
35. $5^{1/3}(5^{-2/3})^2$
36. $3^{3/4}(3^{-3/2})^{-1/2}$
37. $4^{1/3}4^{1/6}$
38. $9^{1/8}9^{3/8}$
39. $1^{-5/6}-4^{-3/2}$
40. $(2^{-1/2})^{-1/2}$

(41–62) Simplify, assuming each variable is positive. Each variable should occur only once and all exponents should be positive in the final simplified form.

41. $x^{1/2}x^{1/6}$
42. $x^{-1/3}x^{-1/9}x^{5/9}$
43. $x^{-5/6}x^{4/3}$
44. $x^{5/8}x^{-3/8}$
45. $(x^{-1/4})^2$
46. $(x^{3/2})^{-4/3}$
47. $(x^{-8}y^4)^{-1/4}$
48. $(x^{-9}y^{-6})^{-2/3}$
49. $\dfrac{x^{5/6}}{x^{-1/6}}$
50. $\dfrac{y^{-5/8}}{y^{3/8}}$

51. $\left(\dfrac{x^{-1/4}}{y^{-1/2}}\right)^4$

52. $\left(\dfrac{x^{-2/3}}{y^{-1/2}}\right)^{-6}$

53. $xy^{5/3}(x^{1/2}y^{2/3})^{-2}$

54. $(xy^{-1/2})^3(x^{-1/2}y^{-1})^{-2}$

55. $\dfrac{x^{1/3}+x^{-1/3}}{x^{-2/3}}$

56. $\dfrac{x^{-1/2}+x^{1/4}}{x^{-4}}$

57. $(x^{1/2}-y^{1/2})(x^{1/2}+y^{1/2})$

58. $(x^{1/2}+y^{1/2})^2$

59. $(x^{1/3}-1)(x^{2/3}+x^{1/3}+1)$

60. $(x^{-1/3}+1)(x^{-2/3}-x^{-1/3}+1)$

61. $\dfrac{(2xy^{-1})^{1/4}}{(32x^5y^{-9})^{1/4}}$

62. $\dfrac{(243x^7y^3)^{1/4}}{(48x^5y^7)^{1/4}}$

7.4 RADICALS

Many radical expressions can be simplified by expressing each radical as a fractional power and then applying the exponent theorems.

A **radical is in simplified form** if:

1. The radicand is integral.
2. The radicand has no factor that is a power whose exponent is greater than or equal to the index of the radical.
3. No radical occurs in a denominator.
4. The index of the radical cannot be reduced.

When a radical is expressed as a fractional power, these statements have the following meaning.

1. The base is integral.
2. The numerator of a fractional exponent is less than its denominator.
3. No fractional exponent occurs in a denominator.
4. The fractional exponent is in simplified form; that is, reduced to lowest terms.

EXAMPLE 1   Simplify $\sqrt[3]{625x^5}$ for $x \geq 0$.

Solution   Changing to a fractional power,
$$(625x^5)^{1/3} = (5^4x^5)^{1/3}$$
Using $(ab)^n = a^nb^n$ and $(b^m)^n = b^{mn}$,
$$(5^4x^5)^{1/3} = 5^{4/3}x^{5/3}$$
$$= 5^15^{1/3}x^1x^{2/3}$$
Rewriting so each fractional exponent is between 0 and 1,
$$5x5^{1/3}x^{2/3}$$
Changing to radical form,
$$5x\sqrt[3]{5x^2}$$

EXAMPLE 2   Simplify $\dfrac{1}{\sqrt[3]{x^2y}}$ for $x > 0$, $y > 0$.

Solution   Changing to a fractional power,

$$\frac{1}{(x^2y)^{1/3}} = \frac{1}{x^{2/3}y^{1/3}}$$

Multiplying numerator and denominator by fractional powers needed to make the exponents in the denominator integers,

$$\frac{x^{1/3}y^{2/3}}{x^{2/3}y^{1/3}x^{1/3}y^{2/3}} = \frac{x^{1/3}y^{2/3}}{xy}$$

Rewriting in radical form,

$$\frac{\sqrt[3]{xy^2}}{xy}$$

EXAMPLE 3   Simplify $\sqrt[6]{81x^4}$ for $x \geq 0$.

Solution   Changing to a fractional power,
$(81x^4)^{1/6} = (3^4x^4)^{1/6}$

Using $(ab)^n = a^nb^n$,
$3^{4/6}x^{4/6}$

Reducing each exponent to lowest terms,
$3^{2/3}x^{2/3} = (3^2x^2)^{1/3}$

Rewriting in radical form,
$\sqrt[3]{9x^2}$

EXAMPLE 4   Simplify $\dfrac{\sqrt[12]{xy^8}}{\sqrt[4]{x^3y^2}}$ for $x > 0$, $y > 0$.

Solution   Changing to fractional powers,
$$\frac{(xy^8)^{1/12}}{(x^3y^2)^{1/4}}$$

Changing exponents to obtain the least common denominator for the exponents,
$$\frac{(xy^8)^{1/12}}{(x^3y^2)^{3/12}}$$

Using exponent theorems,
$$\frac{(xy^8)^{1/12}}{((x^3y^2)^3)^{1/12}} = \left(\frac{xy^8}{x^9y^6}\right)^{1/12}$$

$$= \left(\frac{y^2}{x^8}\right)^{1/12}$$

$$= \left(\frac{y}{x^4}\right)^{2/12}$$

Clearing the fraction of radicals in the denominator,
$$\left(\frac{x^2y}{x^2x^4}\right)^{1/6}$$

Changing to radical form,
$$\frac{\sqrt[6]{x^2y}}{x}$$

EXAMPLE 5 Simplify $\sqrt[3]{6\sqrt{6}}$.

Solution

$\sqrt[3]{6\sqrt{6}} = (6 \cdot 6^{1/2})^{1/3}$

Using $b^m b^n = b^{m+n}$,

$(6^{3/2})^{1/3}$

Using $(b^m)^n = b^{mn}$,

$6^{1/2} = \sqrt{6}$

EXERCISES 7.4

(1–34) Simplify. Assume each variable is a positive real number.

1. $\sqrt[3]{216x^5}$
2. $\sqrt[4]{768y^5}$
3. $\sqrt[5]{224y^{12}}$
4. $\sqrt[3]{32x^7}$
5. $\sqrt[4]{9x^2}$
6. $\sqrt[4]{36x^2y^6}$
7. $\sqrt[6]{144x^2y^2}$
8. $\sqrt[6]{512y^6}$
9. $\sqrt[6]{125a^3b^9}$
10. $\sqrt[6]{36b^4c^2}$
11. $\sqrt[8]{81t^4}$
12. $\sqrt[9]{216t^3}$
13. $\sqrt[3]{4x^2}\sqrt[6]{4x^2}$
14. $\sqrt{8x^3}\sqrt[3]{4x^2}$
15. $\dfrac{\sqrt[6]{10t}}{\sqrt[3]{100t^2}}$
16. $\dfrac{\sqrt{7t}}{\sqrt[4]{7t}}$
17. $\dfrac{\sqrt[12]{6xy}}{\sqrt[4]{216x^3y^3}}$
18. $\dfrac{\sqrt[6]{81x^4y^8}}{\sqrt[3]{3xy}}$
19. $\dfrac{1}{\sqrt[3]{7x^2y}}$
20. $\dfrac{1}{\sqrt[3]{25x^4y^5}}$
21. $\sqrt[4]{\dfrac{1}{25x^2}}$
22. $\sqrt[4]{\dfrac{1}{64x^3y}}$
23. $\sqrt[4]{6t}\sqrt{6t}$
24. $\sqrt[3]{25}\sqrt{5}$
25. $\sqrt[4]{3x}\sqrt[12]{3x}$
26. $\sqrt[6]{2xy}\sqrt[4]{4x^2y^2}$
27. $\sqrt{6}\sqrt[3]{6}\sqrt[6]{6}$
28. $\sqrt{2}\sqrt[3]{4}\sqrt[6]{32}$
29. $\sqrt{10\sqrt{10}}$
30. $\sqrt[3]{7\sqrt{7}}$
31. $\sqrt[3]{2t\sqrt{2t}}$
32. $\sqrt{3xy\sqrt[3]{3xy}}$
33. $\sqrt[5]{\sqrt[3]{32x^{10}}}$
34. $\sqrt[3]{\sqrt{343x^9}}$

7.5 RADICAL EQUATIONS

Equations such as $\sqrt{2x-5}=3$, $\sqrt[3]{x+3}=2$, and $x=6+\sqrt{2x-9}$ are called **radical equations** because a variable occurs in a radicand. Radical equations such as these can be solved by isolating a radical on one side of the equation and raising each side to a power. This process does not produce an equivalent equation but it does produce an equation whose solution set contains the solutions, if any, of the original equation.

Note that the solution set, {2}, of $x=2$ is a subset of the solution set, {2, −2}, of $x^2=2^2$. The set {2} is also a subset of the solution set of $x^3=2^3$; namely of $\{2, -1+i\sqrt{3}, -1-i\sqrt{3}\}$. In general, we have the following theorem.

THEOREM

If n is a positive integer and if A and B are expressions in the variable x, then the solution set of $A=B$ is a subset of the solution set of $A^n=B^n$.

A solution of $A^n=B^n$ that is not a solution of $A=B$ is called an **extraneous root**. Each root of $A^n=B^n$ must be checked in $A=B$ to determine the roots of $A=B$.

EXAMPLE 1 Solve $\sqrt{2x-5}=3$.

Solution   The solution set is a subset of the solution set of
$$(\sqrt{2x-5})^2=3^2$$
$$2x-5=9$$
$$x=7$$
Check   For $x=7$,
$$\sqrt{2x-5}=\sqrt{14-5}=\sqrt{9}=3.$$
The solution set is {7}.

EXAMPLE 2 Solve $\sqrt[3]{x+3}=2$.

Solution   The solution set is a subset of the solution set of
$$(\sqrt[3]{x+3})^3=2^3$$
$$x+3=8$$
$$x=5$$
Check   For $x=5$, $\sqrt[3]{x+3}=\sqrt[3]{5+3}=\sqrt[3]{8}=2$.
The solution set is {5}.

EXAMPLE 3   Solve $x = 6 + \sqrt{2x - 9}$.

Solution   First, isolating the radical,
$$x - 6 = \sqrt{2x - 9}$$
Squaring each side,
$$x^2 - 12x + 36 = 2x - 9$$
Solving,
$$x^2 - 14x + 45 = 0$$
$$(x - 5)(x - 9) = 0$$
$$x = 5 \quad \text{or} \quad x = 9$$
Check   For $x = 5$,
$$5 = 6 + \sqrt{10 - 9}$$
$$5 = 6 + 1 \quad \textit{false}$$
5 is not a solution; 5 is an extraneous root.
For $x = 9$,
$$9 = 6 + \sqrt{18 - 9}$$
$$9 = 6 + 3 \quad \textit{true}$$
9 is a solution.
The solution set is $\{9\}$.

EXAMPLE 4   Solve $\sqrt{3x + 4} - \sqrt{2x - 5} = 2$.

Solution   Isolating one radical,
$$\sqrt{3x + 4} = 2 + \sqrt{2x - 5}$$
Squaring each side and simplifying,
$$3x + 4 = 4 + 4\sqrt{2x - 5} + 2x - 5$$
$$x + 5 = 4\sqrt{2x - 5}$$
Squaring each side,
$$x^2 + 10x + 25 = 16(2x - 5)$$
Solving,
$$x^2 - 22x + 105 = 0$$
$$(x - 7)(x - 15) = 0$$
$$x = 7 \quad \text{or} \quad x = 15$$
Check   For $x = 7$, $\sqrt{21 + 4} - \sqrt{14 - 5}$
$= \sqrt{25} - \sqrt{9} = 5 - 3 = 2.$ *true*
For $x = 15$, $\sqrt{45 + 4} - \sqrt{30 - 5}$
$= \sqrt{49} - \sqrt{25} = 7 - 5 = 2.$ *true*
The solution set is $\{7, 15\}$.

EXERCISES 7.5

(1–30)   Solve. Note that checking is part of the solution process.

1. $\sqrt{3x + 7} = 5$

2. $\sqrt{2x - 4} = 4$

3. $6 + \sqrt{x - 4} = 2$

4. $7 + \sqrt{x + 3} = 3$

5. $\sqrt[3]{5x - 12} = 2$

6. $\sqrt[3]{5 - 4x} + 3 = 0$

7. $\sqrt{x^2 - 6x} = 4$

8. $\sqrt{3x^2 + 25} = 2x$

9. $x + 3 = \sqrt{2x + 9}$

10. $x = 2 + \sqrt{3x - 8}$

11. $x = 3 + \sqrt{2x + 9}$

12. $x = 1 - \sqrt{2x + 1}$

13. $x = 3 - \sqrt{2x + 9}$

14. $x = 3 + \sqrt{x + 3}$

15. $\sqrt{y + 7} = y + 1$

16. $\sqrt{y - 4} = y - 6$

17. $\sqrt{x} + \sqrt{x + 7} = 7$

18. $\sqrt{3 - x} + \sqrt{2 + x} = 3$

19. $\sqrt{3x - 9} - \sqrt{x - 2} = 1$

20. $2\sqrt{x} + \sqrt{x - 3} = 3$

21. $\sqrt{2y + 5} - \sqrt{2y + 1} = 1$

22. $\sqrt{4x + 10} + \sqrt{4x - 2} = 4$

23. $5 + \sqrt[3]{2x + 3} = 0$

24. $7 - \sqrt[3]{3x - 2} = 0$

25. $x - 1 = \sqrt[3]{x^3 - 4x^2 + 3}$

26. $x + 2 = \sqrt[3]{x^3 + 5x^2 + 6x}$

27. $x^{3/2} = 1000$

28. $x^{2/3} = 400$

29. $(2x - 1)^{4/3} = 81$

30. $(5x + 2)^{3/5} = 8$

(31–32) Solve for y.

31. $x^{1/2} + y^{1/2} = a^{1/2}$

32. $x^{2/3} + y^{2/3} = a^{2/3}$

7.6 CHAPTER REVIEW

(1–10) Simplify by using one or more of the exponent theorems. Assume that no variable equals a value that causes a denominator to become zero. (7.1)

1. $(5x^2)^3$

2. $-4x(-x^2)^2$

3. $(5x^3y)^2(3xy^2)^4$

4. $\left(\dfrac{2x^2}{5y}\right)^4$

5. $\dfrac{(4x^4y^2)^3}{(8x^2y^3)^2}$

6. $\dfrac{(243x^5y^4)^5}{(81x^6y^6)^5}$

7. $\dfrac{x^2x^n}{x^{n+1}}$

8. $\dfrac{(xy)^n}{xy^{3n}}$

9. $\dfrac{(10^n)^3}{(2^n5^n)^2}$

10. $\dfrac{(10x^n)^3}{(2x^n)^2(5x^n)^2}$

(11–20) Use the definitions and exponent theorems to express each of the following in simplified form, without using zero or negative exponents. Assume all bases and all denominators to be positive. (7.2)

11. $\left(\dfrac{3}{5}\right)^{-2}$

12. $\dfrac{4}{5^{-3}}$

13. $\left(\dfrac{6}{x}\right)^{-2}\left(\dfrac{x^2}{5}\right)^0$

14. $(-3y^2)^{-3}$

15. $\dfrac{5^3y^{-5}}{5^{-1}y^{-1}}$

16. $(x^{-1}y^2)^{-3}(xy^{-3})^{-2}$

17. $2^{-2}(2^2 \cdot 2^{-4})$

18. $2^{-2}(2^2 + 2^{-4})$

19. $\dfrac{x^{-n}y^{1-n}}{x^{-2n}y^{-n}}$

20. $\dfrac{(x + y)^{-1}}{x^{-1} + y^{-1}}$

(21–28) Simplify, assuming each variable is a positive real number. (7.3)

21. $\dfrac{10^{-8.2}10^{5.6}}{10^{-4.1}}$

22. $\left(\dfrac{8x^{-6}}{y^3}\right)^{-2/3}$

23. $\left(\dfrac{25}{x^2}\right)^{3/2}$

24. $\left(\dfrac{x}{6}\right)^{-1.5}$

25. $\dfrac{5^{3/8}}{5^{-1/8}}$

26. $(xy^{-1/2})^{3/2}(x^{-2}y^{-5})^{-1/4}$

27. $(x^{1/2} - x^{-1/2})^2$

28. $x^{3/4}(x^{-3/2})^{-1/2}$

(29–36) Simplify. Assume each variable is a positive real number. (7.4)

29. $\sqrt[3]{686x^4}$

30. $\sqrt[5]{128y^7}$

31. $\dfrac{1}{\sqrt[3]{18xy^2}}$

32. $\sqrt[4]{\dfrac{2}{5t^3}}$

33. $\sqrt[3]{7n}\,\sqrt[6]{7n}$

34. $\sqrt[3]{4x}\sqrt{4x}$

35. $\sqrt{\sqrt[3]{16x^4}}$

36. $\dfrac{\sqrt[6]{10y}}{\sqrt{10y}}$

(37–42) Solve. (7.5)

37. $10 - \sqrt{5x + 9} = 3$

38. $10 + \sqrt{5x + 9} = 3$

39. $\sqrt[3]{x^2 - 2x - 8} = 3$

40. $x = 5 + \sqrt{3x - 11}$

41. $\sqrt{x + 10} - \sqrt{2x - 3} = 1$

42. $\sqrt{2x - 3} - \sqrt{x + 10} = 1$

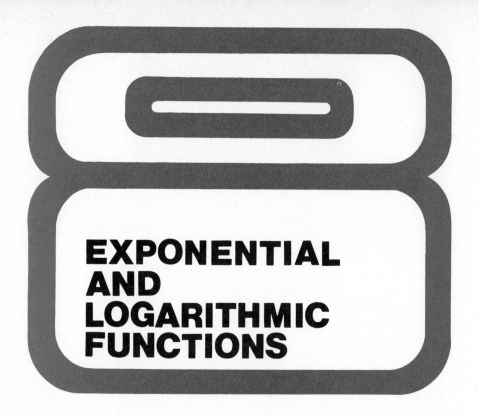

EXPONENTIAL AND LOGARITHMIC FUNCTIONS

In the previous chapter we saw that there is exactly one real number b^x for each rational number x when b is positive. Now we consider the case when the exponent x is an irrational number. For example, we want to assign meanings to expressions such as $2^{\sqrt{3}}$, $5^{\sqrt{2}}$, and $10^{-\sqrt{5}}$.

To help us find a useful definition, we will consider the properties stated in the following theorem. This theorem is proved in more advanced courses in mathematics; however, we accept it here without proof.

THEOREM 1

If b is any positive real number, and if r and s are any rational numbers, then:

1. $b^r = b^s$ if and only if $r = s$, for $b \neq 1$.

2. $b^r > b^s$ if and only if $r > s$, for $b > 1$.

3. $b^r < b^s$ if and only if $r > s$, for $0 < b < 1$.

As examples:

1. $5^{\sqrt{4}} = 5^2$ since $\sqrt{4} = 2$.
2. $2^4 > 2^3$ since $4 > 3$ and $2 > 1$.
3. $\left(\frac{1}{2}\right)^4 < \left(\frac{1}{2}\right)^3$ since $4 > 3$ and $0 < \frac{1}{2} < 1$.

We have seen that an irrational number can be expressed as a decimal that has infinitely many nonrepeating digits. We can approximate an irrational number by a rational number using any finite number of decimal places we wish.

For example,

$$\sqrt{3} = 1.73205...$$

and

$$\sqrt{3} = 1.732 \quad \text{accurate to 3 decimal places}$$

Now, using part 2 of the theorem for the base 2, the following inequalities are true:

$$2^1 < 2^{1.7} < 2^2$$
$$2^{1.7} < 2^{1.73} < 2^{1.8}$$
$$2^{1.73} < 2^{1.732} < 2^{1.74}$$
$$2^{1.7320} < 2^{1.73205} < 2^{1.7321}$$

Now if Theorem 1 is true when the exponent is irrational, then

$$2^{1.7320} < 2^{\sqrt{3}} < 2^{1.7321}$$

Using a hand-held scientific calculator, we find

$$2^{1.7320} \approx 3.3219 \quad \text{and} \quad 2^{1.7321} \approx 3.3221$$

and

$$3.3219 < 2^{\sqrt{3}} < 3.3221$$

Intuitively, we feel it is correct to state $2^{\sqrt{3}} = 3.322$ correct to 3 decimal places.

We note that our approximations for $2^{\sqrt{3}}$ seem to be getting closer and closer to exactly one real number as we improve our approximations for $\sqrt{3}$. Indeed, this is the case and we define $2^{\sqrt{3}}$ to be this exact value.

The above development can be repeated for any irrational exponent. Accordingly, we define b^r for b positive and r irrational to be the exact value that its rational approximations get closer and closer to.

With this definition, which is made more precise in more advanced courses, it can be shown that Theorem 1 and the five exponent theorems remain valid. By accepting these statements, we can now define an exponential function.

DEFINITION   EXPONENTIAL FUNCTION

The exponential function is the function defined by

$$y = b^x, \quad \text{where } b > 0 \quad \text{and} \quad b \neq 1$$

Note that if $b = 1$, then $y = 1^x = 1$ and the function is a linear constant function. We want to exclude this case.

We now restate the exponent theorems in the more general form where the exponent is any real number.

EXPONENT THEOREMS, GENERAL FORM

If a and b are positive real numbers and if x and y are any real numbers, then

1. $b^x b^y = b^{x+y}$
2. $(b^x)^y = b^{xy}$
3. $(ab)^x = a^x b^x$
4. $\left(\dfrac{a}{b}\right)^x = \dfrac{a^x}{b^x}$
5. $\dfrac{b^x}{b^y} = b^{x-y}$

Figure 8.1 shows the basic graph of $y = b^x$ for $b > 1$ and Figure 8.2 shows the basic graph of $y = b^x$ for $0 < b < 1$. Since $\dfrac{1}{b} = b^{-1}$, the graph of $y = b^{-x}$ for $b > 1$ is identical to the graph of $y = \left(\dfrac{1}{b}\right)^x$.

For the exponential function, note that:

1. The domain is the set of real numbers.
2. The range is the set of positive real numbers, $y > 0$.

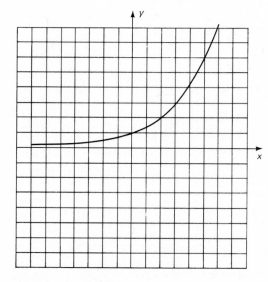

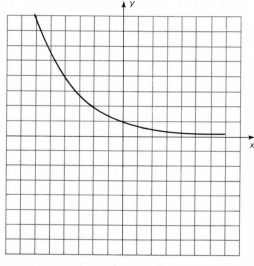

FIG. 8.1 $y = b^x$ for $b > 1$.

FIG. 8.2. $y = b^x$ for $0 < b < 1$.

EXERCISES 8.1

1. Graph $y = 2^x$ by referring to Figure 8.1 and by using the following table of ordered pairs.

x	-3	-2	-1	0	$\frac{1}{2}$	1	2	3
2^x	$\frac{1}{8}$	$\frac{1}{4}$	$\frac{1}{2}$	1	$\sqrt{2}$	2	4	8

2. Graph $y = \left(\frac{1}{2}\right)^x = 2^{-x}$ by referring to Figure 8.2 and by using the following table of ordered pairs.

x	-3	-2	-1	0	1	2	3
2^{-x}	8	4	2	1	$\frac{1}{2}$	$\frac{1}{4}$	$\frac{1}{8}$

3. Graph $y = 4^x$ by referring to Figure 8.1 and plotting ordered pairs for these values of x:
$$-\tfrac{3}{2}, -1, -\tfrac{1}{2}, 0, \tfrac{1}{2}, 1, \tfrac{3}{2}, 2$$

4. Graph $y = 4^{-x}$ by referring to Figure 8.2 and plotting ordered pairs for these values of x:
$$-2, -\tfrac{3}{2}, -1, -\tfrac{1}{2}, 0, \tfrac{1}{2}, 1, \tfrac{3}{2}$$

5. Graph $y = 3^x$.

6. Graph $y = \left(\frac{1}{3}\right)^x$.

7. Graph $y = \left(\frac{2}{3}\right)^x$.

8. Graph $y = \left(\frac{3}{2}\right)^x$.

(9–26) Solve for x.

EXAMPLE 1
$3^x = 81$

Solution   Writing each side as a power having the same base,
$$3^x = 3^4 \quad \text{and} \quad x = 4 \qquad \text{by Theorem 1}$$

EXAMPLE 2
$2^{3x+1} = \frac{1}{32}$

Solution
$$2^{3x+1} = 2^{-5}$$
$$3x + 1 = -5 \qquad \text{by Theorem 1}$$
$$x = -2$$

EXAMPLE 3

$4^{-x} = 8^{x-1}$

Solution

$(2^2)^{-x} = (2^3)^{x-1}$

$2^{-2x} = 2^{3x-3}$

$-2x = 3x - 3$ by Theorem 1

$5x = 3$

$x = \dfrac{3}{5}$

9. $2^x = 16$ **10.** $5^x = 125$

11. $3^{-x} = 243$ **12.** $4^{-x} = 64$

13. $5^x = \dfrac{1}{25}$ **14.** $2^x = \dfrac{1}{32}$

15. $2^x = 4\sqrt{2}$ **16.** $5^x = \dfrac{\sqrt{5}}{5}$

17. $2^{3x} = \dfrac{1}{64}$ **18.** $3^{2x-1} = \dfrac{1}{81}$

19. $4^{1-2x} = 1$ **20.** $6^{3x-4} = 1$

21. $4^x = 2^{x+6}$ **22.** $3^{x-6} = 9^{2x}$

23. $9^x = 3$ **24.** $8^x = 2$

25. $10^{1-x} = (0.01)^x$ **26.** $8^{2x} = 16$

(27–38) Simplify each of the following by using one or more of the five exponent theorems.

EXAMPLE 4

$(5^{\sqrt{3}})^{\sqrt{12}}$

Solution   Using $(b^x)^y = b^{xy}$,

$5^{\sqrt{3}\sqrt{12}} = 5^{\sqrt{36}} = 5^6$

EXAMPLE 5

$2^{\sqrt{18}}2^{\sqrt{8}}$

Solution   Using $b^x b^y = b^{x+y}$,

$2^{\sqrt{18}+\sqrt{8}} = 2^{3\sqrt{2}+2\sqrt{2}} = 2^{5\sqrt{2}}$

This can also be written as $(2^5)^{\sqrt{2}} = 32^{\sqrt{2}}$.

EXAMPLE 6

$\dfrac{6^{\sqrt{45}}}{6^{\sqrt{20}}}$

Solution   Using $\dfrac{b^x}{b^y} = b^{x-y}$,

$6^{\sqrt{45}-\sqrt{20}} = 6^{3\sqrt{5}-2\sqrt{5}} = 6^{\sqrt{5}}$

27. $(2^{\sqrt{5}})^{\sqrt{20}}$ **28.** $(10^{\sqrt{12}})^{\sqrt{3}}$

29. $(5^{\sqrt{2}})^{\sqrt{3}}$ **30.** $(4^{\sqrt{5}})^{\sqrt{2}}$

31. $3^{\sqrt{5}}3^{2-\sqrt{5}}$ **32.** $10^{2-\sqrt{5}}10^{\sqrt{5}-2}$

33. $10^{\sqrt{18}}10^{\sqrt{2}}$ **34.** $2^{\sqrt{3}}2^{\sqrt{75}}$

35. $\dfrac{10^{\sqrt{54}}}{10^{\sqrt{24}}}$ **36.** $\dfrac{5^{\sqrt{90}}}{5^{\sqrt{40}}}$

37. $\dfrac{2^{3+2\sqrt{5}}}{2^{3-2\sqrt{5}}}$ **38.** $\dfrac{3^{2-\sqrt{6}}}{3^{4-\sqrt{6}}}$

8.2 LOGARITHMIC FUNCTIONS

Since $b^r = b^s$ if and only if $r = s$ when $b > 0$ and $b \neq 1$, it follows that the exponential function defined by $y = b^x$ is a one-to-one function. Therefore its inverse is a function, called the logarithmic function.

DEFINITION

The **logarithmic function** with base b is the inverse of the exponential function with base b. It is defined by the equation
$$x = b^y \qquad \text{where } b > 0 \quad \text{and} \quad b \neq 1$$

Since we cannot solve this equation for y by using any of the six algebraic operations, we invent a notation for y instead. We write $y = \log_b x$, which is read "y equals the logarithm of x to the base b." The equation $y = \log_b x$ means "y is the exponent that must be placed on the base b to produce x."

DEFINITION OF $\text{LOG}_b x$

For $b > 0$ and $b \neq 1$, $y = \log_b x$ if and only if $x = b^y$.

Figure 8.3 shows the graph of $y = \log_b x$ for $b > 1$, along with the graph of $y = b^x$ to illustrate their inverse relationship.

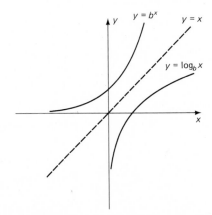

FIG. 8.3. Graph of $y = \log_b x$ and $y = b^x$ for $b > 1$.

The equations $y = \log_b x$ and $x = b^y$ are different forms for stating the same relationship.

The **logarithmic form** of the statement is $y = \log_b x$.

The **exponential form** of the statement is $x = b^y$.

As examples:

$5^3 = 125$ if and only if $\log_5 125 = 3$.

$2^{-4} = \frac{1}{16}$ if and only if $\log_2 \frac{1}{16} = -4$.

$3^0 = 1$ if and only if $\log_3 1 = 0$.

$6^1 = 6$ if and only if $\log_6 6 = 1$.

The last two examples are special cases of a more general result stated in the next theorem.

THEOREM

For $b > 0$ and $b \neq 1$, $\log_b 1 = 0$ and $\log_b b = 1$.

Important properties of logarithms that follow from the exponent theorems are stated in the next theorem.

THEOREM   LOGARITHMIC PROPERTIES

If $b > 0$ and $b \neq 1$ and if r and s are positive real numbers, and k is any real number, then

1. $\log_b rs = \log_b r + \log_b s$

2. $\log_b \dfrac{r}{s} = \log_b r - \log_b s$

3. $\log_b r^k = k \log_b r$

Proof

To prove Statement 1, let $r = b^x$ and $s = b^y$. Then $x = \log_b r$ and $y = \log_b s$ and $rs = b^x b^y = b^{x+y}$. This means

$\log_b rs = x + y$

$\qquad = \log_b r + \log_b s$

For Statement 2,

$$\frac{r}{s} = \frac{b^x}{b^y} = b^{x-y}$$

and

$$\log_b \frac{r}{s} = x - y = \log_b r - \log_b s$$

For Statement 3,

$$r^k = (b^x)^k = b^{kx}$$

and

$\log_b r^k = kx = k \log_b r$

As examples,

$\log_3 10 = \log_3 (2 \cdot 5) = \log_3 2 + \log_3 5$

$\log_3 \frac{5}{2} = \log_3 5 - \log_3 2$

$\log_3 5^4 = 4 \log_3 5$

$\log_3 \sqrt[4]{5} = \log_3 5^{1/4} = \frac{1}{4} \log_3 5$

8.2 LOGARITHMIC FUNCTIONS

Since the exponential function and logarithm function are inverses of each other, the general theorem stating

$$f(f^{-1}(x)) = x \quad \text{and} \quad f^{-1}(f(x)) = x$$

is true for

$$f(x) = b^x \quad \text{and} \quad f^{-1}(x) = \log_b x$$

Using

$$f(\ ) = b^{(\ )}$$

then

$$f(\log_b x) = b^{\log_b x}$$

and

$$f(f^{-1}(x)) = b^{\log_b x} = x$$

Using

$$f^{-1}(\ ) = \log_b(\ )$$

then

$$f^{-1}(b^x) = \log_b(b^x)$$

and

$$f^{-1}(f(x)) = \log_b b^x = x$$

THEOREM

For $b > 0$ and $b \neq 1$, $b^{\log_b x} = x$ for $x > 0$ and $\log_b b^x = x$.

EXERCISES 8.2

(1–16) Write each statement in its logarithmic form. Use $b^y = x$ if and only if $\log_b x = y$.

EXAMPLE 1

$2^5 = 32$

Solution

$\log_2 32 = 5$

EXAMPLE 2

$6^{-2} = \frac{1}{36}$

Solution

$\log_6 \frac{1}{36} = -2$

EXAMPLE 3

$5^{3/2} = 5\sqrt{5}$

Solution

$\log_5 5\sqrt{5} = \frac{3}{2}$

1. $3^4 = 81$
2. $10^3 = 1000$
3. $10^{-3} = 0.001$
4. $2^{-6} = \frac{1}{64}$
5. $9^{1/2} = 3$
6. $4^{3/2} = 8$

7. $10^0 = 1$

8. $10^1 = 10$

9. $8^{-2/3} = \frac{1}{4}$

10. $16^{-3/4} = \frac{1}{8}$

11. $5^1 = 5$

12. $2^0 = 1$

13. $10^4 = 10,000$

14. $10^{-2} = 0.01$

15. $10^{3/2} = 10\sqrt{10}$

16. $6^{4/3} = 6\sqrt[3]{6}$

(17–30) Write each statement in its exponential form.
Use $\log_b x = y$ if and only if $b^y = x$.

EXAMPLE 4

$\log_7 49 = 2$

Solution
$7^2 = 49$

EXAMPLE 5

$\log_5 \frac{1}{125} = -3$

Solution
$5^{-3} = \frac{1}{125}$

EXAMPLE 6

$\log_8 32 = \frac{5}{3}$

Solution
$8^{5/3} = 32$

17. $\log_8 64 = 2$

18. $\log_2 128 = 7$

19. $\log_2 0.125 = -3$

20. $\log_5 0.2 = -1$

21. $\log_3 3\sqrt{3} = \frac{3}{2}$

22. $\log_8 4 = \frac{2}{3}$

23. $\log_{27} \frac{1}{9} = -\frac{2}{3}$

24. $\log_{81} \frac{1}{27} = -\frac{3}{4}$

25. $\log_{25} 0.2 = -0.5$

26. $\log_{32} 0.25 = -0.4$

27. $\log_5 1 = 0$

28. $\log_5 5 = 1$

29. $\log_{10} 10 = 1$

30. $\log_{10} 1 = 0$

(31–50) Find the value of each logarithm.

EXAMPLE 7

$\log_3 9\sqrt{3}$

Solution   Let y equal the logarithm and change
to exponential form.
$$\log_3 9\sqrt{3} = y \quad \text{and} \quad 3^y = 9\sqrt{3}$$
$$3^y = 3^2 3^{1/2}$$
$$3^y = 3^{2+1/2}$$
$$y = 2.5$$

EXAMPLE 8

$\log_{10} 0.00001$

Solution
$$\log_{10} 0.00001 = y \quad \text{and} \quad 10^y = 0.00001$$
$$10^y = 10^{-5}$$
$$y = -5$$

31. $\log_4 16$ **32.** $\log_5 625$

33. $\log_{10} 100{,}000$ **34.** $\log_{10} 1000$

35. $\log_{10} 0.01$ **36.** $\log_{10} 0.001$

37. $\log_3 \frac{1}{81}$ **38.** $\log_2 \frac{1}{64}$

39. $\log_6 6\sqrt{6}$ **40.** $\log_3 9\sqrt{3}$

41. $\log_{125} 25$ **42.** $\log_{81} 27$

43. $\log_8 8$ **44.** $\log_9 9$

45. $\log_9 1$ **46.** $\log_8 1$

47. $\log_8 0.25$ **48.** $\log_{25} 0.20$

49. $\log_{125} 0.04$ **50.** $\log_8 0.125$

(51–72) Write each of the following as a single logarithm.

EXAMPLE 9

$2(\log_b 7 - \log_b 3)$

Solution

$$2(\log_b 7 - \log_b 3) = 2 \log_b \frac{7}{3} \qquad \log_b \frac{r}{s} = \log_b r - \log_b s$$
$$= \log_b \left(\frac{7}{3}\right)^2 \qquad \log_b r^k = k \log_b r$$

EXAMPLE 10

$\log_{10} 23 + \frac{1}{3} \log_{10} 56$

Solution

$$\log_{10} 23 + \frac{1}{3} \log_{10} 56 = \log_{10} 23 + \log_{10} (56)^{1/3} \qquad \log_b r^k = k \log_b r$$
$$= \log_{10} 23 \sqrt[3]{56} \qquad \log_b rs = \log_b r + \log_b s$$

51. $\log_2 5 + \log_2 8$ **52.** $\log_3 60 - \log_3 15$

53. $\log_{10} 96 - \log_{10} 12$ **54.** $\log_{10} 25 + \log_{10} 80$

55. $2 \log_5 31$ **56.** $3 \log_4 40$

57. $\frac{1}{3} \log_{10} 216$ **58.** $\frac{1}{4} \log_{10} 625$

59. $3(\log_2 15 + \log_2 8)$ **60.** $\frac{1}{2}(\log_5 12 + \log_5 17)$

61. $\frac{1}{5}(\log_3 24 - \log_3 35)$ **62.** $4(\log_6 92 - \log_6 37)$

63. $2 \log_5 7 - \log_5 6$ **64.** $\log_4 20 - 3 \log_4 68$

65. $\log_2 52 + \frac{1}{3} \log_2 19$ **66.** $\frac{1}{2} \log_5 83 + \log_5 6$

67. $3(\log_{10} a - \log_{10} b)$ **68.** $\frac{1}{3}(\log_2 a - \log_2 b)$

69. $\frac{1}{2}(\log_5 a + \log_5 b)$ **70.** $4(\log_3 a + \log_3 b)$

71. $2 \log_2 a - \frac{1}{2} \log_2 b$ **72.** $\frac{1}{4} \log_5 a + 3 \log_5 b$

(73–90) Given $\log_{10} 2 = 0.3010$ and $\log_{10} 3 = 0.4771$, find each logarithm.

EXAMPLE 11

$\log_{10} 5$

Solution

$$\log_{10} 5 = \log_{10} \frac{10}{2} = \log_{10} 10 - \log_{10} 2$$
$$= 1 - 0.3010$$
$$= 0.6990$$

EXAMPLE 12

$\log_{10} 18$

Solution

$$\log_{10} 18 = \log_{10} 3^2 \cdot 2$$
$$= 2 \log_{10} 3 + \log_{10} 2$$
$$= 2(0.4771) + 0.3010$$
$$= 1.2552$$

73. $\log_{10} 15$ **74.** $\log_{10} 6$

75. $\log_{10} \frac{10}{3}$ **76.** $\log_{10} \frac{2}{3}$

77. $\log_{10} 8$ **78.** $\log_{10} 9$

79. $\log_{10} \sqrt{2}$ **80.** $\log_{10} \sqrt{3}$

81. $\log_{10} 36$ **82.** $\log_{10} 24$

83. $\log_{10} 90$ **84.** $\log_{10} 400$

85. $\log_{10} \sqrt[3]{6}$ **86.** $\log_{10} \sqrt[3]{15}$

87. $\log_{10} \frac{\sqrt{18}}{12}$ **88.** $\log_{10} \frac{72}{\sqrt[4]{10}}$

89. $\frac{\log_{10} 3}{\log_{10} 2}$ **90.** $\frac{\log_{10} 5}{\log_{10} 2}$

(91–100)   Solve for t by using the theorems $\log_b b^x = x$ and $b^{\log_b x} = x$.

91. $\log_2 2^t = 2.4$ **92.** $\log_5 5^{t-1} = -3.2$

93. $\log_{10} 10^{-t} = 0.05$ **94.** $\log_3 3^{2t+1} = 7$

95. $5^{\log_5 t} = 30$ **96.** $2^{\log_2(t+1)} = 6$

97. $10^{\log_{10}(t^2-4)} = 5$ **98.** $6^{\log_6(8t-t^2)} = 16$

99. $3^{\log_3(5-t)} = 8.2$ **100.** $10^{\log_{10} t(t-1)} = 2$

8.3   COMMON LOGARITHMS

When logarithms are used for calculations, the most convenient base is base 10. A logarithm with base 10 is called a **common logarithm** and the notation $\log x$ is used for $\log_{10} x$:

$$\log x = \log_{10} x$$

The common logarithms of integral powers of 10 can be obtained directly from the definition of the logarithm function.

$\log 1000 = 3$	since	$10^3 = 1000$
$\log 100 = 2$	since	$10^2 = 100$
$\log 10 = 1$	since	$10^1 = 10$
$\log 1 = 0$	since	$10^0 = 1$
$\log 0.1 = -1$	since	$10^{-1} = 0.1$
$\log 0.01 = -2$	since	$10^{-2} = 0.01$
$\log 0.001 = -3$	since	$10^{-3} = 0.001$

Common logarithms of numbers that are not integral powers of 10 are most easily obtained by using a scientific hand-held calculator. Tables, such as the Table of Logarithms in the Endpapers, are also used for this purpose.

TABLE METHOD

To find a common logarithm using tables, we use the property that any positive real number can be written in **scientific notation** as

$d \cdot 10^k$

where d is a decimal, $1 \le d < 10$, and k is an integer.

The following theorem explains how to find a common logarithm.

THEOREM   COMMON LOGARITHM

$\log (d \cdot 10^k) = k + \log d$

Note that

$\log (d \cdot 10^k) = \log d + \log 10^k$
$= k \log 10 + \log d$
$= k + \log d$

If $N = d \cdot 10^k$, a number written in scientific notation, then the **characteristic** of $\log N$ is k and the **mantissa** of $\log N$ is $\log d$.

EXAMPLE 1 Find log 2.34.

Solution   Using Table 2, find the intersection of row 2.3 with column 4.
log 2.34 = 0.3692

EXAMPLE 2 Find log 2340.

Solution
$\log 2340 = \log [2.34(10^3)]$
$= 3 + \log 2.34$
$= 3 + 0.3692$
$= 3.3692$

EXAMPLE 3 Find log 0.0234.

Solution
$\log 0.0234 = \log 2.34(10^{-2})$
$= -2 + \log 2.34$
$= -2 + 0.3692,$ computational form
$= -1.6308,$ formula form

The tables can also be used to find a number whose logarithm is given. This number is called the **antilogarithm,** or **antilog** for short.

DEFINITION ANTILOGARITHM

$y = $ antilog x if and only if $x = \log y$

EXAMPLE 4 Find antilog $(-1 + 0.4518)$.

Solution

$$\log x = -1 + 0.4518$$
$$= -1 + \log 2.83 \quad \text{(from the table)}$$
$$= \log 2.83(10^{-1})$$
$$x = 2.83(10^{-1})$$
$$x = 0.283$$

CALCULATOR METHOD

When a hand-held scientific calculator is used to find logarithms of numbers and numbers whose logarithms are known, the inverse relationship between the logarithmic function and the exponential function is used.

If $f(x) = b^x$, then $f^{-1}(x) = \log_b x$.
If $f(x) = 10^x$, then $f^{-1}(x) = \log x$.
If $f(x) = \log_b x$, then $f^{-1}(x) = b^x$.
If $f(x) = \log x$, then $f^{-1}(x) = 10^x$.

EXAMPLE 5* Find log 46,875 by using a calculator.

Solution
Enter   46875
Press LOG key
Read display and round off to accuracy desired
log 46,875 = 4.6709,   correct to 4 decimal places

EXAMPLE 6 Find log 0.007824 by using a calculator.

Solution
Enter   0.007824
Press LOG key
Read display and round off to 4 decimal places
log 0.007824 = -2.1066

Since antilog $x = y$ if and only if $x = \log y$, there are equivalent expressions for antilog x; namely,

antilog $x = 10^x = $ INV log x

where INV log means the inverse function of the logarithm function; that is, the exponential function.

EXAMPLE 7* Use a calculator to find N correct to 4 significant digits, if log N = 3.2157.

Solution
$N = 10^{3.2157}$ = INV log 3.2157
Method 1
Enter   3.2157
Press INV, log keys
Read display and round off answer
N = 1643
Method 2
Use the y^x key with y = 10.
Enter   10
Press y^x key
Enter   3.2157
Press = key
Read display and round off answer
N = 1643

EXAMPLE 8* Use a calculator to find N correct to 4 significant digits if log N = −2.4896.

Solution
Using Method 1 of Example 7,
Enter   2.4896
Press +/− key (change sign key)
Display will read −2.4896
Press INV, log keys
Read display and round off answer
N = 0.003239

EXERCISES 8.3

(1–32) Find each of the following by using tables or a calculator.

1. log 5.83	**2.** log 8.45
3. log 34.8	**4.** log 13.9
5. log 700	**6.** log 5600
7. log 506,000	**8.** log 1,350,000
9. log 0.985	**10.** log 0.452

*The calculations described in Examples 5–8 were made using the TI–30 calculator, manufactured by Texas Instruments. The methods may vary if a different calculator is used. If necessary, the calculator manual or an instructor should be consulted.

11. log 0.0046   **12.** log 0.00621
13. log 0.0008   **14.** log 0.000479
15. antilog 0.8802   **16.** antilog 0.248
17. antilog 2.936   **18.** antilog 1.1492
19. antilog 3.6911   **20.** antilog 4.871
21. antilog (-0.2277) **22.** antilog (-1.5702)
23. antilog (-2.5935) **24.** antilog (-3.06)
25. $10^{1.48}$ **26.** $10^{0.781}$
27. $10^{2.909}$ **28.** $10^{3.0212}$
29. $10^{-1.684}$ **30.** $10^{-0.4895}$
31. $10^{-3.0214}$ **32.** $10^{-7.15}$

(33–40) Solve for x by using tables or a calculator.

33. $10^x = 35.8$ **34.** $10^x = 2500$
35. $10^x = 0.00743$ **36.** $10^x = 0.00054$
37. $\log x = 4.7686$ **38.** $\log x = 0.4639$
39. $\log x = -0.1570$ **40.** $\log x = -2.091$

8.4   LOGARITHMIC EQUATIONS AND APPLICATIONS

By using the theorems for logarithms, computations involving products, quotients, powers, and roots can be replaced by simpler calculations involving sums, differences, products, and quotients.

EXAMPLE 1   Compute $\sqrt{(0.725)(34.7)}$ by using logarithms.

Solution   Let $N = \sqrt{(0.725)(34.7)}$.

$$\log N = \tfrac{1}{2}(\log 0.725 + \log 34.7)$$

$$\log 0.725 = -1 + 0.8603$$
$$\underline{\log 34.7 =\ \ \ 1 + 0.5403}$$
$$\text{sum} = 1.4006$$

$$\log N = \tfrac{1}{2}\,\text{sum} = 0.7003$$

$$N = 5.015 \quad \text{or} \quad 5.02$$
rounded to 3 significant digits

EXAMPLE 2 Compute $\sqrt[3]{\dfrac{1.380}{24.2}}$ by using logarithms.

Solution Let $N = \sqrt[3]{\dfrac{1.380}{24.2}}$.

$\log N = \frac{1}{3}(\log 1.380 - \log 24.2)$

$\log 1.380 = 0.1399 = 2.1399 - 2$ (2 is added and subtracted so that the subtrac-

$\underline{\log 24.2 = 1.3838 = 1.3838}$ tion will yield a positive decimal)

$\qquad\qquad\text{difference} = 0.7561 - 2$ (1 is added and subtracted so that the division

$\qquad\qquad\qquad\quad = 1.7561 - 3$ by 3 will yield an integer for the characteristic)

$\log N = \frac{1}{3}$ difference $= 0.5854 - 1$

$N = 0.385$ approximated to 3 significant digits

In many areas of application, there are formulas that involve the evaluation of a logarithm. Example 3 illustrates a formula used in chemistry.

EXAMPLE 3 In chemistry, the pH of a solution, a measure of acidity or alkalinity, is defined in terms of its hydrogen ion concentration $[H^+]$ as follows:

$pH = -\log_{10}[H^+]$

(a) Find the pH if $[H^+] = 3.5 \times 10^{-5}$.

(b) Find $[H^+]$ if pH $= 5.4$.

Solution

(a) $pH = -\log(3.5 \times 10^{-5}) = -\log(0.000035)$

Using tables or a calculator,

$\qquad pH = 4.5$

(*Note.* pH values are usually stated correct to 2 significant digits.)

(b) $-\log [H^+] = 5.4$

$\qquad \log [H^+] = -5.4 = -6 + 0.6$

$\qquad\qquad [H^+] = 0.0000040 = 4.0 \times 10^{-6}$

An equation involving logarithms can be solved by using the properties of logarithms and the theorem

$\log_b x = \log_b y$ if and only if $x = y$

It is also helpful to recall that $0 = \log_b 1$ and $1 = \log_b b$.

EXAMPLE 4   Solve $2 \log x - \log(x + 3) + \log 5 = \log 4$.

Solution

$$(\log x^2 + \log 5) - \log(x + 3) = \log 4$$

$$\log \frac{5x^2}{x + 3} = \log 4$$

$$\frac{5x^2}{x + 3} = 4$$

$$5x^2 - 4x - 12 = 0$$

$$(x - 2)(5x + 6) = 0$$

$$x = 2 \quad \text{or} \quad x = -\tfrac{6}{5}$$

Check   For $x = 2$,

$2 \log 2 - \log 5 + \log 5 = 2 \log 2 = \log 4$.

Thus 2 is a solution.

For $x = -\tfrac{6}{5}$, $2 \log x = 2 \log\left(-\tfrac{6}{5}\right)$ which is undefined. Thus $-\tfrac{6}{5}$ is *not* a solution.

Since the domain of the logarithmic function is the set of positive real numbers, the original equation required the restriction that $x > 0$ and $x + 3 > 0$. It is important, then, to check all proposed solutions of a logarithmic equation.

EXERCISES 8.4

(1–16)   Do the following calculations by using logarithms.

1. $87.5\,(96.3)$
2. $0.548\,(6.84)$
3. $\dfrac{1.24}{7.89}$
4. $\dfrac{0.329}{2.47}$
5. $\sqrt{592}$
6. $\sqrt[3]{425}$
7. $\sqrt[5]{1.68}$
8. $\sqrt[4]{83.6}$
9. $(2.13)^{10}$
10. $(1.08)^{12}$
11. $\dfrac{(4.92)(0.0658)}{7.86}$
12. $3.14(0.956)^3$
13. $(0.00729)(2.06)^8$
14. $\dfrac{0.843}{(0.549)^2}$
15. $\sqrt[4]{\dfrac{936}{28.7}}$
16. $\sqrt[3]{\dfrac{24.5}{32.2}}$

17. The period t in seconds for a simple pendulum having length L in feet is given by the approximate formula

$$t = 6.28\sqrt{\frac{L}{32.2}}$$

Use logarithms to find t for $L = 8.50$ feet.

18. The formula for the diameter D of a sphere in terms of its volume V is

$$D = \sqrt[3]{\frac{6V}{\pi}}$$

Find D for $V = 96.45$ cubic centimeters and $\pi = 3.142$.

(19–22) Find the pH for the given hydrogen ion concentration, $[H^+]$. See Example 3.

19. $[H^+] = 5.8 \times 10^{-6}$

20. $[H^+] = 7.4 \times 10^{-3}$

21. $[H^+] = 1.2 \times 10^{-8}$

22. $[H^+] = 3.6 \times 10^{-7}$

(23–26) Find the hydrogen ion concentration for the given pH value. See Example 3.

23. $pH = 4.7$

24. $pH = 8.9$

25. $pH = 7.5$

26. $pH = 3.2$

27. The gain G in decibels (a measure of loudness) due to an increase in power from x watts to y watts is given by

$$G = \log_{10} \frac{y}{x}$$

Find G for a television station that increased its power from 50 to 125 kilowatts.

28. The time t in minutes for a certain substance to cool from $100°$ C to $T°$ C in air having a temperature of $20°$ C is given by

$$t = 5 \log_{10}\left(\frac{100 - 20}{T - 20}\right)$$

Find the time t when $T = 25°$ C.

(29–40) Solve for x. See Example 4.

29. $\log x + \log(x - 4) = \log 12$

30. $\log x + \log(x - 5) - \log(x - 4) = \log 6$

31. $2 \log x - \log(x - 2) = 2 \log 3$

32. $\log(x^2 - 1) - \log(19 - x) = 0$

33. $2 \log(x + 1) - \log 4 - \log(x + 9) = 0$

34. $\log(x + 5) = \log x + \log 5$

35. $\log(x - 2) = \log x - \log 2$

36. $\log(x + 10) + \log(x - 1) - \log(x + 1) = 1$

37. $2 \log(x + 8) - \log(x - 4) - \log(x - 7) = 1$

38. $\frac{1}{2} \log x + \frac{1}{2} \log(x - 5) - \log 6 = 0$

39. $\frac{1}{2} \log(x^2 + 9) - \log(x - 3) - \log 5 = 0$

40. $3 \log(2x - 1) + 2 \log 5 = 2 + \log 2$

8.5 EXPONENTIAL EQUATIONS AND APPLICATIONS

An **exponential equation** is an equation in which a variable occurs in an exponent.

Some exponential equations can be solved by using the property that

$$\log_b x = \log_b y \quad \text{if and only if} \quad x = y$$

EXAMPLE 1   Solve $2^x = 6$.

Solution   Since 6 is not a rational power of 2, we equate the logarithms of both sides.

$\log 2^x = \log 6$

$x \log 2 = \log 6$

$x = \dfrac{\log 6}{\log 2}$,   exact form

To find an approximate value for x, we can either use a hand-held calculator or a table such as Table 2 in the Appendix.

$x = \dfrac{0.7782}{0.3010}$

$x = 2.585$,   correct to 3 decimal places

An exponential equation can be used to find a logarithm of a number to any base.

EXAMPLE 2   Find $\log_5 12$.

Solution   Let $x = \log_5 12$. Then

$5^x = 12$

$\log 5^x = \log 12$

$x \log 5 = \log 12$

$x = \dfrac{\log 12}{\log 5} \approx \dfrac{1.0792}{0.6990} \approx 1.544$

In general, if

$x = \log_b N$

then

$b^x = N$

and

$\log_a b^x = \log_a N$

$x \log_a b = \log_a N$

$x = \dfrac{\log_a N}{\log_a b}$

This result is stated as the following theorem.

THEOREM   CHANGING BASES

$\log_b N = \dfrac{\log_a N}{\log_a b}$

Note that for $N = a$, then

$\log_b a = \dfrac{\log_a a}{\log_a b} = \dfrac{1}{\log_a b}$

For advanced mathematics and applications depending on calculus, logarithms to the base e, called **natural logarithms,** are more useful. The number e is irrational and $e \approx 2.71828$.

Notation. $\ln x = \log_e x$

EXAMPLE 3   Express $\ln x$ in terms of $\log x$.

Solution   Let $y = \ln x$. Then

$$e^y = x$$

and

$$\log e^y = \log x$$
$$y \log e = \log x$$
$$y = \frac{\log x}{\log e}$$

Since

$$\log e \approx \log 2.71828$$
$$\log e \approx 0.43429$$

and

$$\frac{1}{\log e} \approx 2.30259$$

Therefore $\ln x = 2.303 \log x$ for 4-digit accuracy.

Note that $\ln x$ and e^x are inverses of each other.

If $f(x) = e^x$, then $f^{-1}(x) = \ln x$.

If $f(x) = \ln x$, then $f^{-1}(x) = e^x$.

When a calculator is used to find a number whose natural logarithm is known, three different keys might be used; namely, y^x, e^x, or INV $\ln x$.

If $\ln x = N$, then $x = e^N = $ INV $\ln N$.

EXAMPLE 4   Find $\ln 8.72$ by using (a) common logarithm tables, (b) a hand-held calculator.

Solution

(a) Use $\ln x = 2.303 \log x$

$$\ln 8.72 = 2.303 \log 8.72$$
$$= 2.303(0.9405) = 2.166$$

(b) Use $\ln x$ key.

Enter   8.72

Press $\ln x$ key

Read display

$\ln 8.72 = 2.166$

EXAMPLE 5 Find x if ln $x = -1.24$ by using (a) common logarithm tables, (b) a hand-held calculator.

Solution   Note that

ln $x = -1.24$ if and only if $x = e^{-1.24}$

(a) *Table.* Use

$\log x = \log e^{-1.24} = -1.24 \log e$

$\qquad = -1.24(0.4343) = -0.5385$

$\qquad = -1 + 0.4615$

$\qquad x = 0.2894$

(b) *Calculator.* For the TI–30,
use $x = INV$ ln (-1.24).

Enter   1.24

Press $+/-$ (change sign) key

Press INV, ln x keys

Read display

$e^{-1.24} = 0.2894$

If your calculator has an e^x key, then find $e^{-1.24}$.

If your calculator has neither the INV ln x nor the e^x key, the y^x key can be used with $y = e$.

Enter $e = 2.718$

Press y^x key

Enter 1.24

Press "change sign" key

Press $=$ key

Read display

EXAMPLE 6 Radium decomposes according to the equation
$y = y_0 e^{-0.04t}$
where y_0 grams reduces to y grams in t centuries. Find the half-life of radium, the time it takes for radium to decompose to half of its original amount.

Solution   To find the half-life, we need to find t when

$$y = \tfrac{1}{2}y_0$$

$$\tfrac{1}{2}y_0 = y_0 e^{-0.04t}$$

$$\tfrac{1}{2} = e^{-0.04t}$$

$$\log \tfrac{1}{2} = \log e^{-0.04t}$$

$$\log \tfrac{1}{2} = -0.04t \log e$$

$$-0.3010 = -0.04t(0.4343)$$

$$t = \frac{0.3010}{0.04(0.4343)}$$

$$t = 17.3 \text{ centuries}$$

EXAMPLE 7 Find the number of years it takes for $2000 to double itself if it is invested at 8% compound interest converted annually.

Solution The amount A of money that accrues when an amount P is invested at a rate r converted annually for n years is given by

$$A = P(1 + r)^n$$

For this problem, $A = 4000$, $P = 2000$, $r = 0.08$.

$$4000 = 2000(1.08)^n$$
$$(1.08)^n = 2$$
$$\log(1.08)^n = \log 2$$
$$n \log(1.08) = \log 2$$
$$n = \frac{\log 2}{\log 1.08}$$
$$n \approx \frac{0.3010}{0.03342}$$
$$n \approx 9 \text{ years}$$

EXERCISES 8.5

(1–20) Solve for x. Approximate each exact answer to 3 significant digits.

1. $3^x = 20$
2. $5^x = 10$
3. $2^{3x} = 25$
4. $4^{1+x} = 5$
5. $6^{x-3} = 4.5$
6. $2(3^{2x}) = 5$
7. $5^{x-1} = 3^x$
8. $2^{3x-1} = 6^{2-x}$
9. $2^{2x+1} = 9^{x+2}$
10. $8(4^{x-3}) = 5^{x+1}$
11. $(1.05)^x = 2$
12. $(1.06)^{-x} = 0.5$
13. $e^x = 7.5$
14. $e^{2x} = 15$
15. $e^{-x} = 1.54$
16. $e^{-3x} = 0.96$
17. $\ln x = 3.49$
18. $\ln x = 1$
19. $\ln x = -2.36$
20. $\ln x = -0.85$

(21–30) Find each logarithm correct to 4 significant digits.

21. $\log_2 30$
22. $\log_5 93$
23. $\log_3 10$
24. $\log_6 28$
25. $\ln 10$
26. $\ln 6$
27. $\ln 34.2$
28. $\ln 0.76$
29. $\ln 0.434$
30. $\ln 10.9$

31. The population of a certain town of 5000 is increasing by the equation
$$y = 5000\, e^{0.03t}$$
where t is the time in years. In how many years will the population be 8000?

32. Find the half-life of a radioactive substance if its decay equation is given by
$$y = y_0 e^{-0.0123t}$$
where t is in years.

33. Bacteria in a certain culture increase according to the equation
$$y = 200 \, e^{0.5t}$$
where t is measured in hours. Find the number of bacteria y at the end of 15 hours.

34. The population of a certain town is decreasing by the equation
$$y = 6000 \, e^{-0.04t}$$
where t is in years. Find the population in 5 years.

35. Find how many years it takes for \$3000 to double itself if it is invested at 9% compounded annually. (See Example 7.)

36. The formula used for calculating depreciation by the constant percentage method is
$$S = C(1 - r)^n$$
where S is the scrap value after a useful life of n years, C is the original cost, and r is the constant percentage. Find n for a car where $C = \$6000$, $S = \$150$, and $r = 20\%$.

37. The build-up of the current of i amps in an electrical circuit with a condenser of capacitance C farads, a resistance of R ohms, and a source of E volts is given by the equation
$$i = \frac{E}{R} \, e^{-t/CR}$$
Find i for $E = 110$ volts, $R = 250$ ohms, $C = 2 \times 10^{-6}$ farad and $t = 0.001$ second.

38. The concentration y of a dilute sugar solution t hours after the sugar is added to the water is given by
$$y = 0.015 \, e^{-kt}$$
Find k if $y = \frac{1}{250}$ gram/cm³ when $t = 5$ hours.

8.6 CHAPTER REVIEW

(1–2) Graph each of the following. (8.1)

1. $y = \left(\frac{5}{2}\right)^x$ **2.** $y = \left(\frac{5}{2}\right)^{-x}$

(3–8) Solve for x. (8.1)

3. $6^x = 216$ **4.** $36^{-x} = 6$

5. $10^{2x+3} = 0.001$ **6.** $5^{x^2-2x-6} = 1$

7. $8^{-x} = 4^{x-5}$ **8.** $10^x = \dfrac{1}{10\sqrt{10}}$

(9–12) Simplify. (8.1)

9. $5^{\sqrt{10}} 5^{\sqrt{40}}$ **10.** $\left(5^{\sqrt{10}}\right)^{\sqrt{40}}$

11. $\dfrac{6^{\sqrt{48}}}{6^{\sqrt{27}}}$ **12.** $\dfrac{10^{3+\sqrt{5}}}{10^{\sqrt{5}-2}}$

(13–16) Write each statement in its logarithmic form. (8.2)

13. $2^{-3} = 0.125$

14. $10^{1.96} = 91.2$

15. $6^0 = 1$

16. $8^1 = 8$

(17–20) Write each statement in its exponential form. (8.2)

17. $\log_{10} 309 = 2.49$

18. $\log_5 0.0016 = -4$

19. $\log_{2.5} 2.5 = 1$

20. $\log_6 1.00 = 0$

(21–26) Find the value of each logarithm. (82)

21. $\log_2 32$

22. $\log_4 0.125$

23. $\log_{10} 10^8$

24. $\log_7 7$

25. $\log_{1.5} 1$

26. $\log_{10} 0.10$

(27–32) Write each of the following as a single logarithm. (8.2)

27. $3 \log_5 4 - \log_5 10$

28. $\frac{1}{2}(\log_3 7 + \log_3 5)$

29. $\frac{1}{3} \log_b r + 2 \log_b s$

30. $3(\log_b 4 - \frac{1}{2} \log_b s)$

31. $2(\log_{10} a + \log_{10} b) - \frac{1}{5} \log_{10} c$

32. $\frac{1}{4}(\log_2 a - \log_2 b - 2 \log_2 c)$

(33–40) Find the value of each of the following. Use tables or a calculator. (8.3)

33. log 48,200

34. log 0.0235

35. log 0.009674

36. log 7518

37. antilog 2.8609

38. antilog (-1.1500)

39. $10^{-0.3598}$

40. $10^{1.3900}$

(41–42) Do the following calculations by using logarithms. (8.4)

41. $\dfrac{\sqrt[3]{639}}{(4.52)(6.05)}$

42. $\sqrt[4]{\dfrac{12.78}{8.592}}$

(43–44) Using pH $= -\log [H^+]$, find each of the following. (8.4)

43. Find the pH for $[H^+] = 7.8 \times 10^{-6}$.

44. Find $[H^+]$ if pH $= 8.6$.

(45–46) Solve for x. (8.4)

45. $2 \log(x - 2) - \log(x + 8) + \frac{1}{2} \log 9 = 0$

46. $\log x + \log(x - 1) - \log(x - 3) = 1$

(47–54) Solve for x. (8.5)

47. $2^x = 15$

48. $(1.08)^{2x} = 2$

49. $e^{3x} = 8.2$

50. $e^{-2x} = 0.46$

51. $6^{2x-1} = 3^x$

52. $5^{-x} = 2^{x+4}$

53. $\ln x = -2$

54. $\ln x = 1.75$

(55–58) Evaluate to 4 significant digits. (8.5)

55. $\log_4 25$

56. $\log_3 1.24$

56. $\ln 9.2$

58. $\ln 0.32$

POLYNOMIAL FUNCTIONS, THEORY OF EQUATIONS

The theory of equations is concerned with finding the solutions of polynomial equations.

We have seen that the solutions of a quadratic equation may be real or imaginary numbers. In other words, the solutions belong to the set of complex numbers. In general, the solutions of polynomial equations are to be found in the set of complex numbers. This chapter is concerned with methods for finding these solutions.

9.1 DIVISION ALGORITHM, FACTOR AND REMAINDER THEOREMS

In Chapter 3 we learned how to solve certain polynomial equations by the factoring method; however, factoring was restricted to polynomials with integers for coefficients. Now we consider polynomials with complex numbers for coefficients and extend the process of factorization. As examples,

$$x^2 - 2 = (x - \sqrt{2})(x + \sqrt{2})$$
$$x^2 + 9 = (x - 3i)(x + 3i)$$

Solving a polynomial equation is related to factoring the polynomial, and factoring is related to the operation of division. As preparation for the general solution of polynomial equations, in this section we consider some topics related to division and factoring.

DEFINITION POLYNOMIAL FUNCTION

Let $P(x)$ be a polynomial of degree n in the variable x; that is,
$$P(x) = a_n x^n + a_{n-1} x^{n-1} + \cdots + a_1 x + a_0$$
where $a_n \neq 0$ and $a_0, a_1, \cdots, a_n$ are complex numbers. Then a **polynomial function** is a function defined by a polynomial equation that can be expressed in the form $y = P(x)$.
Polynomial function $= \{(x, y) \mid y = P(x)\}$

Recall that a_n is called the **leading coefficient** and a_0 is called the **constant term.** A value of x for which $P(x) = 0$ is called a **zero** of $P(x)$.

Now consider the division of $3x^3 - 13x^2 - 14x + 24$ by $x - 5$:

$$
\begin{array}{r}
3x^2 + 2x - 4 \\
x - 5 \overline{) 3x^3 - 13x^2 - 14x + 24} \\
\underline{3x^3 - 15x^2} \\
2x^2 - 14x + 24 \\
\underline{2x^2 - 10x} \\
-4x + 24 \\
\underline{-4x + 20} \\
4
\end{array}
$$

The quotient polynomial is $3x^2 + 2x - 4$ and the remainder is 4. We can express this result as follows:
$$3x^3 - 13x^2 - 14x + 24 = (x - 5)(3x^2 + 2x - 4) + 4$$
In general we have the following theorem.

THEOREM SPECIAL DIVISION ALGORITHM

If $P(x)$ is a polynomial whose coefficients are complex numbers and if c is any complex number, then there exists a unique polynomial $Q(x)$ whose coefficients are complex numbers, and a unique complex number r such that
$$P(x) = (x - c)Q(x) + r$$
Moreover, if the degree of $P(x)$ is n, then the degree of $Q(x)$ is $n - 1$.

This theorem can be proved by subtracting multiples of $x - c$ from $P(x)$ until a constant is obtained; in other words, the ordinary division process done in a general way. The details of the proof are omitted here.

The preceding division algorithm enables us to prove two important theorems, the **Remainder Theorem** and the **Factor Theorem.**

REMAINDER THEOREM

If $P(x)$, a polynomial with complex numbers for coefficients, is divided by $x - c$ where c is a complex number, then the remainder r is $P(c)$, the value of $P(x)$ for $x = c$;
$r = P(c)$

Proof

$P(x) = (x - c)Q(x) + r$ The Division Algorithm, true for all x
$P(c) = (c - c)Q(c) + r$ Replacing x by c
$P(c) = 0 + r = r$

FACTOR THEOREM

Let $P(x)$ be a polynomial with complex numbers for coefficients and let c be any complex number. Then $x - c$ is a factor of $P(x)$ if and only if $P(c) = 0$.

Proof

An "if and only if" statement infers a two-part proof.

1. If $P(c) = 0$, then since $P(c) = r$, it follows that $r = 0$.
   $P(x) = (x - c)Q(x) + r$
   $P(x) = (x - c)Q(x)$ since $r = 0$.
   Thus $x - c$ is a factor of $P(x)$.
2. If $x - c$ is a factor of $P(x)$, then
   $P(x) = (x - c)Q(x)$
   Thus $r = 0$ and since $P(c) = r$, $P(c) = 0$.

Note that the Factor Theorem tells us that if we know a zero of $P(x)$, then we also know a factor of $P(x)$. Also, if we know a factor $(x - c)$ of $P(x)$, then we know a zero of $P(x)$.

EXERCISES 9.1

(1–16) Find the remainder when the first polynomial is divided by the second polynomial by using (a) synthetic division (see Section 2.2, Chapter 2); (b) the Remainder Theorem.

EXAMPLE 1

$(2x^4 - 5x^3 - 50x - 6) \div (x - 4)$

Solution

(a)

$$
\begin{array}{r}
2 \quad -5 \quad 0 \quad -50 \quad -6 \\
 8 \quad 12 \quad 48 \quad -8 \\
\hline
4)\overline{2 \quad 3 \quad 12 \quad -2 \quad \boxed{-14}}
\end{array}
$$

The remainder is -14.

(b) Remainder $= P(c)$, where $c = 4$

$P(x) = 2x^4 - 5x^3 - 50x - 6$

$P(4) = 2(4^4) - 5(4^3) - 50(4) - 6$

$= 512 - 320 - 200 - 6$

$P(4) = -14$, the remainder

EXAMPLE 2

$(3x^3 + 8x^2 - x + 9) \div (x + 3)$

Solution

(a)

$$
\begin{array}{r}
3 \quad 8 \quad -1 \quad 9 \\
 -9 \quad 3 \quad -6 \\
\hline
-3)\overline{3 \quad -1 \quad 2 \quad \boxed{3}}
\end{array}
$$

The remainder is 3.

(b) $P(x) = 3x^3 + 8x^2 - x + 9$ and

$c = -3$ since $(x + 3 = x - (-3)$

$P(-3) = 3(-27) + 8(9) - (-3) + 9 = 3$, the remainder

EXAMPLE 3

$(x^3 - x^2 + 4x - 4) \div (x - 2i)$

Solution

(a)

$$
\begin{array}{r}
1 \quad -1 \quad 4 \quad -4 \\
 2i \quad -4 - 2i \quad 4 \\
\hline
2i)\overline{1 \quad -1 + 2i \quad -2i \quad \boxed{0}}
\end{array}
$$

The remainder is 0.

(b) $P(x) = x^3 - x^2 + 4x - 4$

$P(2i) = (2i)^3 - (2i)^2 + 4(2i) - 4$

$= -8i + 4 + 8i - 4 = 0$, the remainder

1. $(3x^2 + 2x - 5) \div (x + 2)$
2. $(3x^3 + 2x^2 + 5x - 1) \div (x - 2)$
3. $(x^4 - x^2 + 1) \div (x - 1)$
4. $(x^4 + x^3 - x + 1) \div (x + 1)$
5. $(2x^4 - 3x^3 - 20x^2 - 6) \div (x + 3)$
6. $(2x^4 - 3x^3 - 20x^2 - 6) \div (x - 4)$
7. $(2x^3 - 5x^2 - 28x + 15) \div \left(x - \frac{1}{2}\right)$
8. $(2x^3 + x^2 - 7x - 6) \div \left(x + \frac{3}{2}\right)$

9. $(4x^3 - 5x^2 - 17x + 7) \div \left(x + \frac{7}{4}\right)$

10. $(3x^3 - 5x^2 - 16x + 12) \div \left(x - \frac{2}{3}\right)$

11. $(x^4 + 2x^3 - 10x - 25) \div (x - \sqrt{5})$

12. $(x^4 - 2x^2 - 3) \div (x + \sqrt{3})$

13. $(x^4 + x^3 + 4x - 16) \div (x - 2i)$

14. $(x^4 - 9x^2 + 12x + 10) \div (x - 2 - i)$

15. $(x^4 + 16) \div (x + 2i)$

16. $(x^4 - 9) \div (x - \sqrt{3})$

(17–28) Use the Factor Theorem to determine which of the given numbers is a zero of the given polynomial. Then write the corresponding factor of the polynomial.

EXAMPLE 4

$x^4 - x^3 - 5x^2 - x - 6;$ $(2, -2, 3, -3)$

Solution

(a) $P(2) = 16 - 8 - 5(4) - 2 - 6 = -20$
$P(2) \neq 0$ and   2 is not a zero

(b) $P(-2) = 16 - (-8) - 5(4) - (-2) - 6 = 0$
-2 is a zero   and $x + 2$ is a factor

(c) $P(3) = 81 - 27 - 5(9) - 3 - 6 = 0$
3 is a zero   and $x - 3$ is a factor

(d) $P(-3) = 81 - (-27) - 5(9) - (-3) - 6 = 60$
$P(-3) \neq 0$ and -3 is not a zero

17. $x^3 + 3x^2 - 9x + 5;$ $(1, -1, 5, -5)$

18. $x^3 - 12x - 16;$ $(2, -2, 4, -4)$

19. $x^4 - 8x^2 - 9;$ $(1, -1, 3, -3)$

20. $x^4 - 27x^2 + 50;$ $(2, -2, 5, -5)$

21. $2x^3 - 5x^2 + 2x - 5;$ $\left(1, 5, \frac{5}{2}\right)$

22. $3x^3 - 2x^2 + 3x - 2;$ $\left(2, \frac{2}{3}, \frac{3}{2}\right)$

23. $x^3 + 1;$ $(1, -1, i, -i)$

24. $x^3 - 1;$ $(1, -1, i, -i)$

25. $x^4 - 1;$ $(1, -1, i, -i)$

26. $x^4 + 1;$ $(1, -1, i, -i)$

27. $x^4 + 2x^2 - 8;$ $(\sqrt{2}, 2, 2i)$

28. $x^4 - 4x^2 - 5;$ $(1, 5, \sqrt{5}, i)$

(29–38) (a) Use synthetic division, the Remainder Theorem, and the Factor Theorem to show that the given linear polynomial, $x - c$, is a factor of $P(x)$. Express $P(x)$ as $(x - c)Q(x)$. (b) Check by showing that $P(c) = 0$.

EXAMPLE 5

$P(x) = x^3 + 2x^2 - 2x + 3$; $x + 3$

Solution

$$
\begin{array}{r}
1 \quad\;\; 2 \quad -2 \quad\;\; 3 \\
-3 \quad\;\; 3 \quad -3 \\
\hline
-3\overline{)1} \;\; -1 \quad\;\; 1 \quad\; \textcircled{0}
\end{array}
$$

Since the remainder is 0, by the Remainder Theorem, $P(-3) = r = 0$. Using the Factor Theorem, $x + 3$ is a factor since $P(-3) = 0$.

$x^3 + 2x^2 - 2x + 3 = (x + 3)(x^2 - x + 1)$

Check

$P(x) = x^3 + 2x^2 - 2x + 3$

$P(-3) = -27 + 2(9) - 2(-3) + 3$

$ = -27 + 18 + 6 + 3 = 0$

29. $P(x) = x^4 + 3x - 4$; $x - 1$

30. $P(x) = x^4 - 3x^2 + 2x - 8$; $x - 2$

31. $P(x) = x^3 + 4x^2 + 25$; $x + 5$

32. $P(x) = x^3 - 10x - 3$; $x + 3$

33. $P(x) = 4x^4 - 2x^3 + 6x - 3$; $2x - 1$ $\left[\text{Use } 2x - 1 = 2\left(x - \tfrac{1}{2}\right)\right]$

34. $P(x) = 9x^3 + 6x^2 + 4x + 1$; $3x + 1$ $\left[\text{Use } 3x + 1 = 3\left(x + \tfrac{1}{3}\right)\right]$

35. $P(x) = x^3 - 2x^2 - 7x + 2$; $x - 2 + \sqrt{3}$

36. $P(x) = 2x^3 + 5x^2 - 1$; $x + 1 - \sqrt{2}$

37. $P(x) = x^3 + x^2 - 2$; $x + 1 - i$

38. $P(x) = x^4 + 6x^2 - 27$; $x - 3i$

(39 – 46) Find the value of k so that the given polynomial of degree 1 is a factor of $P(x)$.

EXAMPLE 6

$x + 2$, $P(x) = x^4 + 2x^3 + x + k$

Solution   Using the Factor Theorem with $c = -2$ since

$x + 2 = x - (-2)$

$P(-2) = (-2)^4 + 2(-2)^3 + (-2) + k$

$ = 16 - 16 - 2 + k$

$ = -2 + k$

For $x + 2$ to be a factor, $P(-2) = 0$. Thus $-2 + k = 0$ and $k = 2$.

39. $x - 2$, $P(x) = 3x^2 - 5x + k$

40. $x - 5$, $P(x) = 2x^2 - 4x + k$

41. $x + 3$, $P(x) = x^3 + 2x^2 + k$

42. $x + 4$, $P(x) = x^3 - 5x + k$

43. $x - 3$, $P(x) = 2x^3 + kx + 9$

44. $x + 2$, $P(x) = 3x^3 + kx^2 + 8$
45. $x + 5$, $P(x) = x^3 + kx^2 + kx + 5$
46. $x - 4$, $P(x) = 2x^3 + kx^2 - 13x + 3k$
47. Find n so that the remainder is 23 when $x^3 + n^2x^2 - nx - 4$ is divided by $x - 3$.
48. Find n so that the remainder is 9 when $x^4 + n^2x^2 + nx + 6$ is divided by $x + 1$.

(49–54)   Use the Factor Theorem to determine if $x - c$ is a factor of each of the following:

49. $x^3 - c^3$ **50.** $x^4 - c^4$
51. $x^5 - c^5$ **52.** $x^4 + c^4$
53. $x^5 + c^5$ **54.** $x^6 - c^6$

(55–60)   Use the Factor Theorem to determine if $x + c$ is a factor of each of the following:

55. $x^4 + c^4$ **56.** $x^5 + c^5$
57. $x^6 + c^6$ **58.** $x^4 - c^4$
59. $x^6 - c^6$ **60.** $x^5 - c^5$

9.2 COMPLEX ZEROS OF POLYNOMIAL FUNCTIONS

Finding a zero of a polynomial, in general, is not an easy task. The quadratic formula gives the zeros of polynomials of degree 2.

QUADRATIC FORMULA

If a, b, and c are any complex numbers such that $a \neq 0$ and $ax^2 + bx + c = 0$, then
$$x = \frac{-b \pm \sqrt{b^2 - 4ac}}{2a}$$

There are also formulas for polynomials of degree 3 and 4 but they are quite involved and do not yield the zeros for all cases. For polynomials of degree 5 or greater, it has been shown that it is impossible to derive a formula. However, the Fundamental Theorem of Algebra guarantees that every polynomial of degree greater than zero has a zero which is a complex number.

This theorem was proved by the great German mathematician Carl Friedrich Gauss (1777 – 1855) when he was 20 years old. We accept the theorem without proof in this text. All of the known proofs involve concepts from calculus and are beyond the level of this text.

THEOREM   FUNDAMENTAL THEOREM OF ALGEBRA

If $P(x)$ is a polynomial of degree greater than zero, with complex numbers for coefficients, then there exists a complex number c such that $P(c) = 0$.

The Fundamental Theorem of Algebra and the Factor Theorem enable us to show that a polynomial of degree n can be expressed as a product of n linear factors.

THEOREM   UNIQUE LINEAR FACTORIZATION

If $P(x)$ is a polynomial of degree n where $n \geq 1$, and if the coefficients of $P(x)$ are complex numbers, then there exist n complex numbers, $c_1, c_2, \cdots, c_n$ such that
$$P(x) = a_n(x - c_1)(x - c_2) \cdots (x - c_n)$$
where
$$P(x) = a_n x^n + a_{n-1}x^{n-1} + \cdots + a_1 x + a_0$$

Proof   By the Fundamental Theorem of Algebra, $P(x)$ has a zero, call it c_1, and thus a linear factor $x - c_1$.
$$P(x) = (x - c_1)Q_1(x)$$
and the degree of $Q_1(x)$ is $n - 1$.

Now $Q_1(x)$ has a zero, call it c_2, and thus has a linear factor $x - c_2$.
$$P(x) = (x - c_1)(x - c_2)Q_2(x)$$
and the degree of $Q_2(x)$ is $n - 2$.

We repeat this process until we obtain a polynomial of degree $n - n = 0$; that is, a constant k.
$$P(x) = (x - c_1)(x - c_2) \cdots (x - c_n)k$$
Since k is the coefficient of x^n, $k = a_n$.

The linear factors of $P(x)$ do not have to be distinct. If $x - c$ occurs exactly once, then c is called a **simple zero.** If $x - c$ occurs exactly twice, then c is called a **double zero,** and if $x - c$ occurs exactly three times, then c is called a **triple zero.** In general, we have the following definition.

DEFINITION MULTIPLICITY

c is a zero of $P(x)$ with **multiplicity k** if and only if $x - c$ occurs exactly k times in the linear factorization of $P(x)$.

For example, if
$$P(x) = 3(x - 5)(x - 1)^2(x + 2)^3$$
then:
    5 is a simple zero of $P(x)$.
    1 is a double zero of $P(x)$.
    -2 is a triple zero of $P(x)$.

Now, by agreeing to count a zero of multiplicity k as k zeros, we can state that a polynomial of degree n has exactly n zeros. It follows then that the polynomial equation $P(x) = 0$ has exactly n roots (solutions), when each root is counted with its corresponding multiplicity.

THEOREM

If $P(x)$ is a polynomial having complex coefficients and positive degree, then $P(x) = 0$ has exactly n roots, with a root of multiplicity k counted as k roots.

The next theorem states that the imaginary zeros of a polynomial with real coefficients always occur in pairs.

THEOREM   CONJUGATE IMAGINARY ROOTS

If the coefficients of $P(x)$ are real numbers and if $a + bi$ is an imaginary root of $P(x) = 0$, then $a - bi$ is also a root.

Proof Let
$$F(x) = [x - (a + bi)][x - (a - bi)]$$
$$= x^2 - 2ax + a^2 + b^2$$
Then the coefficients of $F(x)$ are real numbers. When $P(x)$ is divided by $F(x)$, since $F(x)$ is quadratic, the remainder must be linear, one degree less, and
$$P(x) = (x^2 - 2ax + a^2 + b^2)Q(x) + cx + d$$
Moreover, the coefficients of $Q(x)$ and $cx + d$ are real numbers. Now since $a + bi$ is a root,
$$P(a + bi) = 0 \cdot Q(a + bi) + c(a + bi) + d = 0$$
and $ca + d + cbi = 0$. Therefore $ca + d = 0$ and $cb = 0$. Since $b \neq 0$, otherwise $a + bi$ would not be imaginary, $c = 0$ and as a result, $d = 0$. Thus $P(x) = (x^2 - 2ax + a^2 + b^2)Q(x)$ and $P(a - bi) = 0 \cdot Q(x) = 0$ and $a - bi$ is a root.

If a and b are rational numbers and if $a + \sqrt{b}$ is irrational, then $a + \sqrt{b}$ and $a - \sqrt{b}$ are called **conjugate surds.** The conjugate surd zeros of a polynomial with rational coefficients also occur in pairs.

THEOREM   CONJUGATE SURD ROOTS

If $a + \sqrt{b}$ is a root of $P(x) = 0$ where a and b are rational and $\sqrt{b}$ is irrational, and if $P(x)$ has rational coefficients, then $a - \sqrt{b}$ is also a root of $P(x) = 0$.

The proof of this theorem is similar to that for conjugate imaginary roots by letting

$$F(x) = [x - (a + \sqrt{b})][x - (a - \sqrt{b})]$$
$$= x^2 - 2ax + a^2 - b$$

The details of the proof are omitted.

EXERCISES 9.2

(1–10) Factor $P(x)$ and solve $P(x) = 0$.

EXAMPLE 1
$P(x) = 2x^3 - 13x^2 + 18x - 15$ and $P(5) = 0$

Solution

1. Since $P(5) = 0$, $x - 5$ is one factor of $P(x)$. Using synthetic division to find the other factor,

$$
\begin{array}{r}
2 \quad -13 \quad18 \quad -15 \\
10 \quad -15 \quad15 \\
\hline
5)2 \quad-3 \quad3 \quad0
\end{array}
$$

$$P(x) = (x - 5)(2x^2 - 3x + 3)$$

2. Solving $P(x) = 0$,

$$x - 5 = 0 \quad \text{and} \quad x = 5$$

or

$$2x^2 - 3x + 3 = 0$$

Using the Quadratic Formula,

$$x = \frac{-(-3) \pm \sqrt{9 - 4(6)}}{2(2)} = \frac{3 \pm \sqrt{-15}}{4}$$

3. The solution set is $\left\{ 5, \dfrac{3 + i\sqrt{15}}{4}, \dfrac{3 - i\sqrt{15}}{4} \right\}$.

1. $P(x) = x^3 - 8x^2 + 16x - 8$ and $P(2) = 0$
2. $P(x) = x^3 + 5x^2 - 6x - 30$ and $P(-5) = 0$
3. $P(x) = x^3 - 4x^2 - 3x + 18$ and $P(3) = 0$
4. $P(x) = x^3 - 3x^2 + 3x - 1$ and $P(1) = 0$
5. $P(x) = x^3 + 8$ and $P(-2) = 0$
6. $P(x) = x^3 - 64$ and $P(4) = 0$
7. $P(x) = x^4 - 9$ and $P(\sqrt{3}) = 0$ and $P(-\sqrt{3}) = 0$
8. $P(x) = x^4 - x^2 - 2$ and $P(\sqrt{2}) = 0$ and $P(-\sqrt{2}) = 0$
9. $P(x) = x^3 - x^2 + 3x + 5$ and $P(-1) = 0$
10. $P(x) = x^3 - 33x + 18$ and $P(-6) = 0$

(11–18)   Solve each equation using the given information.

EXAMPLE 2 $x^4 + 10x^3 + 20x^2 - 50x - 125 = 0$ where -5 is a double root.

Solution   Using synthetic division to divide $P(x)$ by $x + 5$ two times,

$$\begin{array}{r} 1 \quad 10 \quad\ \ 20 \quad -50 \quad -125 \\ -5 \quad -25 \quad\ \ 25 \quad\ \ 125 \\ \hline -5)\overline{1 \quad\ \ 5 \quad\ -5 \quad -25 \quad\quad\ 0} \end{array}$$

$P(x) = (x + 5)(x^3 + 5x^2 - 5x - 25)$

Now dividing $x^3 + 5x^2 - 5x - 25$ by $x + 5$,

$$\begin{array}{r} 1 \quad\ \ 5 \quad -5 \quad -25 \\ -5 \quad\ \ 0 \quad\ \ 25 \\ \hline -5)\overline{1 \quad\ \ 0 \quad -5 \quad\quad\ 0} \end{array}$$

$P(x) = (x + 5)^2(x^2 - 5)$

Solving $P(x) = 0$,

$(x + 5)^2 = 0$ or $x^2 - 5 = 0$

$x = -5$ or $x = -5$ or $x = \sqrt{5}$ or $x = -\sqrt{5}$

The solution set is $\{-5, \sqrt{5}, -\sqrt{5}\}$, where -5 is a double root.

EXAMPLE 3 $x^4 - 8x^3 + 18x^2 - 27 = 0$ where 3 is a triple root.

Solution   Using synthetic division to divide $P(x)$ by $x - 3$ three times,

$$\begin{array}{r} 1 \quad -8 \quad\ \ 18 \quad 0 \quad -27 \\ 3 \quad -15 \quad 9 \quad\ \ 27 \\ \hline 3)\overline{1 \quad -5 \quad\quad 3 \quad 9 \quad\quad\ 0} \end{array}$$

Repeating the division a second time,

$$\begin{array}{r} 1 \quad -5 \quad\ \ 3 \quad\ \ 9 \\ 3 \quad -6 \quad -9 \\ \hline 3)\overline{1 \quad -2 \quad -3 \quad\ \ 0} \end{array}$$

Repeating the division a third time,

$$\begin{array}{r} 1 \quad -2 \quad -3 \\ 3 \quad\ \ 3 \\ \hline 3)\overline{1 \quad\ \ 1 \quad\ \ 0} \end{array}$$

$x^4 - 8x^3 + 18x^2 - 27 = (x - 3)^3(x + 1) = 0$

$(x - 3)^3 = 0$ or $x + 1 = 0$

$x = 3$ or $x = -1$

The solution set is $\{3, -1\}$, where 3 is a triple root.

11. $x^4 - 6x^3 + 13x^2 - 24x + 36 = 0$, where 3 is a double root.

12. $x^4 - 13x^2 + 20x - 4 = 0$, where 2 is a double root.

13. $2x^4 - 5x^3 + 3x^2 + x - 1 = 0$, where 1 is a triple root.

14. $x^4 + 3x^3 - 6x^2 - 28x - 24 = 0$, where -2 is a triple root.

15. $x^4 + 6x^3 - 2x^2 - 48x - 32 = 0$, where -4 is a double root.

16. $x^4 - 24x^2 - 36x + 27 = 0$, where -3 is a double root.

17. $x^5 + x^4 - x^3 + x^2 + 4x + 2 = 0$, where -1 is a triple root.

18. $x^5 - 12x^4 + 49x^3 - 76x^2 + 48x - 64 = 0$, where 4 is a triple root.

(19–28) Solve each equation using the given information.

EXAMPLE 4 $x^4 - 16x^2 - 8x - 1 = 0$, where $2 + \sqrt{5}$ is a root.

Solution   Since the coefficients of the polynomial are rational, the conjugate $2 - \sqrt{5}$ is also a root. Using synthetic division to first divide by $x - (2 + \sqrt{5})$ and then by $x - (2 - \sqrt{5})$,

$$
\begin{array}{r|rrrrr}
 & 1 & 0 & -16 & -8 & -1 \\
 & & 2+\sqrt{5} & 9+4\sqrt{5} & 6+\sqrt{5} & 1 \\
\hline
2+\sqrt{5})1 & 2+\sqrt{5} & -7+4\sqrt{5} & -2+\sqrt{5} & 0 \\
 & & 2-\sqrt{5} & 8-4\sqrt{5} & 2-\sqrt{5} \\
\hline
2-\sqrt{5})1 & 4 & 1 & 0 \\
\end{array}
$$

The quotient polynomial is $x^2 + 4x + 1$.

Solving,

$x^2 + 4x + 1 = 0$

Completing the square,

$x^2 + 4x + 4 = -1 + 4$

$(x + 2)^2 = 3$

$x = -2 \pm \sqrt{3}$

The solution set is $\{2 + \sqrt{5}, 2 - \sqrt{5}, -2 + \sqrt{3}, -2 - \sqrt{3}\}$.

EXAMPLE 5 $x^4 - 3x^3 - 10x^2 + 69x - 65 = 0$, where $3 - 2i$ is a root.

Solution   Since the coefficients of the polynomial are real, $3 + 2i$ is also a root. Using synthetic division to first divide by $x - (3 - 2i)$ and then by $x - (3 + 2i)$,

$$
\begin{array}{r|rrrrr}
 & 1 & -3 & -10 & +69 & -65 \\
 & & 3-2i & -4-6i & -54+10i & 65 \\
\hline
3-2i)1 & & -2i & -14-6i & 15+10i & 0 \\
 & & 3+2i & 9+6i & -15-10i \\
\hline
3+2i)1 & 3 & -5 & 0 \\
\end{array}
$$

The quotient is $x^2 + 3x - 5 = 0$.

Solving $x^2 + 3x - 5 = 0$ by using the Quadratic Formula,

$$x = \frac{-3 \pm \sqrt{9 + 20}}{2} = \frac{-3 \pm \sqrt{29}}{2}$$

The solution set is

$$\left\{ 3 - 2i, \; 3 + 2i, \; \frac{-3 + \sqrt{29}}{2}, \; \frac{-3 - \sqrt{29}}{2} \right\}.$$

19. $x^4 - 6x^3 + 5x^2 + 12x - 14 = 0$, where $3 + \sqrt{2}$ is a root.
20. $x^4 + 2x^3 + 2x - 1 = 0$, where $-1 + \sqrt{2}$ is a root.
21. $x^4 - 18x^2 + 1 = 0$, where $2 - \sqrt{5}$ is a root.
22. $x^4 + 6x^3 + 8x^2 + 24x + 16 = 0$, where $-3 - \sqrt{5}$ is a root.
23. $x^4 - 2x^3 + 4x - 4 = 0$, where $1 - i$ is a root.
24. $x^4 - 9x^2 + 12x + 10 = 0$, where $2 + i$ is a root.

25. $x^4 - 12x - 5 = 0$, where $-1 - 2i$ is a root.

26. $x^4 + 4x^3 + 10x^2 - 12x - 39 = 0$, where $-2 + 3i$ is a root.

27. $x^4 + x^3 + 4x - 16 = 0$, where $2i$ is a root.

28. $x^4 + x^3 - 6x^2 - 5x + 5 = 0$, where $\sqrt{5}$ is a root.

(29–40) Find a polynomial having the given zeros and satisfying the given conditions.

EXAMPLE 6 Degree 3, real coefficients, two of its zeros are 6 and $3 + 4i$.

Solution The three zeros are
$6, 3 + 4i$, and $3 - 4i$.

The corresponding factors are
$x - 6$, $x - (3 + 4i)$, and $x - (3 - 4i)$.

$$P(x) = (x - 6)(x - 3 - 4i)(x - 3 + 4i)$$
$$= (x - 6)(x^2 - 6x + 25)$$
$$= x^3 - 12x^2 + 61x - 150$$

EXAMPLE 7 Degree 4, rational coefficients, one zero is $4 - \sqrt{2}$ and -3 is a double zero.

Solution The four zeros are
$-3, -3, 4 - \sqrt{2}, 4 + \sqrt{2}$.

$$P(x) = (x - 3)^2(x - 4 + \sqrt{2})(x - 4 - \sqrt{2})$$
$$= (x^2 - 6x + 9)(x^2 - 8x + 14)$$
$$= x^4 - 14x^3 + 71x^2 - 156x + 126$$

29. Degree 3, real coefficients, two zeros are 1 and $4 - 3i$.

30. Degree 3, real coefficients, two zeros are -2 and $2 + 5i$.

31. Degree 3, rational coefficients, two zeros are -3 and $3 + \sqrt{5}$.

32. Degree 3, rational coefficients, two zeros are 5 and $1 - \sqrt{2}$.

33. Degree 4, rational coefficients, two zeros are $i\sqrt{5}$ and $3 - \sqrt{7}$.

34. Degree 4, rational coefficients, two zeros are $1 + \sqrt{3}$ and $-2 + 3i$.

35. Degree 4, rational coefficients, one zero is $3 + \sqrt{6}$ and 4 is a double zero.

36. Degree 4, real coefficients, one zero is $3 + i$ and -2 is a double zero.

37. Degree 3, -5 is a triple zero.

38. Degree 3, 6 is a triple zero.

39. Degree 4, rational coefficients, $1 - \sqrt{3}$ is a double root.

40. Degree 4, rational coefficients, $2 + i\sqrt{3}$ is a double root.

41. If $2 + \sqrt{3}$ is a zero of $4x^3 - 15x^2 + cx + d$ where c and d are rational numbers, find the other two roots, and also find c and d.

42. If $1 + i\sqrt{5}$ is a zero of $x^3 - 6x^2 + cx + d$, where c and d are real numbers, find the other two roots, and also find c and d.

.3 RATIONAL ZEROS OF POLYNOMIAL FUNCTIONS

If the coefficients of a polynomial are integers, then the rational zeros of the polynomial can be found by using the following theorem.

THEOREM   RATIONAL ZEROS

If the coefficients of $P(x) = a_n x^n + \cdots + a_1 x + a_0$ are integers and $a_n \neq 0$, and if $\dfrac{r}{s}$ is a rational number in simplified form (r and s are integers with no common factor except 1 and -1) such that $P\left(\dfrac{r}{s}\right) = 0$, then the numerator r is a factor of a_0 and the denominator s is a factor of a_n.

Proof

The number $\dfrac{r}{s}$ is a zero if and only if

$$a_n \left(\frac{r}{s}\right)^n + \cdots + a_1 \left(\frac{r}{s}\right) + a_0 = 0$$

Multiplying by s^n,

$$a_n r^n + a_{n-1} r^{n-1} s + \cdots + a_1 r s^{n-1} + a_0 s^n = 0$$

Subtracting $a_0 s^n$ from both sides and then factoring the left side,

$$r(a_n r^{n-1} + a_{n-1} r^{n-2} s + \cdots + a_1 s^{n-1}) = -a_0 s^n$$

Since r is a factor of the left side, r must also be a factor of $a_0 s^n$. However, r is not a factor of s since $\dfrac{r}{s}$ is in simplified form and therefore r is a factor of a_0.

Similarly,

$$a_n r^n = -s(a_{n-1} r^{n-1} + \cdots + a_1 r s^{n-2} + a_0 s^{n-1})$$

and since s is not a factor of r, therefore s is a factor of a_n.

In solving $P(x) = 0$, if c is a zero of $P(x)$, then

$$P(x) = (x - c)Q(x)$$

and $P(x) = 0$ is equivalent to

$$x - c = 0 \quad \text{or} \quad Q(x) = 0$$

The equation $Q(x) = 0$ is called the **depressed equation.**

EXAMPLE 1 Find the rational roots of $3x^4 + 4x^3 - x^2 + 4x - 4 = 0$ and solve.

Solution

Possible numerators: 1, −1, 2, −2, 4, −4

Possible denominators: 1, 3

Possible rational roots: $\pm 1, \pm 2, \pm 4, \pm\frac{1}{3}, \pm\frac{2}{3}, \pm\frac{4}{3}$

These possible roots are tested until it is seen that

$$
\begin{array}{r}
3 \quad\;\; 4 \;\; -1 \quad\;\; 4 \;\; -4 \\
-6 \quad\;\; 4 \;\; -6 \quad\;\; 4 \\
\hline
-2)\overline{3 \;\; -2 \quad\;\; 3 \;\; -2 \quad\;\; 0}
\end{array}
$$

$3x^4 + 4x^3 - x^2 + 4x - 4 = (x + 2)(3x^3 - 2x^2 + 3x - 2)$

Eliminating the possibilities that failed the first time $(\pm 1, +2)$, the possible rational roots of the depressed equation are $-2, \pm\frac{1}{3}, \pm\frac{2}{3}$. ($-2$ must be tried again in case it is a multiple root. Also ± 4 and $\pm\frac{4}{3}$ are not possible roots since 4 is not a factor of the constant term -2 of the depressed equation.) Finally, after several trial and error synthetic divisions,

$$
\begin{array}{r}
\phantom{\frac{2}{3})}3 \quad -2 \quad\;\; 3 \quad -2 \\
2 \quad\;\; 0 \quad\;\; 2 \\
\hline
\tfrac{2}{3})\overline{3 \quad\;\; 0 \quad\;\; 3 \quad\;\; 0}
\end{array}
$$

$3x^4 + 4x^3 - x^2 + 4x - 4 = (x + 2)\left(x - \tfrac{2}{3}\right)(3x^2 + 3)$

$\qquad\qquad\qquad\qquad\qquad = 3(x + 2)\left(x - \tfrac{2}{3}\right)(x^2 + 1)$

$\qquad\qquad\qquad\qquad\qquad = 3(x + 2)\left(x - \tfrac{2}{3}\right)(x + i)(x - i)$

Thus the rational roots are -2 and $\frac{2}{3}$ and the imaginary roots are i and $-i$.

EXAMPLE 2 Find the rational roots of $4x^5 - 3x^3 - 3x^2 + 4x - 1 = 0$.

Solution The possible rational roots are $1, -1,$ $\frac{1}{2}, -\frac{1}{2}, \frac{1}{4}, -\frac{1}{4}$. Finally, after testing 1 and -1,

$$
\begin{array}{r}
\phantom{\frac{1}{2})}4 \quad\;\; 0 \;\; -3 \;\; -3 \quad\;\; 4 \;\; -1 \\
2 \quad\;\; 1 \;\; -1 \;\; -2 \quad\;\; 1 \\
\hline
\tfrac{1}{2})\overline{4 \quad\;\; 2 \;\; -2 \;\; -4 \quad\;\; 2 \quad\;\; 0}
\end{array}
$$

Therefore $\frac{1}{2}$ is a root and the depressed equation is

$4x^4 + 2x^3 - 2x^2 - 4x + 2 = 2(2x^4 + x^3 - x^2 - 2x + 1) = 0$

Since 1 and -1 were ruled out from the original equation, the remaining possibilities for $2x^4 + x^3 - x^2 - 2x + 1$ are $\frac{1}{2}$ and $-\frac{1}{2}$. Trying $\frac{1}{2}$,

$$
\begin{array}{r}
\phantom{\frac{1}{2})}2 \quad\;\; 1 \;\; -1 \;\; -2 \quad\;\; 1 \\
1 \quad\;\; 1 \quad\;\; 0 \;\; -1 \\
\hline
\tfrac{1}{2})\overline{2 \quad\;\; 2 \quad\;\; 0 \;\; -2 \quad\;\; 0}
\end{array}
$$

Thus $\frac{1}{2}$ is a double root, and

$4x^5 - 3x^3 - 3x^2 + 4x - 1$

$= \left(x - \frac{1}{2}\right)(2)\left(x - \frac{1}{2}\right)(2x^3 + 2x^2 - 2)$

$= 4\left(x - \frac{1}{2}\right)^2 (x^3 + x^2 - 1) = 0$

The only possibilities for $x^3 + x^2 - 1 = 0$ are 1 and -1, and since these have been eliminated, the only rational roots are $\frac{1}{2}$ and $\frac{1}{2}$, or $\frac{1}{2}$ is a double root.

Finding the rational zeros of a polynomial can be a tedious task, especially when the leading coefficient or the constant term has many factors. However, it is possible to reduce the number of trials by finding an upper bound and a lower bound for the real zeros.

DEFINITION

The real number U is an **upper bound** for the real zeros of $P(x)$ if and only if all real zeros of $P(x)$ are less than or equal to U.

DEFINITION

The real number L is a **lower bound** for the real zeros of $P(x)$ if and only if all real zeros of $P(x)$ are greater than or equal to L.

Now if r is a real zero of $P(x)$, then $L \le r \le U$. Consequently the rational zeros of $P(x)$ must also lie in this interval.

The following theorem provides a method for finding an upper bound and a lower bound for the real zeros of a polynomial. The proof is omitted.

THEOREM   UPPER AND LOWER BOUNDS

Let
$P(x) = a_n x^n + \cdots + a_1 x + a_0$
where the coefficients are real numbers, $a_n > 0$, and $n \ge 1$. When $P(x)$ is divided by $x - c$ using synthetic division:

1. If $c > 0$ and if there are no negative numbers in the bottom row of the synthetic division, then c is an upper bound for the real zeros of $P(x)$.
2. If $c < 0$ and if the numbers in the bottom row of the synthetic division have alternate plus and minus signs (with 0 assigned a plus or minus sign as convenient), then c is a lower bound for the real zeros of $P(x)$.

EXAMPLE 3 Find an upper and lower bound for the real zeros of
$P(x) = x^5 - x^4 - 8x^3 + 6x^2 + 8x - 24$
Then solve $P(x) = 0$.

Solution The possible rational zeros are ± 1, ± 2, ± 3, ± 4, ± 6, ± 8, ± 12, ± 24. Trying the possible positive integral zeros in the listed order, finally,

$$
\begin{array}{r|rrrrrr}
 & 1 & -1 & -8 & 6 & 8 & -24 \\
 & & 3 & 6 & -6 & 0 & 24 \\
\hline
3)\, & 1 & 2 & -2 & 0 & 8 & 0 \\
\end{array}
$$

Therefore 3 is a zero, but since one of the numbers in the third line is negative, it cannot be concluded that 3 is an upper bound. Trying 4,

$$
\begin{array}{r|rrrrrr}
 & 1 & -1 & -8 & 6 & 8 & -24 \\
 & & 4 & 12 & 16 & 88 & 384 \\
\hline
4)\, & 1 & 3 & 4 & 22 & 96 & 360 \\
\end{array}
$$

Since all the numbers in the bottom row are positive, 4 is an upper bound, and 4, 6, 8, 12, and 24 cannot be zeros. Also, 1 and 2 were eliminated by the first two trials. Thus 3 is the only positive rational zero.

Now the negative zeros are investigated; trying -1, -2, -3, -4, -6, -8, -12, and -24 in the order listed,
$P(-1) = -1 - 1 + 8 + 6 - 8 - 24 = -20 \neq 0$
and -1 is not a zero.
Trying -2,

$$
\begin{array}{r|rrrrrr}
 & 1 & -1 & -8 & 6 & 8 & -24 \\
 & & -2 & 6 & 4 & -20 & 24 \\
\hline
-2)\, & 1 & -3 & -2 & 10 & -12 & 0 \\
\end{array}
$$

Therefore -2 is a zero. However, it cannot be concluded that -2 is a lower bound since the signs in the bottom row do not alternate.
Trying -3,

$$
\begin{array}{r|rrrrrr}
 & 1 & -1 & -8 & 6 & 8 & -24 \\
 & & -3 & 12 & -12 & 18 & -78 \\
\hline
-3)\, & 1 & -4 & 4 & -6 & 26 & -102 \\
\end{array}
$$

This time the signs in the bottom row alternate and thus -3 is a lower bound.

For all real zeros r of $P(z)$, $-3 < r < 4$.
Now
$P(x) = (x - 3)(x^4 + 2x^3 - 2x^2 + 8)$
and
$P(x) = (x - 3)(x + 2)(x^3 - 2x + 4)$
Considering the bounds and the trials already made, the only possible rational zero of $x^3 - 2x + 4$ is -2. Trying -2,

$$\begin{array}{r} 1 \quad\ \ 0 \quad -2 \quad\ \ 4 \\ -2 \quad\ \ 4 \quad -4 \\ \hline -2)\,\overline{1 \quad -2 \quad\ \ 2 \quad\ \ 0} \end{array}$$

and -2 is a root.

$P(x) = (x - 3)(x + 2)^2(x^2 - 2x + 2)$

Solving $P(x) = 0$,

$(x - 3)(x + 2)^2 = 0$ or $x^2 - 2x + 2 = 0$

$x = 3,\ \ x = -2,\ \ x = -2,\ \ x^2 - 2x + 1 = -2 + 1$

$(x - 1)^2 = -1$

$x = 1 \pm i$

The solution set is $\{3, -2, 1 + i, 1 - i\}$, where -2 is a double root.

EXERCISES 9.3

(1–18) Find all the rational zeros of each polynomial.

1. $x^4 - 2x^2 - 3x - 2$
2. $x^3 - 7x^2 + 13x - 3$
3. $2x^3 + x^2 - 7x - 6$
4. $2x^4 - x^3 - 3x^2 - 31x - 15$
5. $4x^4 - 13x^3 - 7x^2 + 41x - 14$
6. $2x^3 + 3x^2 - 14x - 21$
7. $2x^4 + x^2 + 2x - 4$
8. $2x^3 - 7x^2 + 10x - 6$
9. $2x^3 - 10x^2 + 13x - 20$
10. $x^3 - x^2 + 2x - 14$
11. $2x^3 - 5x^2 - 28x + 15$
12. $3x^3 + 5x^2 - 16x - 12$
13. $x^4 - x^2 - 2$
14. $x^3 - 5x^2 + 7x + 13$
15. $4x^4 - x^3 - 8x^2 + 18x - 4$
16. $x^5 - 12x^3 + x^2 - 4$
17. $3x^4 - x^3 - 6x^2 + 14x - 4$
18. $5x^4 - 2x^3 - 25x^2 + 70x - 24$

(19–30) Solve by first finding the rational roots.

19. $4x^3 - 3x + 1 = 0$
20. $9x^3 + 2x - 4 = 0$
21. $36x^4 - 65x^2 - 36 = 0$
22. $x^4 + x^2 - 370x - 1200 = 0$
23. $x^4 + x^3 + 10x^2 + 12x - 24 = 0$
24. $x^4 - 2x^3 - 15x^2 + 24x + 36 = 0$
25. $x^4 - 11x^3 + 32x^2 - 22x + 60 = 0$
26. $x^4 + 6x^3 + 17x^2 + 54x + 72 = 0$
27. $x^4 - 5x^3 + 3x^2 - 11x + 60 = 0$
28. $x^4 + 13x^3 + 52x^2 + 84x + 144 = 0$
29. $27x^3 + 27x^2 + 9x + 1 = 0$
30. $64x^3 - 144x^2 + 108x - 27 = 0$
31. Show that $\sqrt{2}$ is irrational by showing that $x^2 - 2 = 0$ has no rational roots.
32. Show that $\sqrt{5}$ is irrational by showing that $x^2 - 5 = 0$ has no rational roots.

9.4   REAL ZEROS OF POLYNOMIAL FUNCTIONS

While the rational zeros of a polynomial can always be found by using the methods of the preceding section, finding the irrational zeros is much more difficult and, in general, it is impossible to obtain them in an exact form involving a finite number of algebraic operations.

However, there are methods for obtaining rational approximations for the irrational zeros. This topic is discussed in this section.

First we consider a theorem called Descartes' Rule of Signs which provides information about the number of real and imaginary zeros of a polynomial.

If the terms of a polynomial are written in descending order (or ascending order), then the polynomial is said to have a variation in sign if the coefficients of two consecutive terms have opposite signs, the missing terms being ignored. For example,

$$x^5 - 3x^3 - 2x + 4$$

has two variations in sign. (Here the missing x^2 term is ignored.) The number of variations in sign provides information regarding the number of positive and negative real zeros of a polynomial with real coefficients.

THEOREM   DESCARTES' RULE OF SIGNS

If $P(x)$ is a polynomial with real coefficients and if $v(p) =$ the number of variations in sign of $P(x)$, then p, the number of positive real zeros of $P(x)$, is either equal to the number of variations or differs from it by an even positive integer; that is,

$p = v(p) - 2k,$ where $k = 0$ or a positive integer

If $v(n)$ is the number of variations in sign of $P(-x)$, then n, the number of negative real zeros of $P(x)$, is either equal to $v(n)$ or differs from it by an even positive integer; that is,

$n = v(n) - 2k,$ where $k = 0$ or a positive integer
The proof of this theorem is omitted.

EXAMPLE 1   Determine the nature of the roots of $x^3 - 3x^2 - 5 = 0$ by using Descartes' Rule of Signs.

Solution
$$P(x) = x^3 - 3x^2 - 5 \quad \text{and} \quad v(p) = 1$$

Therefore $P(x)$ has exactly one positive real zero. (Note that $v(p) - 2 = 1 - 2 = -1$ is impossible since the number of roots must be positive or 0.)

$P(-x) = -x^3 - 3x^2 - 5$ and $v(n) = 0$

Therefore $P(x)$ has no negative real zeros. Finally, $P(x) = 0$ has one positive real root and two conjugate imaginary roots.

EXAMPLE 2 Determine the nature of the roots of $x^3 - 3x + 5 = 0$ by using Descartes' Rule of Signs.

Solution

$P(x) = x^3 - 3x + 5$ and $v(p) = 2$

Therefore $P(x)$ has either two positive real zeros or none at all.

$P(-x) = -x^3 + 3x + 5$ and $v(n) = 1$

Therefore $P(x)$ has exactly one negative real zero. Finally, $P(x) = 0$ has either (one negative real root and two imaginary roots) or (one negative real root and two positive real roots).

EXAMPLE 3 Find the real roots of $25x^5 + 21x^3 + 11x - 6 = 0$ making use of Descartes' Rule of Signs.

Solution

$v(p) = 1$ and there is exactly one positive real root.

$P(-x) = -25x^5 - 21x^3 - 11x - 6$

$v(n) = 0$ and there are no negative real roots.

The possible rational roots are now

$1, 2, 3, 6, \frac{1}{5}, \frac{2}{5}, \frac{3}{5}, \frac{6}{5}, \frac{1}{25}, \frac{2}{25}, \frac{3}{25}, \frac{6}{25}$

Trying 1,

$$
\begin{array}{r|rrrrrr}
 & 25 & 0 & 21 & 0 & 11 & -6 \\
 &   & 25 & 25 & 46 & 46 & 57 \\
\hline
1) & 25 & 25 & 46 & 46 & 57 & 51 \\
\end{array}
$$

Therefore 1 is an upper bound.

Trying $\frac{1}{5}$, it is seen that $\frac{1}{5}$ is not a root.

Trying $\frac{2}{5}$,

$$
\begin{array}{r|rrrrrr}
 & 25 & 0 & 21 & 0 & 11 & -6 \\
 &   & 10 & 4 & 10 & 4 & 6 \\
\hline
\frac{2}{5}) & 25 & 10 & 25 & 10 & 15 & 0 \\
\end{array}
$$

Thus $\frac{2}{5}$ is the only real root. No further trials are necessary since there is exactly one positive real root and no negative real roots.

The Weierstrass Zero Theorem helps us locate a real zero between two rational numbers.

THE WEIERSTRASS ZERO THEOREM

If $P(x)$ is a polynomial with real coefficients and if a and b are real numbers so that $P(a) < 0$ and $P(b) > 0$, then $P(x)$ has a real zero between a and b.

Proof   Since the imaginary zeros of a polynomial with real coefficients occur in conjugate pairs, such as $c + di$ and $c - di$, and since the product

$$(x - c - di)(x - c + di) = x^2 - 2cx + c^2 + d^2$$

is a polynomial with real coefficients, it follows from the unique factorization of polynomials that every polynomial with real coefficients can be expressed as a product of linear factors and quadratic factors, where each linear factor and each quadratic factor have real coefficients.

The quadratic factors have the form

$$(x^2 - 2cx + c^2) + d^2 = (x - c)^2 + d^2$$

which is always positive for all real x and $d \neq 0$.

Now if $P(a) < 0$ and $P(b) > 0$ for the real numbers a and b, then the sign change must result from a sign change of one of the linear factors, say $L(x) = x - r$ where r is a real zero of $P(x)$.

There are two possibilities.

Case 1 $L(a) = a - r < 0$ and $r > a$
           $L(b) = b - r > 0$ and $r < b$

Then $a < r < b$.

Case 2 $L(a) = a - r > 0$ and $r < a$
           $L(b) = b - r < 0$ and $r > b$

Then $b < r < a$.

In either case, the real zero is between a and b.

EXAMPLE 4   Show that $x^3 - 5x - 2 = 0$ has a real root between 2 and 3.

Solution
$P(2) = 8 - 10 - 2 = -4$
$P(3) = 27 - 15 - 2 = 10$
Since $P(2)$ is negative and $P(3)$ is positive, by the Weierstrass Zero Theorem, there is a real root r between 2 and 3, $2 < r < 3$.

EXAMPLE 5   Show that $x^3 - 5x^2 + 3 = 0$ has a real root between 0 and 1.

Solution
$P(0) = 3$ and $P(1) = 1 - 5 + 3 = -1$
Therefore there is a real root between 0 and 1 since $P(0)$ is positive and $P(1)$ is negative.

EXAMPLE 6 For
$$P(x) = 4x^3 - 24x^2 + 45x - 27$$
$$= (2x - 3)^2(x - 3)$$
show that $P(1) < 0$ and $P(2) < 0$ but $P(x)$ has a real zero between 1 and 2.

Solution
$$P(1) = -2 \quad \text{and} \quad P(2) = -1$$
However
$$P\left(\tfrac{3}{2}\right) = 0$$
Therefore $\tfrac{3}{2}$ is a zero and $1 < \tfrac{3}{2} < 2$.

Example 6 shows that the converse of the Weierstrass Zero Theorem is not valid. In other words, $P(x)$ may have a real zero between a and b even though $P(a)$ and $P(b)$ do not differ in sign.

After locating a zero of $P(x)$ between two consecutive integers, further approximations to an irrational zero can be obtained by repeated applications of the Zero Theorem and by using synthetic division. This is shown in Example 7.

EXAMPLE 7 Find, correct to the nearest hundredth, the real root of
$$x^3 - 5x - 2 = 0 \quad \text{between} \quad 2 \text{ and } 3$$

Solution   Note that
$$P(2) = 8 - 10 - 2 = -4$$
and
$$P(3) = 27 - 15 - 2 = 10$$
Thus
$$2 < r < 3$$
We first try 2.5, using synthetic division:

	1	0	−5	−2
		2.5	6.25	3.125
2.5)	1	2.5	1.25	1.125

Now we know that $2 < r < 2.5$ since $P(2) < 0$ and $P(2.5) > 0$. We next try 2.4:

	1	0	−5	−2
		2.4	5.76	1.824
2.4)	1	2.4	0.76	−0.176

Therefore $2.4 < r < 2.5$ since $P(2.4) < 0$ and $P(2.5) > 0$.

Now we continue, using hundredths. Since the absolute value of the remainder for 2.4 is quite a bit less than that for 2.5, let us first try 2.42:

	1	0	−5	−2
		2.42	5.8564	2.0725
2.42)	1	2.42	0.8564	0.0725

Thus $2.40 < r < 2.42$ and we try 2.41:

$$\begin{array}{r|rrr} & 1 & 0 & -5 & -2 \\ & & 2.41 & 5.8081 & 1.9475 \\ \hline 2.41)1 & 2.41 & 0.8081 & -0.0525 \end{array}$$

Now we know $2.41 < r < 2.42$ since $P(2.41) < 0$ and $P(2.42) > 0$. Repeating the process using thousandths, we find the following:

$$\begin{array}{r|rrr} & 1 & 0 & -5 & -2 \\ & & 2.414 & 5.8274 & 1.9973 \\ \hline 2.414)1 & 2.414 & 0.8274 & -0.0027 \end{array}$$

$$\begin{array}{r|rrr} & 1 & 0 & -5 & -2 \\ & & 2.415 & 5.8322 & 2.0098 \\ \hline 2.415)1 & 2.415 & 0.8322 & 0.0098 \end{array}$$

Therefore $2.414 < r < 2.415$.

Correct to the nearest hundredth, $r = 2.41$.

Other approximation techniques exist but the one shown above is comparatively simple to use and easy to remember. Moreover, it is self-checking while providing whatever accuracy is desired.

EXERCISES 9.4

(1–10) Use Descartes' Rule of Signs to determine the nature of the roots of each of the following. See Examples 1, 2, and 3.

1. $2x^3 - x - 5 = 0$
2. $x^6 - 3x^5 + 2x^2 - 3x + 1 = 0$
3. $x^5 + 3x^3 - 5x - 16 = 0$
4. $x^4 + 3x^2 - 5x - 1 = 0$
5. $x^5 + 32 = 0$
6. $x^6 - 64 = 0$
7. $x^6 + 2x^4 + 5x^2 + 3 = 0$
8. $3x^5 - 2x^3 + 5x - 8 = 0$
9. $x^6 - 3x^2 + 4x - 1 = 0$
10. $x^8 - 2x^4 + 1 = 0$

(11–16) Show that each equation has a real root r in the indicated interval. See Examples 4 and 5.

11. $x^3 + x - 15 = 0$, $2 < r < 3$
12. $x^3 + 3x^2 - 12 = 0$, $1 < r < 2$
13. $x^3 + 3x^2 - 4x - 5 = 0$, $-4 < r < -3$
14. $x^3 - x + 4 = 0$, $-2 < r < -1$
15. $2x^4 - 3x^3 - 7x^2 - 8x + 6 = 0$, $0 < r < 1$
16. $2x^4 - 5x^2 - 6x - 2 = 0$, $-1 < r < 0$

(17–22) Find consecutive integers a and b so that $a < r < b$ for each real root r of the given equation.

17. $x^3 + 3x + 8 = 0$
18. $x^3 + 2x - 7 = 0$

19. $x^3 + 5x^2 - 3 = 0$ **20.** $x^3 - 5x^2 + 4x + 5 = 0$

21. $2x^4 - 3x^3 + x - 8 = 0$ **22.** $2x^4 + x^3 - 3x + 1 = 0$

(23–30) Find the indicated root correct to the nearest hundredth for each equation. See Example 7.

23. $x^3 + x^2 - 4x - 15 = 0$, between 2 and 3

24. $x^3 + x - 5 = 0$, between 1 and 2

25. $x^3 - 9x - 5 = 0$, between 3 and 4

26. $2x^3 + x^2 + 3x - 4 = 0$, between 0 and 1

27. $x^4 - 3x^3 + 7x - 8 = 0$, between -2 and -1

28. $x^3 - 3x^2 - 3x + 18 = 0$, root is negative

29. $x^3 + 6x - 2 = 0$, the real root

30. $x^3 + 3x^2 - 5 = 0$, the real root

31. Approximate $\sqrt[3]{5}$ by finding the real root of $x^3 = 5$ correct to the nearest hundredth.

32. Approximate $\sqrt[4]{20}$ by finding the positive real root of $x^4 = 20$ correct to the nearest hundredth.

33. Show that $x^4 + x^3 - 2x^2 - 4 = 0$ has exactly two irrational roots.

34. Show that $2x^4 - 15x + 3 = 0$ has exactly two real and two imaginary roots.

35. Show that $x^n - 1 = 0$ has exactly one real root if n is odd and exactly two real roots if n is even.

36. Show that $x^n + 1 = 0$ has exactly one real root if n is odd and no real roots if n is even.

9.5   CHAPTER REVIEW

(1–2) Use the Remainder Theorem to find the remainder when the first polynomial is divided by the second. (9.1)

1. $(6x^9 - 4x^7 - 3) \div (x - 1)$

2. $(x^6 + x^3 - 3x^2 + 2) \div (x + 2)$

3. Use the Factor Theorem to determine which of the given numbers is a zero of the given polynomial. State the corresponding factor. (9.1)
$x^3 - 12x - 16;$ $(2, -2, 4, -4)$

4. Find k so that $x - 6$ is a factor of $x^3 - 38x + k$. (9.1)

(5–6) Find the quotient and remainder by using synthetic division. (9.1)

5. $(2x^4 - 15x^2 - 6x - 3) \div (x - 3)$

6. $(x^3 + 6x^2 + 5x - 4) \div (x + 4)$

(7–8) Solve $P(x) = 0$. (9.2)

7. $P(x) = x^3 - 8x^2 + 15x + 4$ and $P(4) = 0$
8. $P(x) = x^3 + 2x^2 + 9x + 18$ and $P(-2) = 0$

(9–12) Solve each equation using the given information. (9.2)

9. $x^4 - 2x^3 + x^2 - 12x + 20 = 0$, where 2 is a double root.
10. $x^4 + 7x^3 + 9x^2 - 27x - 54 = 0$, where -3 is a triple root.
11. $x^4 - 8x^3 + 9x^2 + 40x - 70 = 0$, where $4 + \sqrt{2}$ is a root.
12. $x^4 - 6x^3 + 12x^2 + 6x - 13 = 0$, where $3 - 2i$ is a root.

(13–14) Find a polynomial satisfying the given conditions. (9.2)

13. Degree 4, rational coefficients, two zeros are $1 + 2i$ and $2 - \sqrt{5}$.
14. Degree 3, where 5 is a simple zero and 4 is a double zero.

(15–18) Find the rational roots and solve. (9.3)

15. $x^4 - x^3 - 15x^2 + 23x + 12 = 0$
16. $3x^3 + 17x^2 - 2 = 0$
17. $4x^4 - 67x^2 + 85x + 50 = 0$
18. $x^5 - 9x^3 - 8x^2 + 72 = 0$

(19–21) Use Descartes' Rule of Signs to determine the nature of the roots of each of the following. (9.4)

19. $6x^3 + 5x - 12 = 0$
20. $2x^4 - x^3 + 4x^2 - 3x - 2 = 0$
21. $3x^4 + 2x^3 - 7x^2 - 8 = 0$

(22–23) Find consecutive integers a and b so that $a < r < b$ for each real root r of the given equation. (9.4)

22. $x^3 - 6x^2 + 14x - 19 = 0$ **23.** $3x^4 + 2x^3 - 7x^2 - 8 = 0$

(24–25) Find the real roots correct to the nearest hundredth. (9.4)

24. $2x^3 + 5x - 12 = 0$ **25.** $x^4 + 3x - 2 = 0$

FUNCTIONS OF INTEGERS

There are many important applications of functions whose domain is a subset of the set of positive integers. In this chapter we examine such functions.

An ordered set of numbers is called a **sequence.** One familiar example of a sequence is the ordered set of positive integers, 1, 2, 3, 4, 5, ..., used for counting. Stated more formally, we have the following definitions.

DEFINITIONS

A **finite sequence function** is a function whose domain is the set of the first n positive integers, $\{1, 2, 3, \cdots, n\}$.

An **infinite sequence function** is a function whose domain is the set of all positive integers.

The **terms of a sequence** are the numbers in the range of the sequence function.

The terms of a sequence function, listed in order, is called the **sequence.**

EXAMPLE 1   Write the sequence defined by $f(n) = 4n - 1$ having the domain $\{1, 2, 3, 4, 5\}$.

Solution

$f(1) = 4(1) - 1 = 3$
$f(2) = 4(2) - 1 = 7$
$f(3) = 4(3) - 1 = 11$
$f(4) = 4(4) - 1 = 15$
$f(5) = 4(5) - 1 = 19$

The sequence is 3, 7, 11, 15, 19.

The notation c_n is used to designate the **nth term,** or **general term,** of a sequence. For the sequence in Example 1, $c_1 = 3$, $c_2 = 7$, $c_3 = 11$, $c_4 = 15$, $c_5 = 19$, and $c_n = 4n - 1$.

EXAMPLE 2   Write the first five terms of the sequence whose general term is $c_n = 2^{n-1}$.

Solution

$c_1 = 2^{1-1} = 2^0 = 1$
$c_2 = 2^{2-1} = 2^1 = 2$
$c_3 = 2^{3-1} = 2^2 = 4$
$c_4 = 2^{4-1} = 2^3 = 8$
$c_5 = 2^{5-1} = 2^4 = 16$

The first five terms are 1, 2, 4, 8, 16.

EXAMPLE 3   Write the first five terms of the sequence whose general term is $c_n = \dfrac{(-1)^n}{n^2}$.

Solution

$c_1 = \dfrac{(-1)^1}{1^2} = -1$

$c_2 = \dfrac{(-1)^2}{2^2} = \dfrac{1}{4}$

$c_3 = \dfrac{(-1)^3}{3^2} = -\dfrac{1}{9}$

$c_4 = \dfrac{(-1)^4}{4^2} = \dfrac{1}{16}$

$c_5 = \dfrac{(-1)^5}{5^2} = -\dfrac{1}{25}$

The first five terms are $-1, \frac{1}{4}, -\frac{1}{9}, \frac{1}{16}, -\frac{1}{25}$.

DEFINITION

A **series** is the indicated sum of the terms of a sequence. The **nth term of a series** is the nth term of its corresponding sequence.

The finite series $c_1 + c_2 + c_3 + c_4 + \cdots + c_n$ corresponds to the finite sequence

$c_1, c_2, c_3, c_4, \cdots, c_n$

The infinite series

$c_1 + c_2 + c_3 + \cdots + c_n + \cdots$

corresponds to the infinite sequence

$c_1, c_2, c_3, \cdots, c_n, \cdots$

As examples, the series $1 + 2 + 4 + 8 + 16$ corresponds to the sequence 1, 2, 4, 8, 16.

The series $1 - \dfrac{1}{2} + \dfrac{1}{3} - \dfrac{1}{4} + \cdots + \dfrac{(-1)^{n+1}}{n} + \cdots$ corresponds to the sequence

$1, -\dfrac{1}{2}, \dfrac{1}{3}, -\dfrac{1}{4}, \cdots, \dfrac{(-1)^{n+1}}{n}, \cdots$

For a finite series, there is always a finite number which is the sum obtained by adding the terms of the series. Thus the sum of the series $1 + 3 + 9 + 27 + 81 + 243$ is 364. On the other hand, it is important that this sum, a number, is not confused with the series which indicates the addition. This is especially important for infinite series. For example, there is no finite number that can be designated as the sum of the infinite series $3 + 7 + 11 + 15 + 19 + \cdots$. However, there are some infinite series that can be assigned a number, called the **sum.** This shall be seen later.

The Greek capital letter **sigma,** Σ, which corresponds to our letter S, is used in describing series. This **sigma notation** greatly reduces the amount of writing that is necessary. For example, the finite series $3 + 7 + 11 + 15 + 19$ can be written as

$$\sum_{k=1}^{5} (4k - 1)$$

where it is understood that k is to be replaced by each of the numbers 1, 2, 3, 4, and 5 in the expression $4k - 1$, and then the sum of the resulting values is to be indicated.

If $k = 1$,   then $4k - 1 = 3$
  $k = 2$, $4k - 1 = 7$
  $k = 3$, $4k - 1 = 11$
  $k = 4$, $4k - 1 = 15$
  $k = 5$, $4k - 1 = 19$

Thus

$$\sum_{k=1}^{5} (4k - 1) = 3 + 7 + 11 + 15 + 19$$

In general, if the kth term of the series is c_k, then

$$\sum_{k=1}^{n} c_k = c_1 + c_2 + c_3 + \cdots + c_n$$

The symbol Σ is called the **summation symbol,** the letter k is called the **index of summation,** and the replacement set of integers for k is called the **range of summation.** The letter k is often referred to as a **dummy variable** since any other letter could be used and still the same sum would be indicated. Thus

$$\sum_{k=1}^{5} (4k - 1) = \sum_{i=1}^{5} (4i - 1) = 3 + 7 + 11 + 15 + 19$$

An infinite series is designated by the notation,

$$\sum_{k=1}^{\infty} c_k = c_1 + c_2 + \cdots + c_n + \cdots$$

EXAMPLE 4   Write the series

$$\sum_{k=3}^{7} \frac{k(k - 1)}{2}$$

in expanded form.

Solution

$$\frac{3(3 - 1)}{2} + \frac{4(4 - 1)}{2} + \frac{5(5 - 1)}{2} + \frac{6(6 - 1)}{2} + \frac{7(7 - 1)}{2}$$
$$= 3 + 6 + 10 + 15 + 21$$

EXAMPLE 5   Write the series

$$\sum_{i=1}^{\infty} (-1)^{n+1} \left(\tfrac{1}{5}\right)^n$$

in expanded form.

Solution

$$\frac{(-1)^2}{5} + \frac{(-1)^3}{5^2} + \frac{(-1)^4}{5^3} + \frac{(-1)^5}{5^4} + \cdots$$
$$= \frac{1}{5} - \frac{1}{25} + \frac{1}{125} - \frac{1}{625} + \cdots$$

EXERCISES 10.1

(1–20)   Write the first five terms of the sequence whose general term is given. See Examples 2 and 3.

1. $c_n = 2n - 3$ 2. $c_n = n^2$

3. $c_n = \left(\tfrac{1}{3}\right)^n$ 4. $c_n = \tfrac{1}{n}$

5. $c_n = \dfrac{(-1)^{n+1}}{2n + 1}$ 6. $c_n = \dfrac{(-1)^{n+1}}{2^n}$

7. $c_n = 3(-2)^n$ 8. $c_n = 5(10)^{-n}$

9. $c_n = \dfrac{n - 1}{n + 1}$ 10. $c_n = \dfrac{(-1)^n}{n^2 + 1}$

11. $c_n = 43(0.01)^n$ 12. $c_n = 100(1.01)^n$

13. $c_n = \dfrac{(-1)^{n-1}}{n^3}$ 14. $c_n = \dfrac{(-1)^{n+1}}{n(n + 1)}$

15. $c_n = \dfrac{n(n + 1)(n + 2)}{3}$ 16. $c_n = \dfrac{n(n + 1)(2n + 1)}{6}$

17. $c_n = (-1)^{n+1}(3)^{-n}$

18. $c_n = \dfrac{1}{n} - \dfrac{1}{n+1}$

19. $c_n = \dfrac{(-x)^n}{n}$

20. $c_n = \dfrac{(-2)^{n-1}}{x^n}$

(21–40) Write the given series in expanded form. See Examples 4 and 5.

21. $\displaystyle\sum_{k=1}^{5} (3k - 4)$

22. $\displaystyle\sum_{k=1}^{6} (10 - 2k)$

23. $\displaystyle\sum_{k=1}^{4} (k^2 + k)$

24. $\displaystyle\sum_{k=2}^{7} (k^3 - k)$

25. $\displaystyle\sum_{i=2}^{6} 3^{-i}$

26. $\displaystyle\sum_{i=3}^{8} \dfrac{i}{i^2 - 1}$

27. $\displaystyle\sum_{k=1}^{\infty} 4\left(\dfrac{1}{5}\right)^k$

28. $\displaystyle\sum_{k=1}^{\infty} \dfrac{1}{k^2}$

29. $\displaystyle\sum_{k=1}^{\infty} 35(10)^{-2k}$

30. $\displaystyle\sum_{k=1}^{\infty} (-1)^k k^{-1/2}$

31. $\displaystyle\sum_{i=1}^{\infty} \dfrac{(-1)^i}{(i+1)(i+2)}$

32. $\displaystyle\sum_{i=1}^{\infty} 7(10)^{-i}$

33. $\displaystyle\sum_{j=2}^{6} (-1)^{j+1} j^{3/2}$

34. $\displaystyle\sum_{j=3}^{7} [(j+1)^2 - j^2]$

35. $\displaystyle\sum_{k=1}^{6} \dfrac{x^{2k}}{2k}$

36. $\displaystyle\sum_{k=1}^{5} \dfrac{x^k}{k(k+1)}$

37. $\displaystyle\sum_{k=1}^{\infty} \dfrac{(-1)^{k+1} x^k}{k^2}$

38. $\displaystyle\sum_{k=1}^{\infty} \dfrac{(-1)^{k+1} x^{3k}}{k^3}$

39. $\displaystyle\sum_{k=1}^{\infty} \dfrac{kx^k}{5^k}$

40. $\displaystyle\sum_{k=1}^{\infty} \dfrac{x^{-2k}}{k^2 + k}$

(41–54) Express each of the following in sigma notation. (More than one answer is possible. Check with an instructor if in doubt.)

41. $6 + 12 + 18 + 24 + 30$

42. $10 + 9 + 8 + 7 + 6 + 5$

43. $2 + 4 + 6 + 8 + 10 + \cdots$

44. $5 + 10 + 15 + 20 + 25 + \cdots$

45. $1 + \dfrac{1}{2} + \dfrac{1}{3} + \dfrac{1}{4} + \dfrac{1}{5} + \cdots$

46. $1 + 4 + 9 + 16 + 25 + \cdots$

47. $1 - \dfrac{1}{2} + \dfrac{1}{4} - \dfrac{1}{8} + \dfrac{1}{16} - \dfrac{1}{32} + \cdots$

48. $1 - \dfrac{1}{4} + \dfrac{1}{9} - \dfrac{1}{16} + \dfrac{1}{25} - \dfrac{1}{36} + \cdots$

49. $6 - 6^2 + 6^3 - 6^4 + \cdots$

50. $\dfrac{1}{2} + \dfrac{2}{3} + \dfrac{3}{4} + \dfrac{4}{5} + \dfrac{5}{6} + \dfrac{6}{7}$

51. $1 + \dfrac{1}{5} + \dfrac{1}{5^2} + \dfrac{1}{5^3} + \dfrac{1}{5^4}$

52. $1 - \sqrt[3]{4} + \sqrt[3]{9} - \sqrt[3]{16} + \sqrt[3]{25}$

53. $1 - \dfrac{1}{2} + \dfrac{3}{4} - \dfrac{7}{8} + \dfrac{15}{16} - \dfrac{31}{32}$

54. $5 - \dfrac{5}{3} + \dfrac{5}{9} - \dfrac{5}{27} + \dfrac{5}{81} - \dfrac{5}{243}$

10.2 MATHEMATICAL INDUCTION

An important property of the set of positive integers states that if 1 is in a set of positive integers and if $n + 1$ is in this set whenever n is, then the set consists of all the positive integers. Mathematical induction is a method of proof based on this property. It is especially useful for showing that an open statement is true for all values of a variable that are positive integers.

AXIOM OF MATHEMATICAL INDUCTION

Let S be a statement in the variable n; then S is true for all positive integral values of n if the following two conditions are satisfied:

1. S is true for $n = 1$.
2. If S is true for $n = k$ where k is any integer, then S is true for $n = k + 1$.

The so-called "Domino Principle" illustrates the Axiom of Mathematical Induction. Consider a row of dominos standing on edge as shown in Figure 10.1. All the dominos in the row will be knocked down if:

1. The first domino is knocked down, and
2. The distance d between any two adjacent dominos is less than the height h of a domino so that when one domino falls, it causes the adjacent domino to fall.

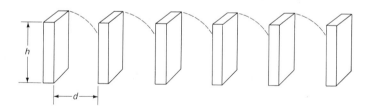

FIG. 10.1. Domino Principle.

It is important to note that a proof by mathematical induction consists of two parts: Part 1, the proof for $n = 1$, and Part 2, the proof for $n = k + 1$ assuming the statement is true for $n = k$.

EXAMPLE 1 Prove by mathematical induction that for all positive integers n,

$$1 + 2 + 3 + \cdots + n = \frac{n(n + 1)}{2}$$

(This statement provides a formula for the sum of the first n positive integers.)

Solution

Part 1 If $n = 1$, the statement becomes

$$1 = \frac{1(1 + 1)}{2} = \frac{2}{2} = 1$$

which is true.

Part 2 Assume for $n = k$ that

$$1 + 2 + 3 + \cdots + k = \frac{k(k + 1)}{2}$$

We need to prove that the given statement is true for $n = k + 1$; that is,

$$1 + 2 + 3 + \cdots + k + (k + 1) = \frac{(k + 1)(k + 2)}{2}$$

Adding $k + 1$ to each side of our assumed statement,

$$1 + 2 + 3 + \cdots + k + (k + 1) = \frac{k(k + 1)}{2} + (k + 1)$$

$$= \frac{k(k + 1) + 2(k + 1)}{2}$$

$$= \frac{k^2 + 3k + 2}{2}$$

$$= \frac{(k + 1)(k + 2)}{2}$$

Since the given statement is true for $n = 1$ and also true for $n = k + 1$ whenever it is true for $n = k$, the statement is true for all positive integers n.

EXAMPLE 2 Prove by mathematical induction that for all positive integers n,

$$a^n - b^n = (a - b)(a^{n-1} + a^{n-2}b + \cdots + ab^{n-2} + b^{n-1})$$

(In other words, show that $a - b$ is a factor of $a^n - b^n$ for all positive integers n.)

Solution

Part 1 If $n = 1$, the statement becomes $a - b = a - b$ which is true.

Before proceeding to Part 2, let us examine the special cases for $n = 2$, 3, and 4 in order to gain a better understanding of the statement.

If $n = 2$, $a^2 - b^2 = (a - b)(a + b)$.

If $n = 3$, $a^3 - b^3 = (a - b)(a^2 + ab + b^2)$.

If $n = 4$, $a^4 - b^4 = (a - b)(a^3 + a^2b + ab^2 + b^3)$.

Part 2 Assume for $n = k$ the statement is true and thus
$$a^k - b^k = (a - b)(a^{k-1} + a^{k-2}b + \cdots + ab^{k-2} + b^{k-1})$$
We now need to show that the given statement is true for $n = k + 1$. We start by adding and subtracting $a^k b$.
$$a^{k+1} - b^{k+1} = a^{k+1} - a^k b + a^k b - b^{k+1}$$
$$= a^k(a - b) + b(a^k - b^k)$$
Now using the statement we assumed true,
$$a^{k+1} - b^{k+1} = a^k(a - b) + b(a - b)(a^{k-1} + a^{k-2}b + \cdots + ab^{k-2} + b^{k-1})$$
$$= (a - b)(a^k + a^{k-1}b + \cdots + ab^{k-1} + b^k)$$
This result is the given statement for $n = k + 1$.

Combining the results of Part 1 and Part 2 we conclude the given statement is true for all positive integers n.

EXAMPLE 3 Prove by mathematical induction that for all positive integers n, and for $r \neq 1$,
$$\sum_{k=1}^{n} r^{k-1} = \frac{1 - r^n}{1 - r}$$
Solution
$$\sum_{k=1}^{n} r^{k-1} = 1 + r + r^2 + \cdots + r^{n-1}$$
Thus we want to prove
$$1 + r + r^2 + \cdots + r^{n-1} = \frac{1 - r^n}{1 - r}$$
Part 1 If $n = 1$, the statement becomes
$$1 = \frac{1 - r}{1 - r} = 1$$
which is true.
Part 2 Assuming the statement is true for $n = k$, then
$$1 + r + r^2 + \cdots + r^{k-1} = \frac{1 - r^k}{1 - r}$$
Adding r^k to each side,
$$1 + r + r^2 + \cdots + r^{k-1} + r^k = \frac{1 - r^k}{1 - r} + r^k$$
$$= \frac{1 - r^k + r^k(1 - r)}{1 - r}$$
$$= \frac{1 - r^{k+1}}{1 - r}$$
This is the form of the statement for $n = k + 1$.

Since Part 1 and Part 2 have been verified, the statement is true for all positive integers n.

EXERCISES 10.2

(1–18) Prove by mathematical induction that each statement is true for all integers n.

1. $1 + 3 + 5 + \cdots + (2n - 1) = n^2$

2. $2 + 4 + 6 + \cdots + 2n = n(n + 1)$

3. $1^2 + 2^2 + 3^2 + \cdots + n^2 = \dfrac{n(n + 1)(2n + 1)}{6}$

4. $1^3 + 2^3 + 3^3 + \cdots + n^3 = \dfrac{n^2(n + 1)^2}{4}$

5. $1 + 2 + 4 + 8 + \cdots + 2^{n-1} = 2^n - 1$

6. $1 + r + r^2 + \cdots + r^{n-1} = \dfrac{r^n - 1}{r - 1}$

7. $\dfrac{1}{1 \cdot 3} + \dfrac{1}{3 \cdot 5} + \dfrac{1}{5 \cdot 7} + \cdots + \dfrac{1}{(2n - 1)(2n + 1)} = \dfrac{n}{2n + 1}$

8. $\dfrac{1}{1 \cdot 2} + \dfrac{1}{2 \cdot 3} + \dfrac{1}{3 \cdot 4} + \cdots + \dfrac{1}{n(n + 1)} = \dfrac{n}{n + 1}$

9. $1 \cdot 2 + 2 \cdot 3 + 3 \cdot 4 + \cdots + n(n + 1) = \dfrac{n(n + 1)(n + 2)}{3}$

10. $1 \cdot 3 + 2 \cdot 4 + 3 \cdot 5 + \cdots + n(n + 2) = \dfrac{n(n + 1)(2n + 7)}{6}$

11. $\displaystyle\sum_{j=1}^{n} 3j = \dfrac{3n(n + 1)}{2}$

12. $\displaystyle\sum_{j=1}^{n} cj = \dfrac{cn(n + 1)}{2}$

13. $\displaystyle\sum_{k=1}^{n} (2k)^2 = \dfrac{2n(n + 1)(2n + 1)}{3}$

14. $\displaystyle\sum_{k=1}^{n} \left(\tfrac{1}{2}\right)^k = \dfrac{2n - 1}{2^n}$

15. $\displaystyle\sum_{k=1}^{n} 5^{-k} = \dfrac{5^n - 1}{4(5^n)}$

16. $\displaystyle\sum_{k=1}^{n} (a + kd) = \dfrac{n}{2}[2a + (n + 1)d]$

17. $\displaystyle\sum_{k=1}^{n} ar^{k-1} = \dfrac{a(1 - r^n)}{1 - r}$

18. $a + b$ is a factor of $a^{2n} - b^{2n}$
[Use $a^{2n+2} - b^{2n+2} = a^{2n}(a^2 - b^2) + b^2(a^{2n} - b^{2n})$.]

(19–23) Prove each of the listed exponent theorems where m and n are natural numbers and x and y are real numbers. Use the definition $x^1 = x$ and $x^n = xx^{n-1}$ for $n > 1$.

19. $x^m x^n = x^{m+n}$ **20.** $(x^m)^n = x^{mn}$ **21.** $(xy)^n = x^n y^n$

22. $\dfrac{x^m}{x^n} = x^{m-n}$ for $m > n$ and $x \neq 0$

23. $\left(\dfrac{x}{y}\right)^n = \dfrac{x^n}{y^n}$ for $y \neq 0$

(24–27) Prove each of these basic summation formulas.

24. $\displaystyle\sum_{k=1}^{n} 1 = n$

25. $\displaystyle\sum_{k=1}^{n} k = \frac{n(n+1)}{2}$

26. $\displaystyle\sum_{k=1}^{n} k^2 = \frac{n(n+1)(2n+1)}{6}$

27. $\displaystyle\sum_{k=1}^{n} k^3 = \frac{n^2(n+1)^2}{4}$

(28–35) Use one or more of the basic summation formulas in Exercises 24–27 and the properties

$$\sum_{k=1}^{n} (f(k) + g(k)) = \sum_{k=1}^{n} f(k) + \sum_{k=1}^{n} g(k)$$

$$\sum_{k=1}^{n} cf(k) = c \sum_{k=1}^{n} f(k)$$

to express each of the following in terms of n only.

EXAMPLE 4

$$\sum_{k=1}^{n} (2k^2 - k + 5)$$

Solution

$$\sum_{k=1}^{n} (2k^2 - k + 5) = 2 \sum_{k=1}^{n} k^2 - \sum_{k=1}^{n} k + 5 \sum_{k=1}^{n} 1$$

$$= \frac{2n(n+1)(2n+1)}{6} - \frac{n(n+1)}{2} + 5n$$

$$= \frac{4n^3 + 3n^2 + 29n}{6}$$

28. $\displaystyle\sum_{k=1}^{n} (k^2 + k)$

29. $\displaystyle\sum_{k=1}^{n} (3k^2 - 4)$

30. $\displaystyle\sum_{k=1}^{n} (k^2 - 2k + 3)$

31. $\displaystyle\sum_{k=1}^{n} (6k^2 + 5k - 4)$

32. $\displaystyle\sum_{k=1}^{n} (2k^2 - k - 1)$

33. $\displaystyle\sum_{k=1}^{n} (k^3 - k)$

34. $\displaystyle\sum_{k=1}^{n} (k^3 + k^2)$

35. $\displaystyle\sum_{k=1}^{n} (k^3 + k^2 + k)$

10.3 ARITHMETIC PROGRESSIONS

The sequence 5, 9, 13, 17, 21, 25 is an example of an **arithmetic sequence,** also called an **arithmetic progression.** It has the property that each term, except the first, can be obtained from the previous term by adding the constant 4.

0.3 ARITHMETIC PROGRESSIONS

DEFINITION

An **arithmetic progression** is a sequence in which each term, except the first, can be obtained from the previous term by adding a constant d, called the **common difference.**

If we let a represent the first term, then an arithmetic progression can be written as follows:

$a, a + d, a + 2d, a + 3d, a + 4d, \cdots$

The nth term, or general term, is c_n where

$c_n = a + (n - 1)d$

This statement can be verified by using mathematical induction. Note for $n = 1$, $c_1 = a$. Now if $c_k = a + (k - 1)d$, then

$c_{k+1} = a + (k - 1)d + d = a + (k + 1 - 1)d$

Thus the statement is true for all positive integers n.

EXAMPLE 1 Find the 35th term of the arithmetic progression 2, 5, 8, 11, 14, ...

Solution

$a = 2$ and $n = 35$

The common difference d is obtained by subtracting any term from the next term:

$d = 5 - 2 = 3$

Also, $d = 8 - 5 = 11 - 8$, and so on.

Now using

$c_n = a + (n - 1)d$

$c_{35} = 2 + (34)(3) = 104$

DEFINITION

The **arithmetic means** between two given numbers are the terms of an arithmetic progression for which the given numbers are the first and last terms.

EXAMPLE 2 Insert three arithmetic means between 4 and 20.

Solution   The arithmetic progression has the form

$4, c_2, c_3, c_4, 20$

Using $c_n = a + (n - 1)d$, where $c_5 = 20$,

$20 = 4 + (5 - 1)d$ and $d = 4$

Then

$c_2 = 4 + 4 = 8$, $c_3 = 8 + 4 = 12$, $c_4 = 12 + 4 = 16$

The arithmetic means are 8, 12, 16.

If one arithmetic mean is inserted between two numbers *a* and *b,* then this mean is called the **average** or the **arithmetic mean** of the two numbers.

$$\text{Arithmetic mean of } (a \text{ and } b) = \frac{a + b}{2}.$$

Using $c_n = a + (n - 1)d$, note that

$$b = a + (3 - 1)d \quad \text{and} \quad d = \frac{b - a}{2}$$

$$a + \frac{b - a}{2} = \frac{2a + b - a}{2} = \frac{a + b}{2}$$

DEFINITION

The **sum** S_n **of an arithmetic progression** having *n* terms is the sum of the *n* terms of its corresponding series.

In symbols,

$$S_n = a + (a + d) + (a + 2d) + \cdots + [a + (n - 1)d]$$

To obtain a formula for S_n, we express S_n in an alternate form:

$$S_n = c_n + (c_n - d) + (c_n - 2d) + \cdots + [c_n - (n - 1)d]$$

Adding the two equations expressing S_n,

$$2S_n = (a + c_n) + (a + c_n) + \cdots + (a + c_n)$$

where there are *n* terms of the form $a + c_n$.

$$2S_n = n(a + c_n)$$

$$S_n = \frac{n(a + c_n)}{2} \qquad (1)$$

Replacing c_n by $a + (n - 1)d$, then

$$S_n = \frac{n}{2}[2a + (n - 1)d] \qquad (2)$$

Either Equation (1) or Equation (2) can be used to find the sum.

To verify that the formulas for the sum are valid, we will prove, by using mathematical induction, that

$$S_n = \frac{n}{2}[2a + (n - 1)d]$$

> *Part 1* $n = 1$.
>
> $$S_1 = \frac{1}{2}(2a + 0d) = a$$
>
> *Part 2* Assume
>
> $$S_k = \frac{k}{2}[2a + (k - 1)d]$$
>
> and prove that
>
> $$S_{k+1} = \frac{k + 1}{2}(2a + kd)$$

Adding
$$c_{k+1} = a + kd \text{ to } S_k,$$

$$S_{k+1} = c_{k+1} + S_k = a + kd + \frac{k}{2}[2a + (k-1)d]$$

$$= a + kd + \frac{2ka + k(k-1)d}{2}$$

$$= \frac{2a + 2kd + 2ka + k^2d - kd}{2}$$

$$= \frac{2ak + 2a}{2} + \frac{k^2d + kd}{2}$$

$$= \frac{(k+1)2a}{2} + \frac{(k+1)kd}{2}$$

$$= \frac{k+1}{2}(2a + kd)$$

EXAMPLE 3 Find the sum of the first 12 terms of the arithmetic progression 1, 3, 5, 7, ...

Solution
$$a = 1, \quad d = 2, \quad \text{and} \quad n = 12$$
$$c_n = a + (n-1)d$$
$$c_{12} = 1 + (12-1)(2) = 23$$
$$S_n = \frac{n}{2}(a + c_n)$$
$$S_{12} = \frac{12}{2}(1 + 23) = 144$$

EXAMPLE 4 Find the sum of the first 100 positive integers.

Solution $1 + 2 + 3 + \cdots + 99 + 100$ is an arithmetic series with $a = 1$, $d = 1$, $n = 100$, $c_n = 100$.
$$S_n = \frac{n}{2}(a + c_n)$$
$$S_{100} = \frac{100}{2}(1 + 100) = 50(101) = 5050$$

EXAMPLE 5 Find the sum of the arithmetic series
$$1 + 3 + 5 + \cdots + 39$$

Solution
$$a = 1, d = 2, \text{ and } c_n = 39$$
$$c_n = a + (n-1)d$$
$$39 = 1 + (n-1)2, \qquad 2(n-1) = 38, \qquad n = 20$$
$$S_n = \frac{n(a + c_n)}{2}$$
$$= \frac{20(1 + 39)}{2} = 400$$

EXERCISES 10.3

(1–8) Write the terms of the arithmetic progression for the given values where a is the first term, d is the common difference, and n is the number of terms.

1. $a = 3$, $d = 5$, $n = 4$
2. $a = 6$, $d = 4$, $n = 5$
3. $a = -10$, $d = 6$, $n = 6$
4. $a = -8$, $d = 2$, $n = 6$
5. $a = 12$, $d = -2$, $n = 5$
6. $a = 15$, $d = -3$, $n = 4$
7. $a = 19$, $d = -4$, $n = 5$
8. $a = -9$, $d = -5$, $n = 6$

(9–18) Find the indicated term of each of the following arithmetic progressions.

9. 40th term of 1, 3, 5, 7, ...
10. 20th term of 3, 8, 13, 18, ...
11. 51st term of 87, 81, 75, ...
12. 12th term of 9, 3, −3, ...
13. 39th term of 2, 2.5, 3, ...
14. 100th term of $\frac{1}{4}, \frac{1}{2}, \frac{3}{4}$, ...
15. 17th term of $10 + \sqrt{2}$, 10, $10 - \sqrt{2}$, ...
16. 16th term of $2\sqrt{5}, 4\sqrt{5}, 6\sqrt{5}$, ...
17. 15th term of $x - k$, x, $x + k$, ...
18. 21st term of $3x + k$, $3x$, $3x - k$, ...

(19–28) Insert the indicated number of arithmetic means between the two given numbers.

19. One between 75 and 83
20. One between 87 and 95
21. Three between 10 and −10
22. Three between 80 and 90
23. Two between 45 and 54
24. Two between −8 and 4
25. Four between 1 and 2
26. Five between 24 and −6
27. Seven between 0 and 1
28. Six between 2 and 16

(29–46) Find the sum of each arithmetic series.

29. $2 + 4 + 6 + \cdots + 100$
30. $1 + 2 + 3 + \cdots + 200$
31. $12 + 8 + 4 + \cdots + (-20)$
32. $30 + 29.5 + 29 + \cdots + 5$
33. $3 + 6 + 9 + \cdots + 3n$
34. $1 + 4 + 10 + \cdots + (3n - 2)$
35. $6 + 10 + 14 + \cdots + (4n + 2)$
36. $5 + 10 + 15 + \cdots + 5n$

37. $\sum\limits_{k=1}^{50} \dfrac{k+1}{2}$

38. $\sum\limits_{k=1}^{10} (5k-3)$

39. $\sum\limits_{i=1}^{20} (2i-1)$

40. $\sum\limits_{i=1}^{40} (i+10)$

41. $\sum\limits_{j=51}^{250} j$

42. $\sum\limits_{j=15}^{60} 2j$

43. $\sum\limits_{i=1}^{n} (2-3i)$

44. $\sum\limits_{i=1}^{n} (3i+2)$

45. $\sum\limits_{k=1}^{n} k$

46. $\sum\limits_{k=1}^{n} (2k-1)$

47. How many multiples of 7 are between 10 and 100? (*Hint:* $a = 14$, $d = 7$. Find c_n and n.)

48. How many multiples of 11 are between 100 and 200? (*Hint:* $a = 110$ and $d = 11$. Find c_n and n.)

49. A man contributes to his savings fund by putting in $1000 the first year and by increasing his contribution by $50 each year thereafter. How much has he saved at the end of 10 years?

50. A debt of $20,000 is paid by paying $100 at the end of each month plus 1% interest on the amount unpaid at the end of that month. What is the total payment?

51. The fare charged by a certain taxicab company is 60 cents for the first $\frac{1}{4}$ mile and 30 cents for each $\frac{1}{4}$ mile thereafter. What is the fare from the center of a city to an airport 10 miles away?

52. When a body falls from rest in a vacuum, it falls 16 feet the first second and 32 feet each second thereafter. How far does the body fall in 20 seconds? How far in k seconds?

53. A person is offered two jobs. Job 1 has a beginning salary of $9000 per year and an $800 raise at the end of each year. Job 2 has a beginning salary of $4500 for the first 6 months and a raise of $200 at the end of each 6 months. Find the salary for each job for the sixth year. Which job pays the most?

54. A grocer wants to display some cans in the shape of a pyramid with one can on top, 3 cans in the row beneath, 5 cans in the next row, and so on. How many cans should be in the bottom row if 64 cans are to be displayed?

55. A certain automobile depreciated in value by $300 a year. If the original cost was $4000 when new, find the value of the automobile when it was 8 years old.

56. In 1970 the value of a certain house was $30,000. If the value of the house since 1970 has been increasing by $500 a year, find the value of the house in 1990.

57. If the total number of a certain species of bird is 348 today, and if the number is decreasing by 15 a year, what will be the total number of these birds 20 years from now?

58. If the temperature decreases by 4°F for each 1000-foot increase in elevation, find the temperature at 9000 feet if it is 70°F at sea level.

59. The cost of digging a well is $5 for the first foot and then each foot thereafter costs 25 cents more than the previous foot.
    (a) Find the cost of digging a well 250 feet deep.
    (b) How deep is a well that cost $1767.50 to dig?
60. A trellis is to be made consisting of 15 slats of wood, varying uniformly from 16 inches at the bottom to 9 inches at the top. What length of wood will be needed to make the slats?

10.4 GEOMETRIC PROGRESSIONS

The sequence
    4, 20, 100, 500, 2500
is an example of a geometric sequence, also called a geometric progression. It has the property that each term after the first can be obtained from the preceding term by multiplying by the constant 5.

DEFINITION

A **geometric progression** is a sequence in which each term after the first can be obtained from the preceding term by multiplying by a constant r, called the **common ratio.**

If we let a represent the first term, then a geometric progression can be written as follows:
    $a, ar, ar^2, ar^3, ar^4, \ldots$

THEOREM

The **nth term,** or **general term** of a geometric progression, is c_n where
$c_n = ar^{n-1}$

This statement can be verified by using mathematical induction. Note for $n = 1$, $c_1 = ar^{1-1} = ar^0 = a$. Now if $c_k = ar^{k-1}$, then $c_{k+1} = ar^{k-1}r$ and $c_{k+1} = ar^k$. For $n = k + 1$, the formula $c_n = ar^{n-1}$ becomes $c_{k+1} = ar^k$. So the formula is verified.

EXAMPLE 1 Find the sixth term of the geometric progression 3, 15, 75, ...

Solution

$a = 3$ and $r = \frac{15}{3} = \frac{75}{15} = 5$

Using $c_n = ar^{n-1}$ with $n = 6$,
$c_6 = 3(5^5) = 9375$

DEFINITION

The **geometric means** between two numbers are the terms of a geometric progression for which the given numbers are the first and last terms.

EXAMPLE 2   Insert two geometric means between 1 and 64.

Solution   The geometric progression has the form

$$1, c_2, c_3, 64, \quad \text{where } a = 1$$

Using

$$c_n = ar^{n-1}$$
$$64 = 1 \cdot r^{4-1}, \quad r^3 = 64, \quad r = 4$$

then

$$c_2 = 1 \cdot 4 = 4 \quad \text{and} \quad c_3 = 4 \cdot 4 = 16$$

The two geometric means are 4 and 16.

EXAMPLE 3   Insert three real geometric means between 3 and 48.

Solution   The geometric progression is

$$3, c_2, c_3, c_4, 48.$$
$$a = 3 \quad \text{and} \quad c_5 = 3r^4 = 48, \quad \text{or} \quad r^4 = 16$$

Thus

$$r^2 = \pm 4 \quad \text{and} \quad r = \pm 2, \quad \text{or} \quad r = \pm 2i$$

For the means to be real, $r = 2$ or $r = -2$. There are two possibilities: The real means are 6, 12, 24 or -6, 12, -24.

If one geometric mean m is to be inserted between two numbers a and b, then the sequence is

$$a, m, b$$

Equating ratios,

$$\frac{m}{a} = \frac{b}{m}$$

and

$$m^2 = ab$$

and

$$m = \sqrt{ab} \quad \text{or} \quad m = -\sqrt{ab}$$

If a and b are positive numbers, we want their geometric mean to be positive so that m is between a and b. Also if a and b are negative, we want m to be negative. Therefore we make the following definition.

DEFINITION

The **geometric mean** of the real numbers a and b is $\sqrt{ab}$ when a and b are positive and $-\sqrt{ab}$ when a and b are negative.

EXAMPLE 4   Find the geometric mean of
(a) 4 and 9
(b) -5 and -20

Solution
(a) $m = \sqrt{4 \cdot 9} = \sqrt{36} = 6$
(b) $m = -\sqrt{(-5)(-20)} = -\sqrt{100} = -10$

DEFINITION

The **sum S_n of a geometric progression** is the sum of the terms of the series associated with the geometric sequence. In symbols,
$$S_n = a + ar + ar^2 + ar^3 + \cdots + ar^{n-1}$$

If each term of the indicated sum is multiplied by $-r$, then
$$-rS_n = -ar - ar^2 - ar^3 - \cdots - ar^{n-1} - ar^n$$
Adding these two equations,
$$S_n - rS_n = a - ar^n$$
$$S_n(1 - r) = a(1 - r^n)$$
Now, if $r \neq 1$, then
$$S_n = \frac{a(1 - r^n)}{1 - r} \quad \text{and} \quad S_n = \frac{a(r^n - 1)}{r - 1}$$

THEOREM

The sum S_n of a geometric progression, where $r \neq 1$, is given by
$$S_n = \frac{a(r^n - 1)}{r - 1}$$

Proof by mathematical induction on n.
Part 1   For $n = 1$,
$$S_1 = \frac{a(r^1 - 1)}{r - 1} = a$$
Part 2   Assume, for $n = k$,
$$S_k = \frac{a(r^k - 1)}{r - 1}$$
We have to prove, for $n = k + 1$,
$$S_{k+1} = \frac{a(r^{k+1} - 1)}{r - 1}$$
Now
$$\begin{aligned}
S_{k+1} &= c_{k+1} + S_k \\
&= ar^k + \frac{a(r^k - 1)}{r - 1} \\
&= \frac{ar^k(r - 1) + a(r^k - 1)}{r - 1} \\
&= \frac{ar^{k+1} - a}{r - 1} \\
&= \frac{a(r^{k+1} - 1)}{r - 1}
\end{aligned}$$

EXAMPLE 5 Find the sum of the geometric progression
5, 10, $\cdots$, 320

Solution

$a = 5$ and $r = \frac{10}{5} = 2$

$c_n = ar^{n-1} = 5(2)^{n-1} = 320$

$2^{n-1} = 64 = 2^6$

$n - 1 = 6$ and $n = 7$

$S_7 = \dfrac{a(r^7 - 1)}{r - 1} = \dfrac{5(2^7 - 1)}{2 - 1} = 5(128 - 1) = 5(127) = 635$

An **infinite geometric series** is one that has the form

$a + ar + ar^2 + ar^3 + \cdots$

which can also be written as

$$\sum_{k=1}^{\infty} ar^{k-1}$$

Now the sum of a finite geometric series,

$S_n = \dfrac{a - ar^n}{1 - r}$

can also be written as

$S_n = \dfrac{a}{1 - r}(1 - r^n)$

If $|r| < 1$ (or $-1 < r < 1$), then r^n becomes smaller and smaller as n becomes larger and larger. For example, let $r = \frac{1}{10} = 0.1$. Then

$r^2 = (0.1)^2 = 0.01$

$r^3 = (0.1)^3 = 0.001$

$r^4 = (0.1)^4 = 0.0001$

$\vdots$

$r^9 = (0.1)^9 = 0.000000001$

By taking n large enough, r^n can be made as close to 0 as one wants. Then $1 - r^n$ will be as close to 1 as one wants and, as a result, S_n can be made as close to $a/(1 - r)$ as one wants. Therefore S, the sum of an infinite geometric series with $|r| < 1$ is defined as this value, $a/(1 - r)$.

DEFINITION

If $|r| < 1$ and $S = a + ar + ar^2 + \cdots$, then

$S = \dfrac{a}{1 - r}$

A more precise way of stating this result symbolically is as follows:

$\lim_{n \to \infty} S_n = \dfrac{a}{1 - r}$

which is read "the limit of S_n as n increases without bound is $a/(1 - r)$."

EXAMPLE 6 Find the sum of the infinite geometric series
$12, 3, \frac{3}{4}, \ldots$

Solution

$a = 12, \qquad r = \frac{1}{4}$

$S = \dfrac{a}{1 - r} = \dfrac{12}{1 - \frac{1}{4}} = \dfrac{12(4)}{4 - 1} = \dfrac{48}{3} = 16$

EXAMPLE 7 Express the repeating decimal fraction $2.363636\ldots$ as a common fraction.

Solution $2.363636\ldots$
$= 2 + 0.36 + 0.0036 + 0.000036 + \cdots$
Note that $0.36 + 0.0036 + 0.000036 + \cdots$ is an infinite geometric series with $a = 0.36$ and $r = 0.01$.

$S = \dfrac{a}{1 - r} = \dfrac{0.36}{1 - 0.01} = \dfrac{36}{100 - 1} = \dfrac{36}{99} = \dfrac{4}{11}$

Thus $2.363636\ldots = 2 + \frac{4}{11} = \frac{26}{11}$.

EXERCISES 10.4

(1–8) Write the terms of the geometric progression for the given values where a is the first term, r is the common ratio, and n is the number of terms.

1. $a = 5, r = 2, n = 6$ **2.** $a = 4, r = 3, n = 5$

3. $a = 6, r = \frac{1}{3}, n = 5$ **4.** $a = 1, r = \frac{2}{5}, n = 6$

5. $a = 1, r = -4, n = 4$ **6.** $a = 3, r = -5, n = 4$

7. $a = 3, r = -\frac{1}{2}, n = 6$ **8.** $a = 5, r = -\frac{1}{4}, n = 5$

(9–18) Find the indicated term of each of the following geometric progressions.

9. Fifth term of $1, 7, 49 \ldots$
10. Eighth term of $6, 12, 24, \ldots$
11. Tenth term of $-1, 2, -4, \ldots$
12. Ninth term of $2, -10, 50, \ldots$
13. Fifteenth term of $9, 0.9, 0.09, \ldots$
14. Sixth term of $0.25, 0.0025, 0.000025, \ldots$
15. Eighth term of $18, 12, 8, \ldots$
16. Tenth term of $50, 20, 8, \ldots$
17. Sixth term of $4, 2\sqrt{2}, 2, \ldots$
18. Seventh term of $2, -\sqrt{6}, 3\sqrt{2}$

(19–24) Insert the indicated number of geometric means between the two given numbers.

19. Two between 54 and 16
20. Two between 3 and 34
21. Three between 2 and 162
22. Three between 81 and 16
23. Two between 4 and 20
24. Two between 5 and 10

(25–32) Find the geometric mean of the two given numbers.

25. 3 and 75
27. -8 and -50
29. 20 and 30
31. -4 and -24

26. 6 and 24
28. -7 and -63
30. 15 and 20
32. -2 and -45

(33–44) Find the sum of each geometric series.

33. $1 + 7 + \cdots + 7^5$
35. $3 + 6 + 12 + \cdots + 96$
37. $75 - 15 + 3 + \cdots + 0.0048$
38. $2 - 12 + \cdots + 2592$

34. $10 + 5 + \cdots + 10(0.5)^6$
36. $25 + 10 + \cdots + 0.256$

39. $\displaystyle\sum_{k=1}^{8} 2(3^k)$

40. $\displaystyle\sum_{k=1}^{6} 5(3^{-k})$

41. $\displaystyle\sum_{i=2}^{7} 6\left(-\frac{2}{3}\right)^i$

42. $\displaystyle\sum_{i=2}^{8} 3(-2)^i$

43. $\displaystyle\sum_{j=1}^{5} 100(1.01)^j$

44. $\displaystyle\sum_{j=1}^{6} 100(1.01)^{-j}$

(45–56) Find the sum of each infinite geometric series.

45. $1 + \frac{1}{2} + \frac{1}{4} + \cdots$

46. $1 + \frac{2}{5} + \frac{4}{25} + \cdots$

47. $1 - \frac{2}{3} + \frac{4}{9} - \frac{8}{27} + \cdots$

48. $1 - \frac{3}{4} + \frac{9}{16} - \frac{81}{64} + \cdots$

49. $10 + 2 + 0.4 + \cdots$

50. $-72 + 12 - 2 + \cdots$

51. $\frac{1}{75} - \frac{1}{90} + \frac{1}{108} - \cdots$

52. $\frac{1}{54} - \frac{1}{36} + \frac{1}{24} - \cdots$

53. $\displaystyle\sum_{k=1}^{\infty} 100\left(\frac{3}{4}\right)^k$

54. $\displaystyle\sum_{k=1}^{\infty} 32\left(-\frac{5}{8}\right)^k$

55. $\displaystyle\sum_{j=1}^{\infty} (1.04)^{-j}$

56. $\displaystyle\sum_{j=1}^{\infty} 1000(1.05)^{-j}$

(57–66) Express each repeating decimal as a common fraction.

57. 0.444...
59. 2.5454...
61. 0.8080...
63. 1.2555...
65. 6.132132...

58. 5.666...
60. 24.2525...
62. 0.0909...
64. 0.5222...
66. 0.4522522...

67. Each path length of the bob of a swinging pendulum is 95% of the preceding arc length. If its initial arc length is 18 inches, find the total distance the bob travels before the pendulum comes to rest.

68. On each rebound, a certain ball bounces back $\frac{5}{8}$ of its preceding height. If the ball is dropped from a height of 4 feet, find the total distance it travels before coming to rest.

69. How many ancestors has an individual had in the twelve generations preceding him if it is assumed that there were no intermarriages?

70. When the oil used in operating certain machinery is refined so that it can be used again, 25% is lost each time it is refined. If there is originally 200 gallons of oil, which is refined each time it becomes dirty, find approximately the total amount used in operating the machinery before all of the oil is lost.

71. A piece of equipment costing $12,500 depreciates by 20% of its value each year. At the end of each year the amount the equipment has depreciated is placed in a fund. In how many years will the fund contain $8404?

72. If $100 is deposited at the end of each month in a savings fund and if $\frac{1}{2}$% interest is paid on the money in the fund each month, how much money is in the fund at the end of one year?

73. If an air pump removes 60% of the air in a container with each stroke, find the percentage of air left in the container after four strokes. How many strokes would it require to reduce the air to less than $\frac{1}{2}$%?

74. In a certain culture, a bacterium divides into two bacteria every 30 minutes. If there were 50 bacteria in the culture at the beginning, how many bacteria are there at the end of 3 hours?

75. In the "double-up" system of betting, a person doubles his next bet each time he loses. If, using this system, a person's bets have been $1, $2, $4, and so on, how much must his next bet be if he loses 8 consecutive times?

76. If the inflation rate is 5% per year, what will be the rental price of an apartment 5 years from now if the rental price now is $200 per month?

10.5 CHAPTER REVIEW

(1–4) Write the first five terms of the sequence whose general term is given. (10.1)

1. $c_n = n^2 - n$

2. $c_n = (-1)^{n-1}\left(\frac{3}{2}\right)^n$

3. $c_n = 2n^{-3/2}$

4. $c_n = (1 + 2 + 3 + \cdots + n)$

(5–6) Write each series in expanded form. (10.1)

5. $\displaystyle\sum_{k=1}^{6} \frac{3}{2k^2 - 1}$

6. $\displaystyle\sum_{i=1}^{\infty} \frac{(-1)^{i+1}x^{2i}}{(2i)^2}$

(7–8) Prove by mathematical induction that each statement is true for all integers n. (10.2)

7. $1 + 4 + 7 + \cdots + (3n - 2) = \dfrac{n(3n - 1)}{2}$

8. $\displaystyle\sum_{k=1}^{n} (k^2 - k) = \frac{n(n^2 - 1)}{3}$

(9–10) Find the indicated term of the arithmetic progression. (10.3)

9. 26th term of 4, 7, 10, 13, …

10. 41st term of 0.125, 0.250, 0.375, …

(11–13) Insert the indicated number of arithmetic means. (10.3)

11. One between 85 and 97

12. Two between 0 and 1

13. Three between -6 and 4

(14–15) Find the sum of the arithmetic series. (10.3)

14. $51 + 55 + 59 + \cdots + 99$

15. $6 + 4 + 2 + \cdots + (8 - 2n)$

(16–17) Find the indicated term of the geometric progression. (10.4)

16. Seventh term of 16, 24, 36, …

17. Tenth term of 81, -54, 36, …

(18–19) Insert the indicated number of geometric means. (10.4)

18. Two between 500 and 4

19. Three between 16 and 625

(20–21) Find the geometric mean between the two given numbers. (10.4)

20. 5 and 30

21. -8 and -98

(22–25) Find the sum of each geometric series. (10.4)

22. $2 + 12 + 72 + \cdots 2(6^7)$

23. $8 + 4\sqrt{2} + 4 + \cdots + 1$

24. $1 + \frac{3}{4} + \frac{9}{16} + \cdots$

25. $9 - 6 + 4 - \cdots$

26. Express 5.303030 … as a common fraction. (10.4)

11 COUNTING, BINOMIAL THEOREM, PROBABILITY

There are many applications in business, economics, science, and in other areas involving statistics and probability, where it is necessary to determine the number of ways one or more events can occur.

First let us consider the number of ways a student can select one science course and one foreign language course if the student can select any of the 3 science courses of chemistry, biology, physics, and any of the 4 language courses of French, Spanish, German, and Chinese. We can construct a diagram such as that shown in Figure 11.1.

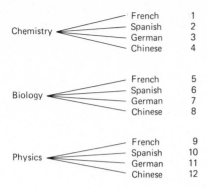

Chemistry	French	1
	Spanish	2
	German	3
	Chinese	4
Biology	French	5
	Spanish	6
	German	7
	Chinese	8
Physics	French	9
	Spanish	10
	German	11
	Chinese	12

FIG. 11.1

By counting, we see there are 12 ways in which the student can make the selection. Note also that $12 = 3 \cdot 4$. This special case illustrates the general principle stated as the following axiom.

AXIOM FUNDAMENTAL PRINCIPLE OF COUNTING

If one event can happen in n ways, and if, after this event has happened, a second event can happen in k ways, then nk is the number of ways both events can happen, in the indicated order.

EXAMPLE 1 In how many ways can 1 shirt and 1 pair of slacks be selected from 6 shirts and 5 pairs of slacks?

Solution Using the Fundamental Principle of Counting, the number of ways is the product of 6 and 5, $6 \cdot 5 = 30$. There are 30 ways.

The Fundamental Principle of Counting can be extended to more than two events. If the first event can happen in n_1 ways, the second in n_2 ways, ..., the kth in n_k ways, then the number of ways in which the k events can occur in this order is

$n_1 n_2 \ldots n_k$

EXAMPLE 2 In how many ways can a dinner consisting of 1 soup, 1 salad, 1 entree, and 1 dessert be selected from 3 soups, 4 salads, 8 entrees, and 5 desserts?

Solution
$3 \cdot 4 \cdot 8 \cdot 5 = 480$
There are 480 ways.

EXAMPLE 3   In how many different orders can the 6 teams of the Eastern division of the National League complete the baseball season?

Solution   If we use 6 horizontal lines,

_____, _____ _____ _____ _____ _____, to indicate from left to right, first place, then second place, and so on, we see there are 6 different ways for a team to be in first place, then 5 different ways for second place, and so on. Thus the number of different orders is

$6 \cdot 5 \cdot 4 \cdot 3 \cdot 2 \cdot 1 = 720$

There is a special symbol for the product of the first n natural numbers, namely, $n!$, which is read "n factorial."

DEFINITION OF n FACTORIAL, $n!$

$1! = 1, \qquad n! = n(n-1)!, \qquad 0! = 1$

We define 0! as 1 for convenience in using formulas. Note that

$5! = 5(4)!$
$\quad = 5 \cdot 4 \cdot (3!)$
$\quad = 5 \cdot 4 \cdot 3 \cdot (2!)$
$\quad = 5 \cdot 4 \cdot 3 \cdot 2 \cdot 1 = 120$

EXAMPLE 4   How many 3-digit numbers can be formed from the digits 1, 3, 5, 7, 9 if no two of the digits selected are equal?

Solution   There are 5 ways for the first digit to be selected, then 4 ways for the second, and then 3 ways for the third. The number of 3-digit numbers is

$5 \cdot 4 \cdot 3 = 60$

Example 3 illustrates the number of permutations of 6 things taken 6 at a time and Example 4 illustrates the number of permutations of 5 things taken 3 at a time.

DEFINITION

A **permutation** is an ordered arrangement of all or some of the elements of a set.

In general, we have the following theorem.

THEOREM

Let $P(n, r)$ be the number of permutations of n elements taken r at a time. Then

$P(n, r) = \dfrac{n!}{(n-r)!}$

and

$P(n, n) = n!$

Proof

$$P(n, r) = n(n - 1)(n - 2) \ldots (n - r + 1)$$
$$= \frac{n(n - 1)(n - 2) \ldots (n - r + 1) \cdot (n - r)!}{(n - r)!}$$
$$= \frac{n!}{(n - r)!}$$

Other notations for $P(n, r)$ are $_nP_r$, $P_{n,r}$, and $P_r{}^n$.

EXAMPLE 5   In how many ways can 5 different prizes be given to 100 people if no person can receive more than one prize?

> Solution   This is the number of permutations of 100 elements taken 5 at a time.
> $$P(100, 5) = \frac{100!}{(100 - 5)!} = \frac{100!}{95!}$$
> $$= 100(99)(98)(97)(96)$$
> $$= 9,034,502,400$$

EXAMPLE 6   In how many ways can 6 persons be seated around a circular table?

> Solution   Having placed 1 person, then the number of ways the remaining 5 persons can be seated relative to the first is $P(5, 5) = 5! = 120$.

In general, the number of ways in which n elements can be arranged in a circle is $(n - 1)!$.

EXERCISES 11.1

(1–16)   Find the value of each of the following.

1. $4!$
2. $5!$
3. $6!$
4. $7!$
5. $\dfrac{10!}{6!}$
6. $\dfrac{12!}{9!}$
7. $\dfrac{8!}{3!}$
8. $\dfrac{20!}{15!}$
9. $P(6, 6)$
10. $P(8, 8)$
11. $P(6, 2)$
12. $P(8, 4)$
13. $P(9, 2)$
14. $P(16, 1)$
15. $P(30, 2)$
16. $P(20, 3)$
17. In how many ways can 1 beverage and 1 sandwich be selected from 5 different beverages and 8 different sandwiches?
18. In how many ways can 1 magazine and 1 newspaper be selected from 20 magazines and 4 newspapers?
19. How many different looking homes are possible if there is a choice of 5 different roofs, 4 different garage arrangements, and 3 different types of entrances?
20. If there are 3 candidates for president, 4 candidates for secretary, and 6 candidates for treasurer, in how many ways can these offices be filled?

21. A set of signal flags consists of 6 flags, each a different color. How many displays of 4 flags can be made on a vertical staff?

22. In how many ways can 5 pictures be displayed in a row on a wall if the selection is to be made from 10 different pictures?

23. Five novels and 3 textbooks are to be arranged on a shelf so that the novels are together and the textbooks are together. How many arrangements are possible?

24. Four adults and 5 children are to be seated in a row so that each adult is between 2 children. How many arrangements are possible?

25. A newspaper awards a prize each week during football season to the person who has selected the winner or tie game in each of 6 listed games. In how many different ways can the entry blank be filled in?

26. A committee of 4 members is to be formed from 5 Americans, 4 Russians, 4 Englishmen, and 3 Frenchmen. The committee is to consist of 1 American, 1 Russian, 1 Englishman, and 1 Frenchman. How many different committees can be formed?

27. In an artificial language, 3-letter words are to be formed from the consonants T, R, N, S, D, L and from the vowels A, E, I, O. How many words are possible if a vowel must be placed between two consonants?

28. If 6 horses are entered in a race, in how many ways can 3 horses finish first, second, and third?

29. In how many ways can 7 persons be seated around a circular dining table?

30. In how many ways can 6 persons be seated around a circular table if 2 persons insist on sitting together?

31. In how many ways can 8 different plants be arranged around the border of a circular plot of land?

32. In how many ways can 8 dancers, 4 men and 4 women, be arranged in a circle if each woman is between two men?

33. In how many ways can a tourist visit 4 of 10 foreign countries?

34. How many batting orders are possible for a baseball team of 9 men if the pitcher bats last?

11.2 COMBINATIONS

In the previous section we dealt with permutations, ordered arrangements of some or all of the elements of a set. For some problems, the order is not important and we are concerned with only what elements are in the set.

For example, how many committees of 3 members each can be formed from 8 persons? If we consider the order in which the members are chosen, then the number of ways the selections can be made is

$$P(8, 3) = 8 \cdot 7 \cdot 6 = 336$$

However, we consider the committee $\{A, B, C\}$ to be the same as each of the following:

$$\{A, C, B\}, \{B, C, A\}, \{B, A, C\}, \{C, A, B\}, \{C, B, A\}$$

In other words, the same committee has been counted 6 times where 6 is the number of permutations of 3 elements taken 3 at a time.

Then the number of committees possible is

$$\frac{P(8, 3)}{P(3, 3)} = \frac{336}{6} = 56$$

The number 56 is the number of combinations of 8 elements taken 3 at a time.

DEFINITION

A **combination** of r objects of a set S is a subset of S that contains r distinct objects.

DEFINITION

The **number of combinations of n elements taken r at a time** is the number of sets of objects containing exactly r elements that can be formed from a set containing exactly n elements. The **number of combinations** of n elements taken r at a time is expressed in symbols as $C(n, r)$. Other notations used are $_nC_r$, $C_{n,r}$, C_r^n, and $\left(\dfrac{n}{r}\right)$.

Since there are $r!$ permutations for a combination of r elements,

$$C(n, r) = \frac{P(n, r)}{r!}$$

Since

$$P(n, r) = \frac{n!}{(n - r)!}$$

we can also write

$$C(n, r) = \frac{n!}{r!(n - r)!}$$

THEOREM

For $n \geq r$,

$$C(n, r) = \frac{n!}{r!(n - r)!}$$

where $C(n, r)$ is the number of combinations of n elements taken r at a time.

EXAMPLE 1   Find $C(9, 5)$.

Solution   Using

$$C(n, r) = \frac{n!}{r!(n - r)!}$$

$$C(9, 5) = \frac{9!}{5!(9 - 5)!} = \frac{9!}{5!\,4!} = \frac{9(8)(7)(6)5!}{5!(4)(3)(2)}$$

$$= 126$$

EXAMPLE 2   Find $C(6, 6)$.

Solution

$$C(6, 6) = \frac{6!}{6!(6 - 6)!} = \frac{1}{0!} = \frac{1}{1} = 1$$

$$C(n, n - r) = \frac{n!}{(n - r)![n - (n - r)]!} = \frac{n!}{(n - r)!\,r!}$$
$$= C(n, r)$$

THEOREMS

$C(n, n) = 1$
$C(n, r) = C(n, n - r)$

EXAMPLE 3   How many sets of 4 different magazines can be selected from 12 different magazines?

Solution   Since the order in which the magazines are selected is not important, we want the number of combinations of 12 elements taken 4 at a time.

$$C(12, 4) = \frac{12!}{4!\,8!} \left(\frac{n!}{r!(n - r)!} \right)$$

$$= \frac{12(11)(10)(9)8!}{4!\,8!}$$

$$= \frac{12(11)(10)(9)}{4(3)(2)(1)}$$

$$= 495$$

EXAMPLE 4   In how many ways can an automobile dealer select 5 different models of cars from 15 models?

Solution

$$C(15, 5) = \frac{15!}{5!\,10!}$$

$$= \frac{15(14)(13)(12)(11)10!}{5!\,10!}$$

$$= \frac{15(14)(13)(12)(11)}{5(4)(3)(2)(1)}$$

$$= 3003$$

EXAMPLE 5 A committee of 5 persons is to be selected from 10 Democrats and 7 Republicans. Each committee is to consist of 3 Democrats and 2 Republicans. How many committees are possible?

Solution   The number of ways the Democrats can be selected is $C(10, 3)$. The number of ways the Republicans can be selected is $C(7, 2)$.

Using the Fundamental Principle of Counting, the number of possible committees is the product

$$C(10, 3) \cdot C(7, 2) = \frac{10!}{3!\,7!} \cdot \frac{7!}{5!\,2!}$$

$$= \frac{10!}{3!\,5!\,2!}$$

$$= \frac{10(9)(8)(7)(6)}{3(2)(2)}$$

$$= 2520$$

EXERCISES 11.2

(1–12)   Find the value of each of the following.

1. $C(9, 3)$
2. $C(9, 6)$
3. $C(10, 4)$
4. $C(14, 10)$
5. $C(12, 12)$
6. $C(18, 18)$
7. $C(8, 1)$
8. $C(7, 1)$
9. $C(7, 2)$
10. $C(20, 19)$
11. $C(15, 13)$
12. $C(16, 12)$

13. In a certain survey each voter was asked to check the 3 most important issues out of 10 that were listed. In how many ways could this be done?

14. In how many ways could a group of 12 persons select 2 persons as delegates to a conference?

15. In how many ways can 3 different flavors of ice cream be selected from 21 flavors?

16. A grocer has a choice of 8 different brands of canned vegetables. How many ways can she select 3 different brands?

17. On a certain examination the instructions are to answer any 5 of 8 questions. In how many ways can this be done?

18. In how many ways can 6 similar scholarships be awarded to 10 applicants?

19. In how many ways can 4 models of refrigerators and 3 models of stoves be selected from 9 models of refrigerators and 6 models of stoves?

20. How many committees containing 3 men and 4 women can be formed from a group of 8 men and 10 women?

21. In how many ways can an entertainment program consisting of 3 comedy acts and 4 musical acts be formed from 7 comedy acts and 9 musical acts?

22. In the manufacture of 100 machine parts, 8 of the parts were defective. In how many sets of 5 of these parts will exactly 1 part be defective?

23. A standard bridge deck of 52 cards contains 4 aces. How many ways can 5 cards be selected, exactly 3 of which are aces?

24. A standard bridge deck of 52 cards contains 13 cards called hearts. (The other cards are called spades, clubs, and diamonds.) How many ways can 5 cards be selected, exactly 4 of which are hearts?

25. In how many ways can a student select 1 science course out of 4 offered, 1 language course out of 5 offered, and 3 other courses out of 12 offered?

26. In the Western Division of the American League of Baseball, there are 7 teams. How many games will be played if each team plays every other team 3 times?

11.3 THE BINOMIAL THEOREM

In this section we develop a formula for expressing a power of a binomial as a polynomial, called the **expansion of the power of the binomial.** The following expansions can be obtained by performing the indicated multiplications.

$$(a + b)^1 = a + b$$
$$(a + b)^2 = a^2 + 2ab + b^2$$
$$(a + b)^3 = a^3 + 3a^2b + 3ab^2 + b^3$$
$$(a + b)^4 = a^4 + 4a^3b + 6a^2b^2 + 4ab^3 + b^4$$
$$(a + b)^5 = a^5 + 5a^4b + 10a^3b^2 + 10a^2b^3 + 5ab^4 + b^5$$

By examining these special cases, the following properties can be observed for $(a + b)^n$ when $n = 1, 2, 3, 4,$ or 5.

1. Each expansion has $n + 1$ terms.
2. The first term is a^n and the last term is b^n.
3. The sum of the exponents of a and b in each term is n.
4. The exponent of a is one less than that of the preceding term.
5. The exponent of b is one more than that of the preceding term.
6. The coefficient of the second term and the next to last term is n. Morever,

$$n = C(n, 1) = \frac{n!}{1!(n - 1)!} = n$$

7. The coefficient of $a^{n-r}b^r$ is $C(n, r)$.

Statement 7 follows by observing the following:

$C(3, 1) = 3$ and $C(3, 2) = \dfrac{3!}{2!1!} = 3$

$C(4, 1) = 4$ and $C(4, 2) = \dfrac{4!}{2!2!} = \dfrac{4 \cdot 3}{2} = 6$ and $C(4, 3) = \dfrac{4!}{3!1!} = 4$

$C(5, 1) = 5$ and $C(5, 2) = \dfrac{5!}{2!3!} = \dfrac{5 \cdot 4}{2} = 10$ and $C(5, 3) = 10$ and $C(5, 4) = 5$

We now form a conjecture for $(a + b)^n$:
$$(a + b)^n = a^n + C(n, 1)a^{n-1}b + C(n, 2)a^{n-2}b^2 + \cdots$$
$$+ C(n, r)a^{n-r}b^r + \cdots + b^n$$

The Binomial Theorem states that this formula is true for all positive integers n.

THE BINOMIAL THEOREM

For all positive integers n,
$$(a + b)^n = a^n + C(n, 1)a^{n-1}b + \cdots + C(n, r)a^{n-r}b^r + \cdots + b^n$$

ALTERNATE FORM

$$(a + b)^n = a^n + na^{n-1}b + \frac{n(n - 1)}{2}a^{n-2}b^2 + \cdots$$

$$+ \frac{n!}{r!(n - r)!}a^{n-r}b^r + \cdots + b^n$$

Proof By mathematical induction.
Part 1
$n = 1$, $(a + b)^1 = a + b$, true.
Part 2 Assume statement true for $n = k$,
$$(a + b)^k = a^k + ka^{k-1}b + \cdots + \frac{k!}{r!(k - r)!}a^{k-r}b^r + \cdots + b^k$$

We will verify this statement for $n = k + 1$.
$$(a + b)^{k+1} = (a + b)(a + b)^k = a(a + b)^k + b(a + b)^k$$
Then

$$a(a + b)^k = a^{k+1} + ka^k b + \cdots + \frac{k!}{r!(k - r)!}a^{k-r+1}b^r + \cdots + ab^k \tag{1}$$

$$b(a + b)^k = a^k b + \cdots + \frac{k!}{(r - 1)!(k - r + 1)!}a^{k-r+1}b^r + \cdots + kab^k + b^{k+1} \tag{2}$$

Now adding the two coefficients of $a^{k-r+1}b^r$,
$$\frac{k!}{r!(k - r)!} + \frac{k!}{(r - 1)!(k - r + 1)!} = \frac{k!(k - r + 1 + r)}{r!(k - r + 1)!} = \frac{(k + 1)!}{r!(k + 1 - r)!} = C(k + 1, r)$$

Now adding Equations (1) and (2) we obtain the statement of the Binomial Theorem for $n = k + 1$:
$$(a + b)^{k+1} = a^{k+1} + (k + 1)a^k b + \cdots + C(k + 1, r)a^{k+1-r}b^r + \cdots + b^{k+1}$$
Thus the theorem is proved.

EXAMPLE 1   Use the Binomial Theorem to write the expansion of $(x - 2)^6$. Then simplify.

Solution

$$(x - 2)^6 = x^6 + 6x^5(-2) + C(6, 2)x^4(-2)^2 + C(6, 3)x^3(-2)^3 + C(6, 4)x^2(-2)^4 + 6x(-2)^5 + (-2)^6$$

$$= x^6 - 12x^5 + \frac{6(5)}{2}x^4(4) + \frac{6(5)(4)}{1(2)(3)}x^3(-8) + \frac{6(5)(4)(3)}{1(2)(3)(4)}x^2(16) + 6x(-32) + 64$$

$$= x^6 - 12x^5 + 60x^4 - 160x^3 + 240x^2 - 192x + 64$$

EXAMPLE 2   Use the Binomial Theorem to expand $(x + y)^7$.

Solution

$$(x + y)^7 = x^7 + 7x^6y + C(7, 2)x^5y^2 + C(7, 3)x^4y^3 + C(7, 4)x^3y^4 + C(7, 5)x^2y^5 + 7xy^6 + y^7$$

$$= x^7 + 7x^6y + \frac{7(6)}{2}x^5y^2 + \frac{7(6)(5)}{1(2)(3)}x^4y^3 + \frac{7(6)(5)(4)}{1(2)(3)(4)}x^3y^4 + \frac{7(6)(5)(4)(3)}{1(2)(3)(4)(5)}x^2y^5 + 7xy^6 + y^7$$

$$= x^7 + 7x^6y + 21x^5y^2 + 35x^4y^3 + 35x^3y^4 + 21x^2y^5 + 7xy^6 + y^7$$

Now $C(n, r)a^{n-r}b^r$ is the $(r + 1)$th term of $(a + b)^n$. Therefore

the rth term of $(a + b)^n$ is

$$C(n, r - 1)a^{n-r+1}b^{r-1}$$

EXAMPLE 3   Find the sixth term of $(5y + 0.2)^{10}$.

Solution   Using $n = 10$, $r = 6$, $a = 5y$, $b = 0.2$,

$$C(n, r - 1)a^{n-r+1}b^{r-1} = C(10, 5)a^5b^5$$

$$C(10, 5)a^5b^5 = \frac{10!}{5!5!}(5y)^5(0.2)^5$$

$$= \frac{10(9)(8)(7)(6)}{5(4)(3)(2)(1)}y^5$$

$$= 252y^5$$

EXAMPLE 4   Approximate $(1.05)^{10}$ correct to the nearest hundredth by using the Binomial Theorem.

Solution   Using $(1 + 0.05)^{10}$,

$$(1 + 0.05)^{10} = 1^{10} + 10(1)^9(0.05) + C(10, 2)1^8(0.05)^2 + C(10, 3)1^7(0.05)^3 + C(10, 4)1^6(0.05)^4 + \cdots$$

$$= 1 + 0.5 + \frac{10(9)}{2}(0.0025) + \frac{10(9)(8)}{1(2)(3)}(0.000125) + \frac{10(9)(8)(7)}{1(2)(3)(4)}(0.00000625) + \cdots$$

$$= 1.5 + 0.1125 + 0.015 + 0.0013125 + \cdots$$

Since the terms are decreasing, we note that the terms that have been omitted will not contribute to the hundredths' position. Therefore

$(1.05)^{10} = 1.63$ correct to the nearest hundredth

EXERCISES 11.3

(1–14)   Expand by using the Bionomial Theorem and simplify. See Examples 1 and 2.

1. $(x + 4)^3$　　　　　　　　　**2.** $(x - 5y)^3$

3. $(2x - 3y)^4$　　　　　　　　**4.** $(3x + 2y)^4$

5. $(t^2 + 2)^5$ **6.** $(2t^2 - 3)^6$

7. $(y + 1)^7$ **8.** $(2y - 1)^7$

9. $(x - 1)^8$ **10.** $\left(2x + \dfrac{y}{2}\right)^8$

11. $\left(x + \dfrac{1}{x}\right)^6$ **12.** $(5x - 0.2)^5$

13. $(x - y)^9$ **14.** $(x + y)^9$

(15–22) Write, in simplified form, the specified term in the binomial expansion. See Example 3.

15. Fifth, $(x - 3)^{10}$ **16.** Fourth, $\left(\dfrac{x}{2} + 1\right)^{16}$

17. Fourth, $(y + 4)^{12}$ **18.** Sixth, $(1 - 0.02y)^8$

19. Sixth, $\left(x + \dfrac{1}{x}\right)^9$ **20.** Fifth, $\left(r - \dfrac{2}{s}\right)^7$

21. Third, $(2x - y)^{15}$ **22.** Eighth, $\left(x + \dfrac{1}{x}\right)^{12}$

(23–30) Approximate each of the following by finding the sum of the first four terms of a binomial expansion. See Example 4.

23. $(1.02)^{12}$ **24.** $(1.01)^{10}$

25. $(1.05)^{11}$ **26.** $(2.02)^8$

27. $(0.97)^{10}$; use $(1 - 0.03)^{10}$

28. $(0.98)^8$; use $(1 - 0.02)^8$

29. $(1.95)^7$; use $(2 - 0.05)^7$

30. $(1.98)^9$; use $(2 - 0.02)^9$

31. Evaluate $(1 - i)^8$ where $i^2 = -1$.

32. Evaluate $\left(\dfrac{1}{2} + \dfrac{i\sqrt{3}}{2}\right)^9$ where $i^2 = -1$.

1.4 BINOMIAL SERIES

If k is any rational number and if $|x| < 1$, then it can be shown, by using methods of calculus, that it is valid to equate $(1 + x)^k$ with an infinite series; that is,

$$(1 + x)^k = 1 + kx = \frac{k(k - 1)}{2} x^2 + \cdots + \frac{k(k - 1) \cdots (k - n + 1)}{1(2)(3) \cdots (n)} x^n + \cdots$$

Note that the coefficient of x^n in this infinite series is identical in form to that of x^n in the binomial expansion of $(1 + x)^k$ when k is a positive integer. For this reason, the infinite series above is called the **binomial series**.

EXAMPLE 1   Express $(1 + x)^{1/2}$ as an infinite series.

Solution   Using $k = \frac{1}{2}$ in the expansion for $(1 + x)^k$,

$$(1 + x)^{1/2} = 1 + \frac{1}{2}x + \frac{\frac{1}{2}\left(\frac{1}{2} - 1\right)}{2}x^2 + \frac{\frac{1}{2}\left(\frac{1}{2} - 1\right)\left(\frac{1}{2} - 2\right)}{1 \cdot 2 \cdot 3}x^3$$
$$+ \frac{\frac{1}{2}\left(\frac{1}{2} - 1\right)\left(\frac{1}{2} - 2\right)\left(\frac{1}{2} - 3\right)}{1 \cdot 2 \cdot 3 \cdot 4}x^4 + \cdots$$

Simplifying,

$$(1 + x)^{1/2} = 1 + \frac{1}{2}x - \frac{1}{8}x^2 + \frac{1}{16}x^3 - \frac{5}{128}x^4 + \cdots$$

EXAMPLE 2   Approximate $\sqrt{1.1} = (1.1)^{1/2}$ by an appropriate binomial expansion using the first four terms.

Solution   From Example 1, the sum of the first four terms of $(1 + x)^{1/2}$ is

$$1 + \frac{1}{2}x - \frac{1}{8}x^2 + \frac{1}{16}x^3$$

Replacing x by 0.1,

$$(1 + 0.1)^{1/2} \approx 1 + \frac{1}{20} - \frac{1}{800} + \frac{1}{16,000}$$

Simplifying,

$$\sqrt{1.1} \approx 1 + 0.05 - 0.00125 + 0.0000625$$
$$\approx 1.0488125$$

A calculator or a table of square roots gives $\sqrt{1.1} = 1.048809$. Thus the answer to Example 2 is correct to the fifth decimal place.

As Example 2 illustrates, the sum of the first n terms of the binomial series can be used to approximate $(1 + x)^k$, if $|x| < 1$, and this sum will differ from the correct value of $(1 + x)^k$ by a small error if n is taken large enough. For some series, only a few terms need to be evaluated for a useful approximation.

Whenever the terms of the series alternate in sign, the error made by using n terms is between 0 and the value of the $n + 1$st term. Referring to Example 2, the fifth term has value -0.00000390625. Thus the error is in the sixth decimal place. This is verified by comparing the result obtained with the tabular value stated; that is, $1.048809 - 1.0488125 = -0.0000035$.

EXAMPLE 3   Write the first five terms of the binomial expansion of $1/(2 - x)^3$.

Solution

$$\frac{1}{(2 - x)^3} = (2 - x)^{-3} = 2^{-3}\left(1 - \frac{x}{2}\right)^{-3} = \frac{1}{8}\left(1 - \frac{x}{2}\right)^{-3}$$

Replacing x by t and k by -3 in

$$(1 + x)^k \approx 1 + kx + \frac{k(k-1)}{2} x^2 + \frac{k(k-1)(k-2)}{2 \cdot 3} x^3$$
$$+ \frac{k(k-1)(k-2)(k-3)}{2 \cdot 3 \cdot 4} x^4$$

$$(1 + t)^{-3} \approx 1 - 3t + \frac{-3(-4)}{2} t^2 + \frac{-3(-4)(-5)}{2 \cdot 3} t^3 + \frac{-3(-4)(-5)(-6)}{2 \cdot 3 \cdot 4} t^4$$
$$\approx 1 - 3t + 6t^2 - 10t^3 + 15t^4$$

Replacing t by $(-x/2)$ and multiplying by $\frac{1}{8}$,

$$\frac{1}{8}\left(1 - \frac{x}{2}\right)^{-3} \approx \frac{1}{8}\left[1 - 3\left(-\frac{x}{2}\right) + 6\left(-\frac{x}{2}\right)^2 - 10\left(-\frac{x}{2}\right)^3 + 15\left(-\frac{x}{2}\right)^4\right]$$
$$(2 - x)^{-3} \approx \frac{1}{8} + \frac{3}{16}x + \frac{3}{16}x^2 + \frac{5}{32}x^3 + \frac{15}{128}x^4$$

EXAMPLE 4 Approximate $1/(1.95)^3$, using three terms of the binomial expansion. Round off each term to 4 decimal places and round off the final result to 3 decimal places.

Solution

$$\frac{1}{(1.95)^3} = (2 - 0.05)^{-3}$$

Using the result of Example 3 and replacing x by 0.05,

$$(2 - 0.05)^{-3} \approx \frac{1}{8} + \frac{3}{16}\left(\frac{5}{100}\right) + \frac{3}{16}\left(\frac{25}{10,000}\right)$$
$$\approx 0.1250 + 0.0094 + 0.0005$$
$$\approx 0.1349 \approx 0.135$$

EXERCISES 11.4

(1–10) Write the first five terms of the binomial series for each of the following. Assume $|x| < 1$.

1. $(1 + x)^{1/3}$ **2.** $(1 + x)^{1/4}$

3. $(1 - x)^{1/2}$ **4.** $(1 - x^2)^{3/2}$

5. $(1 + x^2)^{-2}$ **6.** $(1 + x^3)^{-4}$

7. $(2 + x)^{-1/2}$ **8.** $(3 + x)^{-1}$

9. $\left(x - \frac{1}{x}\right)^{1/4}$, where $x \neq 0$ **10.** $\left(x - \frac{2}{x}\right)^{1/2}$, where $x \neq 0$

(11–20) Approximate each of the following to 4 significant figures.

11. $\sqrt[3]{1.02}$ **12.** $\sqrt[5]{1.5}$

13. $\sqrt[5]{0.99}$ **14.** $\sqrt[4]{0.98}$

15. $\dfrac{1}{\sqrt{15}}$ **16.** $\dfrac{1}{\sqrt{26}}$

17. $\sqrt[5]{34}$ **18.** $\sqrt[5]{30}$

19. $\dfrac{1}{(1.02)^{10}}$ **20.** $\dfrac{1}{(0.95)^8}$

11.5 PROBABILITY FUNCTIONS

A die (singular of dice) is a cube containing one of the numbers from 1 to 6 on each of its six sides.

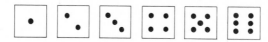

When a die is rolled, there are six ways the die may land, according to the number that appears on top of the die. We say that the probability of rolling a 4 is $\frac{1}{6}$ because there is one way to roll a 4 and 6 ways the die may land. This means that if the die were rolled a great many times, a 4 would appear about $\frac{1}{6}$ of the time. If the die were rolled 600 times, we would expect to roll a 4 about 100 times.

Insurance companies use probability to determine the premiums people pay for insurance. Mortality tables are used for life insurance. In one case, statistics were compiled from 100,000 persons born and the number of those living at different ages up to 85. Suppose a table lists that 66,114 of these people were alive at age 70. We would say, then, the probability of a person at birth living to the age of 70 is $\frac{66,114}{100,000} \approx 66\%$.

Analyzing these special examples, we note three features of a probability situation.

1. An **experiment** that produces a set of specific results or outcomes.
2. A **sample space** S, the set of all possible results or outcomes of the experiment.
3. An **event** E, a subset of the sample space.

The table below illustrates these for our two special examples.

Experiment	Rolling a die	Predicting life expectancy
Sample space S	The 6 ways the die can land	The 100,000 persons
Event E	The 1 way the 4 is rolled	The 66,114 persons still alive at age 70
Probability of E, $p(E)$	$\frac{1}{6}$	$\frac{66,114}{100,000} \approx 66\%$

1.5 PROBABILITY FUNCTIONS

The **mathematical probability,** also called an **a priori** probability, is one that is predicted before an experiment occurs and is based on information about the number of outcomes of an event and the number of possible outcomes. It is assumed that all outcomes are equally likely to occur.

DEFINITION   MATHEMATICAL PROBABILITY

Let
E = a subset of a sample space S.
$n(E)$ = the number of outcomes for event E.
$n(S)$ = the number of all possible outcomes in S.
$p(E)$ = the probability that E occurs.
Then
$$p(E) = \frac{n(E)}{n(S)}$$

The following theorems follow from the definition.

THEOREM 1

$0 \le p(E) \le 1$

THEOREM 2

If it is impossible for event E to happen, then
$p(E) = 0$

THEOREM 3

If event E is certain to happen, then
$p(E) = 1$

THEOREM 4

If $p(\text{not } E)$ is the probability that event E does not happen, then
$p(\text{not } E) = 1 - p(E)$
and
$p(E) + p(\text{not } E) = 1$

For the introductory example of rolling a die, the probability of an event is predicted before the experiment takes place. This example illustrates a mathematical, or a priori, probability.

For the example of predicting a life expectancy, the probability was based on an experiment that had taken place. This involved a sampling procedure; namely, the selection of the 100,000 persons. This type of probability is called an **empirical, or a posteriori, probability.** It is assumed that with a good sampling procedure, the probability determined on the basis of the sample is a good approximation to the mathematical probability.

EXAMPLE 1   A card is drawn at random from a standard deck of cards. What is the probability that the card is (a) an ace? (b) a spade?

> Solution   A standard deck of cards contains 52 cards. There are 4 suits, 2 red suits called hearts and diamonds and 2 black suits called spades and clubs. Each suit contains 13 cards; ace, king, queen, jack, 10, 9, 8, 7, 6, 5, 4, 3, 2.
>
> (a) Since there are 4 ways to draw an ace and 52 ways to draw a card,
>
> $$n(E) = 4 \quad \text{and} \quad n(S) = 52$$
>
> For $p(E)$, the probability of drawing an ace,
>
> $$p(E) = \frac{n(E)}{n(S)} = \frac{4}{52} = \frac{1}{13}$$
>
> (b) Since there are 13 ways to draw a spade and 52 ways to draw a card, for $p(E)$ = the probability of drawing a spade,
>
> $$p(E) = \frac{n(E)}{n(S)} = \frac{13}{52} = \frac{1}{4}$$

EXAMPLE 2   There are 60 pieces of candy in a bag of which 15 are white, 20 are green, and 25 are red. What is the probability that one piece of candy, selected at random, (a) is red? (b) is not red? (c) is yellow? (d) is white or green or red?

> Solution
>
> (a) For selecting a red candy,
>
> $$p = \frac{25}{60} = \frac{5}{12}$$
>
> (b) For selecting a candy not red, the probability is $p(\text{not } E)$ where $p(E)$ is the probability of selecting a red candy.
>
> Since $p(\text{not } E) = 1 - p(E)$,
>
> $$p(\text{not } E) = 1 - \frac{5}{12} = \frac{7}{12}$$
>
> (c) For selecting a yellow candy, since there is no yellow candy in the bag,
>
> $$n(E) = 0 \quad \text{and} \quad p(E) = \frac{0}{60} = 0$$
>
> (d) For selecting a white or green or red candy,
>
> $$n(E) = 60 = n(S) \quad \text{and} \quad p(E) = \frac{60}{60} = 1$$

EXAMPLE 3   If two dice are rolled, what is the probability that the sum of the numbers rolled is 10?

> Solution   The sample space is a set of ordered pairs. The total number of ordered pairs possible is $6 \cdot 6 = 36$.
>
> The event E consists of these ordered pairs whose sum is 10; that is,
>
> $$E = \{(4, 6), (5, 5), (6, 4)\} \quad \text{and} \quad n(E) = 3$$
>
> $$p(E) = \frac{3}{36} = \frac{1}{12}$$

EXAMPLE 4 A pinochle deck of cards consists of 48 cards. There are 4 suits, 2 red suits called hearts and diamonds and 2 black suits called spades and clubs. Each suit contains 12 cards, two of each of the following: ace, king, queen, jack, ten, nine.

If 2 cards are drawn at random, what is the probability that these 2 cards are diamonds?

Solution

Sample space $n(S)$ = the number of ways that 2 cards can be selected from 48. Thus $n(S) = C(48, 2)$, the number of combinations of 48 things taken 2 at a time.

Event $n(E)$ = the number of ways of drawing 2 diamonds from 12 diamonds. Thus $n(E) = C(12, 2)$.

Therefore

$$p(E) = \frac{C(12, 2)}{C(48, 2)} = \frac{12!}{2!10!} \div \frac{48!}{2!46!} = \frac{12(11)}{48(47)} = \frac{11}{188}$$

EXAMPLE 5 Weather statistics for a certain city show that during the month of September there are 12 days of rain. A person is planning to visit this city some day in September. What is the probability that it will rain that day?

Solution Since there are 30 days in September, $n(S) = 30$.

$$p(E) = \frac{12}{30} = \frac{2}{5}$$

EXAMPLE 6 Using the Mortality Table in this section, find the probability that a person whose age is:

(a) 25 will live to be 60.
(b) 50 will live another 5 years.
(c) 50 will die within the next 5 years.

Solution

(a) $n(S)$ = number alive at age 25 = 96,747
   $n(E)$ = number alive at age 60 = 82,282
   $p(E) = \frac{82,282}{96,747} \approx 85\%$

(b) $n(S)$ = number alive at age 50 = 90,747
   $n(E)$ = number alive at age 55 = 87,310
   $p(E) = \frac{87,310}{90,747} \approx 96\%$

(c) The probability of not being alive at age 55 is $p(\text{not } E)$ for E of part (b).
   $p(\text{not } E) = 1 - \frac{87,310}{90,747} \approx 4\%$

A MORTALITY TABLE

AGE	NUMBER LIVING	AGE	NUMBER LIVING
Birth	100,000	45	93,043
5	98,207	50	90,747
10	98,033	55	87,310
15	97,865	60	82,282
20	97,391	65	75,084
25	96,747	70	66,114
30	96,147	75	54,060
35	95,452	80	39,540
40	94,522	85	25,029

Source: Division of Vital Statistics, Public Health Service, 1976.

EXERCISES 11.5

(1–6) A bag contains 12 marbles; 5 are red, 4 are white, and 3 are green. Find the probability of selecting, at random:

1. a white ball

2. a green ball

3. a ball that is not white

4. a ball that is not red

5. a blue ball

6. a ball that is not blue

(7–12) One piece of fruit is placed in each of 100 box lunches. An apple is in each of 45 boxes, an orange in each of 25 boxes, and a banana in each of 30 boxes. Find the probability of selecting at random a box that contains:

7. an orange

8. an apple

9. an apple or a banana

10. an orange or an apple

11. an apple, an orange, or a banana

12. a pear

(13–20) If a card is drawn at random from a standard deck of cards, find the probability that the card is:

13. a king

14. a heart

15. a red card

16. not a red card

17. not a club

18. not a 2

19. an ace or a king

20. a red card or a black card

(21–28) If a card is drawn at random from a pinochle deck of cards (see Example 4), find the probability that the card is:

21. a 10

22. a diamond

23. not a spade

24. an 8

25. a black card

26. not a jack

27. a king of hearts or queen of hearts

28. a spade or a jack of diamonds

(29–34) If two dice are rolled, find the probability that the sum of the numbers rolled is:

29. 6 **30.** 7 **31.** 11 **32.** 15
33. 7 or 11 **34.** 2, 3, 4, 5, 6, 7, 8, 9, 10, 11, or 12

(35–38) Two candies are drawn at random from a bag that contains 3 white candies, 4 green candies, and 5 red candies. Find the probability that:

35. both candies are red **36.** both candies are green
37. both candies are not red **38.** both candies are not green

(39–42) Two cards are drawn at random from a pinochle deck of cards. (See Example 4.) Find the probability that:

39. both cards are spades **40.** both cards are aces
41. both cards are not clubs **42.** both cards are not kings

(43–46) Three cards are drawn at random from a standard deck of cards. Find the probability that:

43. all 3 cards are queens **44.** all 3 cards are hearts
45. all 3 cards are not aces **46.** all 3 cards are not diamonds

(47–52) Using the Mortality Table in this section, find the probability that a person whose age is:

47. 20 will live to be 70.
48. 50 will live to be 65.
49. 40 will live another 5 years.
50. 65 will die within 5 years.
51. 70 will die within 5 years.
52. 75 will live another 5 years.

53. In the game of Keno, 20 numbers are selected from the integers from 1 to 80. Suppose you play this game by selecting 5 of the 80 numbers before the 20 are chosen. You will win if the 5 numbers you selected are among the 20 chosen. Find your probability of winning.

54. A certain machine is rated as 98% effective; that is, for every 100 parts it manufactures, 98 are perfect and 2 are defective. How many defective parts would you expect to find in a random sample of 400 parts?

55. A pair of dice are rolled 900 times. Of these times, a sum of 6 appears exactly 30 times.
 (a) If the outcomes are equally likely, how many times should you expect a sum of 6 in 900 rolls?
 (b) Should one suspect that the dice are loaded (that is, that the outcomes are not equally likely)?

56. A packet of seeds states that 85% of the seeds will germinate. If 2500 seeds are planted, how many should be expected to germinate?

57. If the probability is $\frac{3}{5}$ that a person who enters a certain restaurant will buy the daily special, how many specials should be prepared if 350 persons are expected to enter the restaurant?

11.6   COMBINED PROBABILITIES

In this section we consider the probability that one event *and* another event occur and also the probability that one event *or* another event occurs.

When two events happen, it might be the case that both events happen at the same time. For example, drawing an ace and drawing a king from a standard deck of cards cannot happen in one draw. On the other hand, drawing an ace and drawing a heart can occur on the same draw; namely, drawing the ace of hearts. We say, in the first case, that the events are mutually exclusive, while, in the second case, they are not mutually exclusive. More formally, we have the following definition.

DEFINITION   MUTUALLY EXCLUSIVE EVENTS

Two events, E_1 and E_2, are **mutually exclusive** (or **disjoint**) if and only if their intersection is the empty set, $E_1 \cap E_2 = \varnothing$.

Now an element is in set E_1 or in set E_2 if and only if it is in their union, $E_1 \cup E_2$. If E_1 and E_2 have no elements in common, then $E_1 \cap E_2 = \varnothing$. In this case, the events are mutually exclusive and
$$n(E_1 \cup E_2) = n(E_1) + n(E_2)$$
However, if the intersection is not empty, then $n(E_1) + n(E_2)$ counts each element in the intersection twice. For the number in the union, we have to subtract the number in the intersection. Then
$$n(E_1 \cup E_2) = n(E_1) + n(E_2) - n(E_1 \cap E_2)$$

THEOREM 1   PROBABILITY OF E_1 OR E_2

If E_1 and E_2 are any events in a sample space S, then the probability that E_1 or E_2 occurs is given by
$$p(E_1 \text{ or } E_2) = p(E_1 \cup E_2) = p(E_1) + p(E_2) - p(E_1 \cap E_2)$$

Proof

$$p(E_1 \cup E_2) = \frac{n(E_1 \cup E_2)}{n(S)}$$

$$= \frac{n(E_1) + n(E_2) - n(E_1 \cap E_2)}{n(S)}$$

$$= \frac{n(E_1)}{n(S)} + \frac{n(E_2)}{n(S)} - \frac{n(E_1 \cap E_2)}{n(S)}$$

$$= p(E_1) + p(E_2) - p(E_1 \cap E_2)$$

THEOREM 2 PROBABILITY OF E_1 OR E_2, TWO MUTUALLY EXCLUSIVE EVENTS

$$p(E_1 \text{ or } E_2) = p(E_1) + p(E_2)$$

EXAMPLE 1 Find the probability of drawing an ace or a heart from a standard deck of cards.

Solution Let

E_1 = set of ways of drawing an ace, and $n(E_1) = 4$

E_2 = set of ways of drawing a heart, and $n(E_2) = 13$

Then $E_1 \cap E_2$ = set of ways of drawing an ace that is a heart.

$n(S) = 52$ and $n(E_1 \cap E_2) = 1$

$$p(E_1 \text{ or } E_2) = p(E_1) + p(E_2) - p(E_1 \cap E_2)$$
$$= \frac{4}{52} + \frac{13}{52} - \frac{1}{52} = \frac{16}{52} = \frac{4}{13}$$

EXAMPLE 2 Find the probability of drawing an ace or a king from a standard deck of cards.

Solution Let

E_1 = set of ways of drawing an ace and $n(E_1) = 4$

E_2 = set of ways of drawing a king and $n(E_2) = 4$

Note that E_1 and E_2 are mutually exclusive. Since no card is an ace and a king, $E_1 \cap E_2 = \varnothing$. The probability of drawing an ace or a king is

$$p(E_1 \text{ or } E_2) = p(E_1) + p(E_2) = \frac{4}{52} + \frac{4}{52} = \frac{2}{13}$$

We have seen that the probability of drawing an ace from a standard deck of cards is $\frac{4}{52} = \frac{1}{13}$. Now let us consider the probability of drawing an ace, knowing that one ace has already been drawn. Since there are 3 aces remaining and a total of 51 cards remaining, the probability is $\frac{3}{51} = \frac{1}{17}$. This last probability is called a **conditional probability** because it depends on a given event having occurred.

Notation   Conditional Probability

$p(E_2|E_1)$ denotes the **conditional probability** of E_2 given that E_1 has occurred, where E_1 and E_2 are any two events in a sample space.

Now let us consider the probability that one event occurs and another event occurs. There are two possibilities to consider: the occurrence of one event may depend on the occurrence of the other event or it may not.

First let us recall the Fundamental Principle of Counting, as stated in Section 11.1. If one event E_1 can happen in n_1 ways, and if, after this has happened, a second event E_2 can happen in n_2 ways, then the number of ways both events can happen in this order is $n_1 n_2$.

Now if k_1 is the number of possible outcomes in the sample space for E_1, that is, $n(S_1) = k_1$, and if k_2 is the number of possible outcomes in the sample space S_2 for E_2 after E_1 has occurred, then the number of outcomes in the sample space for E_1 and E_2 is $k_1 k_2$.

Therefore the probability that E_1 and E_2 occur, $p(E_1 \text{ and } E_2)$, is given by

$$p(E_1 \text{ and } E_2) = \frac{n_1 n_2}{k_1 k_2} = \left(\frac{n_1}{k_1}\right)\left(\frac{n_2}{k_2}\right)$$

Now $\dfrac{n_1}{k_1} = p(E_1)$ but $\dfrac{n_2}{k_2}$ is not necessarily equal to $p(E_2)$ because it is based on the fact that E_1 has occurred. In fact, $\dfrac{n_2}{k_2} = p(E_2|E_1)$, the conditional probability of E_2 given E_1. When $p(E_2|E_1) = p(E_2)$, we say that the events are independent.

DEFINITION   INDEPENDENT EVENTS

Let E_1 and E_2 be events in a sample space.
Then E_1 and E_2 are **independent** events if and only if $p(E_2|E_1) = p(E_2)$.
Two events are **dependent** if and only if they are not independent.

The results of the preceding discussion can now be summarized as the following theorem.

THEOREM 3   PROBABILITY OF E_1 AND E_2

Let E_1 and E_2 be events in a sample space. Let $p(E_1 \text{ and } E_2)$ be the probability that E_1 and E_2 occur. Then
$p(E_1 \text{ and } E_2) = p(E_1 \cap E_2) = p(E_1)p(E_2|E_1)$
If E_1 and E_2 are independent, then $p(E_2|E_1) = p(E_2)$ and
$p(E_1 \text{ and } E_2) = p(E_1)p(E_2)$

EXAMPLE 3 Find the probability that two cards drawn from a standard deck are aces if the first card is replaced before the second card is drawn.

Solution   Let

E_1 = set of ways to draw an ace

E_2 = set of ways to draw an ace

The two events are independent since the first card is replaced, and the second drawing is not affected by the first drawing.

$$p(E_1 \text{ and } E_2) = p(E_1)p(E_2) = \left(\tfrac{4}{52}\right)\left(\tfrac{4}{52}\right) = \left(\tfrac{1}{13}\right)\left(\tfrac{1}{13}\right) = \tfrac{1}{169}$$

EXAMPLE 4 Find the probability that two cards drawn from a standard deck of cards are aces if the first card drawn is not replaced.

Solution   Let

E_1 = set of ways to draw an ace

E_2 = set of ways to draw an ace

$p(E_2|E_1)$ = the probability of drawing an ace after an ace has been removed from the deck

$$p(E_1 \text{ and } E_2) = p(E_1)p(E_2|E_1) = \left(\tfrac{4}{52}\right)\left(\tfrac{3}{51}\right) = \tfrac{1}{221}$$

EXAMPLE 5 Box 1 contains 6 red balls and 4 blue balls. Box 2 contains 5 red balls and 7 blue balls. Two balls are drawn, one from each box. Find the probability that (a) both of the balls are red; (b) one of the balls is red.

Solution   Let

E_1 = set of ways of drawing a red ball from Box 1

E_2 = set of ways of drawing a red ball from Box 2

(a) $p(E_1 \text{ and } E_2)$ = probability that the ball from Box 1 is red and the ball from Box 2 is red

Since the events are independent,

$$p(E_1 \text{ and } E_2) = p(E_1)p(E_2) = \left(\tfrac{6}{10}\right)\left(\tfrac{5}{12}\right) = \tfrac{1}{4}$$

(b) $p(E_1 \text{ or } E_2)$ = probability that the ball from Box 1 is red or the ball from Box 2 is red

The events are not mutually exclusive, since both balls can be red. Note from part (a), $E_1 \cap E_2 \neq \varnothing$.

$$p(E_1 \text{ or } E_2) = p(E_1) + p(E_2) - p(E_1 \text{ and } E_2)$$
$$= \tfrac{6}{10} + \tfrac{5}{12} - \tfrac{1}{4} = \tfrac{46}{60} = \tfrac{23}{30}$$

EXERCISES 11.6

1. A card is drawn from a standard deck of 52 cards. Find the probability that:
   (a) It is a queen.
   (b) It is a queen if it is known that one queen has already been drawn and not replaced.
   (c) It is a queen if it is known that 4 cards have already been drawn, none have been replaced, and none of the 4 cards is a queen.

(2–5) A poll was conducted asking a sample of persons how they would vote on a certain proposition in the coming election. The results of the poll are shown in the following table.

Party Affiliation	Yes	No	Undecided
Democrat	60	25	15
Republican	20	30	10
Independent	15	5	20

If a person is selected at random, find the probability that she will vote

2. Yes   **3.** No

4. Yes, if it is known she is a Democrat.

5. No, if it is known she is not a Republican.

(6–9) A card is drawn from a standard deck of 52 cards. Find the probability that it is:

6. a heart or a spade   **7.** a 10 or a black card

8. a club or a jack   **9.** a diamond or a red card

(10–13) Two cards are drawn from a standard deck of 52 cards, the first being replaced before the second is drawn. Find the probability that:

10. Both cards are clubs.   **11.** Both cards are queens.

12. One card is a heart and one card is an ace.

13. One card is a 3 and one card is a red card.

(14–17) Two cards are drawn from a standard deck of 52 cards, but the first card drawn is not replaced. Find the probability that:

14. Both cards are kings.   **15.** Both cards are diamonds.

16. The first card is the ace of hearts and the other card is a heart.

17. The first card is a club and the other card is a black card.

(18–23) Box 1 contains 2 red balls and 4 green balls. Box 2 contains 7 red balls and 3 green balls. Two balls are drawn, one from each box. Find the probability that:

18. Both balls are green.   **19.** Both balls are red.

20. One of the balls is green. **21.** One of the balls is red.

22. One ball is red and the other ball is green.

23. Neither ball is red.

(24–25) Two coins are tossed. Find the probability that:

24. Both coins land heads.

25. One coin lands heads and other lands tails.

(26–29) Three coins are tossed. Find the probability that:

26. All 3 coins land heads.

27. Exactly 2 coins land tails.

28. At least 2 coins land tails.

29. At most 2 coins land tails.

(30–33) It is known there are 3 defective batteries in a box of 16 batteries. After one battery is selected, it is not replaced and a second battery is selected. Find the probability that:

30. Both batteries are defective.

31. Both batteries are not defective.

32. The first is defective and the second is not.

33. The first or the second battery is defective.

(34–36) The probability that a certain married couple will have a child born with a certain trait is $\frac{1}{4}$. Find the probability that of 2 children born to the couple:

34. One has the trait and one does not.

35. Both do not have the trait.

36. Both have the trait.

(37–42) If the probability for the birth of a girl is 52% and the probability for the birth of a boy is 48%, for a couple selected at random, find the probability that:

(37–39) If they have 2 children:

37. Both are boys. **38.** Both are girls.

39. One is a boy and one is a girl.

(40–42) If they have 3 children:

40. All 3 are boys. **41.** All 3 are girls.

42. 2 are girls and 1 is a boy.

(43–46) A pair of dice are tossed twice. Find the probability of getting:

43. A sum of 6 or 8 on the first roll.

44. A sum of 7 or 11 on the first roll.

45. A sum of 6 on the first roll and of 7 on the second roll.

46. A sum of 6 on the first roll and of 6 on the second roll.

(47–50) On a certain multiple choice test, there are 10 questions, each having 5 choices for the answer, with each question having exactly one correct answer.

47. If a student selects an answer at random for 2 of the questions, find the probability that:
(a) Both answers are correct.
(b) One answer is correct and the other is not.

48. If a student selects an answer at random for 3 of these questions, find the probability that:
(a) All 3 answers are correct.
(b) No answer is correct.
(c) At least 1 answer is correct.

49. If a student selects all of the answers at random, what is the probability that all 10 answers are correct?

50. If a student selects the answers to all 10 questions at random, what is the probability that at least 5 answers are correct?

(51–54) In a certain city, weather statistics indicate that it usually rains 12 days in September. Assuming it is equally likely to rain any day of the week, find the probability that:

51. It rains on one day and the next day.
52. It rains on one day or the next day.
53. It rains on 3 consecutive days.
54. It rains on one of 3 consecutive days.

(55–58) The probability of a certain man living to the age of 60 is $\frac{3}{4}$. The probability of a certain woman living to the age of 60 is $\frac{5}{6}$. Find the probability that:

60. Both live to the age of 60.
56. Both do not live to the age of 60.
57. The man lives to be 60 but the woman does not.
58. The man or the woman lives to the age of 60.

(59–61) From a group of 12 persons consisting of 8 men and 4 women, 2 are selected at random to share an office, the others having individual offices. Find the probability that the 2 selected to share an office are:

59. both men
60. both women
61. a man and a woman

(62–65) A committee of 3 persons is selected at random from 5 men and 3 women. Find the probability that the people on the committee are:

62. all women
63. all men
64. 2 men and 1 woman
65. 2 women and 1 man

(66–69) In a certain raffle, 100 tickets are sold for two prizes. After the ticket is drawn for the first prize, it is not replaced for the drawing of the second prize. If a person has bought 5 tickets, find the probability that:

66. He wins the first prize and the second prize.
67. He wins the first prize but not the second prize.
68. He wins the second prize but not the first prize.
69. He wins the first prize or the second prize.

(70–72) In a certain community, the probability of the two Democratic candidates winning the election are $\frac{2}{3}$ for one and $\frac{3}{8}$ for the other. Find the probability that:

70. Both candidates win.
71. Only one candidate wins.
72. At least one candidate wins.

11.7 CHAPTER REVIEW

(1–4) Find the value of each of the following. (11.1)

1. $8!$ **2.** $\dfrac{16!}{12!}$ **3.** $P(6, 6)$ **4.** $P(15, 4)$

(5–7) Answer each question. (11.1)

5. In how many ways can 9 charms be arranged on a charm bracelet?
6. If 12 contestants are entered in the Olympics decathalon, in how many ways can they win the gold medal, the silver medal, and the bronze medal?
7. A color wheel consists of a circle divided into 8 equal sectors, each painted a different color. If 8 colors are used, in how many ways can the sectors be painted?

(8–10) Find the value of each combination. (11.2)
8. $C(8, 3)$ **9.** $C(10, 6)$ **10.** $C(80, 80)$

(11–12) Answer each question. (11.2)

11. A cafeteria offers a dinner special consisting of a meat dish and a choice of 2 side orders. If there are 7 side orders to choose from, in how many ways can the 2 side orders be chosen?
12. In how many ways can 3 different types of daffodil bulbs and 5 different types of tulip bulbs be selected from 8 different types of daffodil bulbs and 12 different types of tulip bulbs?

(13–14) Use the Binomial Theorem to write the simplified expansion. (11.3)

13. $(3x - 1)^6$ **14.** $(2x + 0.5)^5$

(15–16) Write the specified term in the binomial expansion. (11.3)

15. Fourth, $(y + 2)^8$

16. Seventh, $\left(x - \dfrac{1}{x}\right)^{10}$

(17–18) Approximate by finding the sum of the first four terms of a binomial expansion. (11.3)

17. $(2.04)^{12}$ **18.** $(0.95)^{10}$

(19–20) Approximate each of the following by using a binomial series. (11.4)

19. $\sqrt[4]{1.05}$ **20.** $\dfrac{1}{(0.96)^{10}}$

(21–24) Answer each question. (11.5)

21. The English alphabet consists of 26 letters, 5 of which are vowels (a, e, i, o, u) and the rest are consonants. Find the probability that:
 (a) One letter selected at random is a consonant.
 (b) Two letters selected at random are both vowels.
 (c) Two letters selected at random are not both vowels.
 (d) Of two letters selected at random, one is a consonant or a vowel.

22. If a pair of dice are rolled, find the probability that:
 (a) The sum of the numbers rolled is 8.
 (b) The sum of the numbers rolled is 2, 3, or 12.
 (c) The sum of the numbers rolled is not 7.
 (d) The sum of the numbers rolled is 14.

23. The outer edge of a standard Nevada roulette wheel contains 38 compartments. Of these, 36 are painted red and black, alternately, and each has one of the numbers from 1 to 36. The other two compartments are painted green and bear the signs 0 and 00, one on each. The wheel is spun, a ball is placed on the wheel, and finally the ball rolls into one of the compartments. Find the probability that the ball lands on:
 (a) the number 17 (b) an even number (not 0 or 00)
 (c) a number from 1 to 12 (d) the numbers 9 or 12

24. Suppose you select 2 of 10 different numbers and another person selects 4 of these 10 different numbers. Find the probability that the other person has selected the same 2 numbers that you did.

(25–28) Answer each question. (11.6)

25. In an election, three candidates, *R, D.* and *I,* are running for the same office. The probability of *R* winning is $\frac{1}{2}$ and the probability of *D* winning is $\frac{1}{3}$. Find the probability that

(a) R or D wins   (b) D or I wins

26. A survey revealed that out of 800 persons who had lunch at a certain restaurant, 600 ordered the special lunch, 300 ordered dessert, and 200 ordered both the special and the dessert. Find the probability that the next person who enters orders:
    (a) the special lunch or the dessert
    (b) the special lunch and the dessert
    (c) the special lunch but not the dessert
    (d) neither the special nor the dessert

27. A pair of dice are rolled two times. Find the probability of getting:
    (a) a 9 on the first roll and a 7 on the second roll
    (b) a 9 on the first roll and a 9 on the second roll

28. A box contains 8 balls, numbered from 1 to 8. One ball is drawn from the box and is not replaced. Then a second ball is drawn. Find the probability that:
    (a) The numbers on the balls are 5 and 7.
    (b) One of the numbers is 5 or 7.
    (c) Neither number is 5 or 7.
    (d) Both numbers are even.

ANSWERS

EXERCISES 1.1

1. a) $\{1, 3\}$

   b) $\{-3, 0, 1, 3\}$

   c) $\left\{-3, \dfrac{-1}{3}, 0, 1, 3, \dfrac{19}{3}\right\}$

   d) $\{-\sqrt{3}, \sqrt{3}\}$

5. $\{1, 2, 3, 4, 5, 6, 7\}$, finite

9. $\{-3, -2, -1, 0, 1, 2, 3\}$, finite

13. $\{1, 3, 5, 7, 15, 21, 35, 105\}$, finite

17. $\{15, 21, 35, 105\}$, finite

21. $\emptyset$, finite

25. $\{\sqrt{2}, \sqrt{3}, \sqrt{5}, \sqrt{6}, \sqrt{7}, \sqrt{8}, \sqrt{10}, \ldots\}$, infinite

29. Rational, integral

33. Rational, nonintegral

37. Irrational

41. a) $\{2, 5, 7, 10, 14, 70\}$
   b) $\{2\}$

45. a) $\{3, 4, 5, 6, \ldots\}$
   b) $\{6, 7, 8, \ldots\}$

49. a) $\{\ldots, -3, -2, -1, 1, 2, 3, \ldots\}$
   b) $\emptyset$

3. a) $\left\{\dfrac{9}{9}\right\}$

   b) $\left\{-\sqrt{9}, \dfrac{-90}{9}, \dfrac{9}{9}\right\}$

   c) $\left\{-\sqrt{9}, \dfrac{-90}{9}, \dfrac{9}{9}, \sqrt{\dfrac{1}{9}}\right\}$

   d) $\{-\pi, \sqrt{90}\}$

7. $\{-4, -5, -6, \ldots\}$, infinite

11. $\{11, 13, 15, 17, 19\}$, finite

15. $\{3, 5, 7\}$, finite

19. $\{12, 24, 36, \ldots\}$, infinite

23. $\left\{0, \dfrac{1}{2}, \dfrac{-1}{2}, 1, -1, \dfrac{3}{2}, \dfrac{-3}{2}, \ldots\right\}$, infinite

27. The set of integers, infinite

31. Irrational

35. Rational, nonintegral

39. Rational; integral

43. a) $\{1, 2, 5, 10, 25\}$
   b) $\{1, 5\}$

47. a) Set of real numbers
   b) $\emptyset$

ANSWERS

EXERCISES 1.2

1. Reflexive
3. Symmetric
5. Substitution (or transitive)
7. Substitution
9. Substitution (or transitive)
11. Symmetric
13. Reflexive
15. Commutative, addition
17. Inverse, addition
19. Inverse, multiplication
21. Commutative, multiplication
23. Identity, multiplication
25. Associative, multiplication
27. Associative, addition
29. Identity, addition
31. Distributive
33. Closure, multiplication
35. Associative and commutative, addition
37. Associative and commutative, multiplication
39. Associative, Inverse, Identity (addition)
41. Associative, Inverse, Identity (multiplication)
43. Commutative (addition), Distributive
45. Distributive
47. Distributive, Identity (multiplication)
    Commutative, Associative (addition)
49. Distributive, Commutative (addition)
51. (1) Distributive axiom
    (2) Symmetric axiom

53. (1) Distributive axiom
    (2) Distributive axiom
    (3) Substitution axiom

55. (1) Associative axiom, multiplication
    (2) Inverse axiom, multiplication
    (3) Substitution axiom
    (4) Identity axiom, multiplication
    (5) Substitution (or transitive) axiom

57. (1) Definition of subraction
    (2) Identity axiom, addition
    (3) Substitution (or transitive) axiom

59. (1) Identity axiom, addition
    (2) Inverse axiom, addition
    (3) Substitution axiom
    (4) Associative axiom (addition), substitution
    (5) Inverse axiom (addition), substitution
    (6) Identity axiom (addition), substitution

EXERCISES 1.3

1. 84	3. -36	5. -70	7. 1,000
9. 0	11. -5	13. -12	15. 18
17. -7	19. 4	21. 65	23. 16
25. -8	27. -45	29. $\frac{1}{2}$	31. $\frac{-1}{30}$
33. $\frac{2}{9}$	35. $\frac{-1}{2}$	37. -12	39. 80
41. -1	43. -125	45. 3	47. 0

49. $\dfrac{-11}{3}$ **51.** $\dfrac{-32}{9}$ **53.** 26 **55.** -72

57. 144 **59.** 5 **61.** 6 **63.** 0

65. 49 **67.** -25 **69.** -268 **71.** $\dfrac{9}{16}$

73. 3 **75.** 120 **77.** 0 **79.** 2

81. (1) Distributive axiom
    (2) Identity axiom, addition
    (3) Zero Factor Theorem
    (4) Identity axiom, multiplication
    (5) Inverse axiom, addition
    (6) Addition Cancellation Theorem
    (7) Symmetric axiom

83. (1) Definition of subtraction
    (2) Theorem 9
    (3) Theorem 1, $-(-b) = b$
    (4) Commutative axiom, addition
    (5) Definition of subtraction

85. (1) Commutative axiom, multiplication
    (2) Theorem 11
    (3) Commutative axiom, multiplication

87. (1) Definition of subtraction
    (2) Distributive axiom
    (3) Theorem 11
    (4) Definition of subtraction

89. $\dfrac{a}{c} + \dfrac{b}{c} = a\left(\dfrac{1}{c}\right) + b\left(\dfrac{1}{c}\right)$ Definition of division

    $= (a + b)\dfrac{1}{c}$ Distributive axiom

    $= \dfrac{a + b}{c}$ Definition of division

91. $\dfrac{-a}{b} = \dfrac{(-1)(-a)}{(-1)b}$ Fundamental Theorem of Fractions

    $= \dfrac{a}{-b}$ Theorems 13 and 8

and

    $\dfrac{-a}{b} = (-1)a\left(\dfrac{1}{b}\right)$ Definition of division and Theorem 8

    $= (-1)\dfrac{a}{b}$ Definition of division

    $= -\dfrac{a}{b}$ Theorem 8

therefore,

$\dfrac{-a}{b} = \dfrac{a}{-b} = -\dfrac{a}{b}$ Substitution (or transitive) axiom

93. $\dfrac{a}{c} - \dfrac{b}{c} = \dfrac{a}{c} + \left(-\dfrac{b}{c}\right)$ Definition of subtraction

    $= \dfrac{a}{c} + \dfrac{-b}{c}$ Theorem 21

    $= \dfrac{a + (-b)}{c}$ Theorem 19

    $= \dfrac{a - b}{c}$ Definition of subtraction

95. $\dfrac{1}{b} + \dfrac{1}{d} = \dfrac{1(d)}{b(d)} + \dfrac{1(b)}{d(b)}$ Fundamental Theorem of Fractions

$= \dfrac{d}{bd} + \dfrac{b}{db}$ Identity axiom, multiplication

$= \dfrac{d}{bd} + \dfrac{b}{bd}$ Commutative axiom, multiplication

$= \dfrac{d + b}{bd}$ Theorem 19

$= \dfrac{b + d}{bd}$ Commutative axiom, addition

EXERCISES 1.4

1. $10 > 8$

3. $14 < 17$

5. $26 > 0$

7. $0 > -8$

9. $-15 > -20$

11. $-25 < -10$

13. $13 > -9$

15. $-7 < 4$

17. $0.5 > \dfrac{1}{4}$

19. $\dfrac{1}{8} = 0.125$

21. $-0.5 = \dfrac{-1}{2}$

23. $2x < 0$

25. $n - 3 \geq 0$

27. $x + y < 9$

29. $10 \geq x$

31. $-5 < 2x < 5$

33. $-7 \leq y < 2$

35. $3.1415 < \pi < 3.1416$

37. $11 \leq x \leq 19$

39. $6.45 \leq x \leq 6.55$

41. $\{-1, -2, -3, -4, \ldots\}$

43. $\{-4, -3, -2, -1, 0, \ldots\}$

45. $\{-1, 0, 1\}$

47. $\{-6, -5, -4, -3, -2\}$

49. $\{0, 1, 2, 3, \ldots\}$

51.

53.

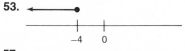

55.

57.

59.

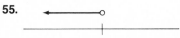

61.

63.

65. (1) Definition of $<$
    (2) Order axiom 4
    (3) Definition of subtraction

67. (1) Order axiom 4
    (2) Definition of Subtraction and Distributive axiom

(4) Identity axiom, addition
(5) Substitution axiom

(3) Associative, Commutative axioms, addition
(4) Identity, Inverse axiom, addition
(5) Definition of subtraction
(6) Substitution axiom
(7) Order axiom 4
(8) Substitution axiom

69. (1) Theorem of Exercise 65
(2) Order axiom 4
(3) Order axiom 3
(4) Theorems: $a(-b) = -ab$, $(-a)(-b) = ab$, Distributive axiom, Substitution axiom.
(5) Commutative axiom, addition
(6) Order axiom 4
(7) Definition of $<$

71. If $a < b$, then $b > a$ Definition of $<$
$b > a$ if and only if $b - a > 0$ Order axiom 4
$(b - a)c > 0$ Order axiom 3
$bc - ac > 0$ Distributive axiom
$bc > ac$ Order axiom 4
$ac < bc$ Definition of $<$

73. If $a < 0$, then $-a > 0$ Theorem of Exercise 65
If $b < 0$, then $-b > 0$ Theorem of Exercise 65
$(-a)(-b) > 0$ Order axiom 3
$ab > 0$ Theorem: $(-a)(-b) = ab$

EXERCISES 1.5

1. 64
3. 125
5. $\frac{9}{100}$

7. $\frac{512}{27}$
9. 16
11. 0.36

13. $\frac{-1}{216}$
15. 243
17. $-100,000$

19. -48
21. 900
23. -180
25. $-5,000$
27. $-1,125$
29. 40

31. 78
33. $\frac{1}{2}$
35. 8.367

37. 3.162
39. 6
41. 20

43. $\frac{7}{8}$
45. 3.893
47. 9.655

49. 8, Theorem 1
51. 9.5, Definition of $\sqrt{a}$
53. 21, Theorem 2
55. 6, Theorem 3
57. $2\sqrt{10}$, Theorem 3
59. $7\sqrt{2}$, Theorem 3

61. $6\sqrt{5}$, Theorem 3

63. $\dfrac{\sqrt{3}}{3}$, Theorem 4

65. $\dfrac{\sqrt{10}}{10}$, Theorem 4

67. $\dfrac{1}{6}$, Theorem 4

69. $\dfrac{\sqrt{14}}{2}$, Theorem 4

71. $2\sqrt{3}$, Theorems 3 and 4

73. $21 - 8\sqrt{5}$

75. $67 + 42\sqrt{2}$

77. 31

79. 63

81. $4 - \sqrt{3}$

83. $\dfrac{-1 + 2\sqrt{2}}{2}$

85. $\dfrac{4 + \sqrt{10}}{6}$

87. $6 - 4\sqrt{2}$

89. $3 - \sqrt{5}$

91. $\sqrt{5} + 1$

93. $7 - 2\sqrt{7}$

95. $8 - 3\sqrt{7}$

97. $\dfrac{\sqrt{5} + \sqrt{3}}{2}$

99. $17 + 12\sqrt{2}$

101. $2i$

103. $5i$

105. $i\sqrt{3}$

107. $i\,2\sqrt{6}$

109. $1 + 9i$

111. 9

113. $-10 + 4i$

115. $-3 + 4i$

117. $16 - 3i$

119. $-16 + 30i$

121. $27 + 36i$

123. $-2 - i\,2\sqrt{3}$

125. 25

127. 2

129. 59

131. $\dfrac{3 + 4i}{25}$

133. $4 + 2i$

135. $\dfrac{-5 - 12i}{13}$

137. (1) Definition of x^2
    (2) Commutative, Associative axioms, multiplication
    (3) Definition of $\sqrt{x}$ for $x > 0$
    (4) Definition of $\sqrt{x}$ for $x > 0$
    (5) Theorem on Existence of Positive Roots (Section 1.3)

139. (1) Definition of addition of complex numbers
    (2) Definition of addition of complex numbers
    (3) Commutative axiom, addition of real numbers
    (4) Substitution axiom

1.6 CHAPTER REVIEW

1. $\{52, 54, 56, 58\}$, finite

2. $\{2, 1, 0, -1, -2, \ldots\}$, infinite

3. $\left\{\dfrac{1}{10}, \dfrac{2}{10}, \dfrac{3}{10}, \dfrac{4}{10}, \dfrac{5}{10}, \dfrac{6}{10}, \dfrac{7}{10}, \dfrac{8}{10}, \dfrac{9}{10}\right\}$, finite

4. $\{\sqrt{2}, -\sqrt{2}, 2\sqrt{2}, -2\sqrt{2}, 3\sqrt{2}, -3\sqrt{2}, \ldots\}$, infinite

5. $\emptyset$, finite

6. irrational

7. rational, nonintegral

8. rational, integral

9. $A \cup B = \{1, 2, 3, 7, 21\}$, $A \cap B = \{3, 7\}$
10. $A \cup B = \{4, -4, 6, -6, 8, -8, 12, -12, \ldots\}$
    $A \cap B = \{12, -12, 24, -24, 36, -36, \ldots\}$
11. Symmetric 　　　　　　　　12. Reflexive 　　　　　　　13. Substitution
14. Associative, multiplication
15. Commutative, addition
16. Inverse, addition
17. Commutative, multiplication
18. Identity, multiplication
19. Associative, addition
20. Inverse, multiplication
21. Identity, addition 　　　　　　　　　　　　　　　　22. Distributive
23. Associative, Inverse, Identity (addition)
24. Associative, Inverse, Identity (multiplication)
25. Associative, Inverse, Identity (multiplication)
26. Distributive, Associative (multiplication)
27. 18 　　　　　　　　　28. -5 　　　　　　　　29. -6
30. 0 　　　　　　　　　　31. -1 　　　　　　　　32. 150
33. 11 　　　　　　　　　　34. 18 　　　　　　　　　35. 19
36. 7 　　　　　　　　　　37. $2x - 8 > 0$ 　　　　38. $y + 4 \le 0$
39. $x - 6 \ge y$ 　　　　　　　　　　　　　40. $-6 < 5n < 8$
41. $\{3, 2, 1, 0, -1, -2, \ldots\}$ 　　　　42. $\{-3, -2, -1, 0, 1, 2, \ldots\}$
43. $\{-4, -3, -2, -1, 0, 1, 2, 3, 4\}$ 　　44. $\{3, 4, 5, 6, 7, 8\}$
45.

46.

47.

48. $6\sqrt{2}$

49. $5\sqrt{3}$ 　　　　　　　　　　50. $\dfrac{\sqrt{6}}{3}$

51. $\dfrac{\sqrt{15}}{10}$ 　　　　　　　　52. $16 - 6\sqrt{7}$

53. 13 　　　　　　　　　　　54. $\dfrac{1 - \sqrt{5}}{3}$

55. $\dfrac{4 - \sqrt{6}}{2}$ 　　　　　　　56. $7i$

57. $-i\sqrt{6}$ 　　　　　　　　　58. $3 - i$
59. $-16 + 30i$ 　　　　　　　60. 20

61. $\dfrac{7 + 4i}{65}$ 　　　　　　　　62. $-i$

EXERCISES 2.1

1. 69   **3.** 4   **5.** -3

7. -128 **9.** -9 **11.** 0

13. $x^2 + x - 6$ **15.** $x^3 - 6x^2 - 4x + 24$

17. $3x^2 - 9x - 4$ **19.** $t^3 + 27$

21. $-y^2 - 16y + 7$ **23.** $3x + 3$

25. $x^2 - 2xy + y^2 - x + y$ **27.** $3x^2 - 18x$

29. $2x^3 + 10x^2$ **31.** $-6y^3 + 42y^2$

33. $8x^3 - 16x^2 + 8x$ **35.** $-x^4y - 6x^3y^2 - 9x^2y^3$

37. $x^2 - 3x - 40$ **39.** $2y^2 - 13y + 21$

41. $x^2 - 2x + 1$ **43.** $36y^2 + 60y + 25$

45. $81x^2 - 16$ **47.** $x^4 - 5x^2 + 4$

49. $16x^4 - y^4$ **51.** $x^3 + 5x^2 - 16x - 80$

53. $35x^2 + 2xy - y^2$ **55.** $36x^2y^2 - 9x^2 - 4y^2 + 1$

57. $2x^2 + 4x - 5$ **59.** $4x^2 - 3xy - y^2$

61. $4x^2 - 4x + 1$ **63.** $x - 3 + \dfrac{-19}{x - 3}$

65. $3x - 2$ **67.** $x^2 + 4x + 4$

69. $y^2 + 1 + \dfrac{-2}{y^2 - 4}$ **71.** $y^2 + 2y - 5 + \dfrac{20}{y^2 + 5}$

73. $x^2 - 4$ **75.** $x^2 - x + 1$

77. $x^2 + 5x + 25$ **79.** $x^3 - x^2 + x - 1$

81. $2x + 3 + \dfrac{-4}{3x - 2}$ **83.** $y^2 + 2y - 3$

85. $x^3 + 4 + \dfrac{10}{x^2 - 2}$ **87.** $x^2 - 4x + 4$

EXERCISES 2.2

1. $x^2 - 2x + 2 + \dfrac{5}{x - 4}$ **3.** $y^2 - 5y - 1 + \dfrac{5}{y + 3}$

5. $x^2 + x + 5 + \dfrac{15}{x - 5}$ **7.** $t^3 - 2t^2 + 3t - 6$

9. $x - 1 + \dfrac{2}{x + 1}$ **11.** $x^2 - 5x + 25$

13. $2x^2 + 6$ **15.** $5x^2 - 25x + 10$

EXERCISES 2.3

1. $(x - 10)(x + 10)$

3. $(x - 2)(x - 5)$

5. $8x(x^2 - 5x + 2)$

7. $(4y + 7)^2$

9. $(3x - y)(9x^2 + 3xy + y^2)$

11. $(t^2 + 1)(4t^2 - 5)$

13. $9xy(x - 4y)$

15. $(1 + 4y)(1 - 4y + 16y^2)$

17. $4x^2(x - 1)(x - 9)$

19. $(x - 5)^2(x + 5)^2$

21. $12x(x - 1)(5x + 3)$

23. $8y^2(2x + y)(4x^2 - 2xy + y^2)$

25. $(x - 2)(x + 2)(x^2 + 2x + 4)(x^2 - 2x + 4)$

27. $x(x - 2)(x + 2)(7x^2 + 1)$

29. $(x + 7 - y)(x + 7 + y)$

31. $(x^2 - x + 1)(x^2 + x - 1)$

33. $(x - 5y - 4xy)(x - 5y + 4xy)$

35. $(x^2 + 2x + 2)(x^2 - 2x + 2)$

37. $(y^2 + 5y + 3)(y^2 - 5y + 3)$

39. $(x + 2)(x^2 + 4)$

41. $(y - 1)(y + 1)(y + 6)$

43. $(x + y)(x - y)^2$

45. $xy(x - 2)(y + 3)$

47. $5(2r - 7)(s + 1)$

49. $(2x + 5)(3x + 4)$

51. $(2x - 9)(5x + 4)$

53. $(2y + 25)(8y - 3)$

55. $2x^2(x + 2)(10x - 3)$

57. $x(x - 2)(x + 2)(x - 5)(x + 5)$

59. $(x + 2)(x + 1)(x^2 - x + 1)$

61. $(xy - x - 1)(xy + x - 1)$

63. $(r^2 - ar + a^2)(r^2 + ar + a^2)$

65. $u(u + 1)(u - 1)^2$

67. $(x^2 - bx + a)(x^2 + bx + a)$

69. $xy(a + b)(x + y)$

EXERCISES 2.4

1. $\dfrac{3}{2}$

3. $\dfrac{x - 9}{x + 9}$

5. -1

7. $\dfrac{y}{y + 3}$

9. $\dfrac{4 + y}{4 - y}$

11. $\dfrac{1}{y^2 - 6y + 36}$

13. $-20x - 4$

15. $\dfrac{x + 4}{x + 2}$

17. $\dfrac{n + 2}{n}$

19. $\dfrac{x^2 - 4}{4x^2}$

21. $\dfrac{2x + 3}{x(x + 3)}$

23. $\dfrac{5x - 4}{20x^2}$

25. $\dfrac{4}{(y - 2)(y + 2)}$

27. 0

29. $\dfrac{x}{2(x - 2)}$

31. $\dfrac{y^2 + 2y - 1}{y^3}$

33. $\dfrac{x^2 - 6x - 6}{(x - 6)^2}$

35. $\dfrac{4xy}{(x - y)(x + y)}$

37. $\dfrac{9}{t - 9}$

39. $\dfrac{x}{(x - 1)(x + 2)(x - 3)}$

41. 0

43. $\dfrac{-1}{20t}$

45. 0

47. $\dfrac{1}{y(x-y)}$

EXERCISES 2.5

1. $\dfrac{1}{4x}$

3. $\dfrac{5x^3}{12y^3z^2}$

5. $\dfrac{x+4}{x+3}$

7. $\dfrac{1}{x-2}$

9. $\dfrac{4x}{(x+4)(x+2)}$

11. $\dfrac{(y-1)(y+5)}{(y+1)(y-5)}$

13. $\dfrac{6(x+1)}{5}$

15. $\dfrac{(x+7)(x-7)}{(x+5)(x-5)}$

17. 1

19. $\dfrac{x}{y}$

21. $\dfrac{3x}{5}$

23. $\dfrac{7}{11}$

25. $\dfrac{-x-1}{x}$

27. 240

29. $\dfrac{y-x}{y+x}$

31. $\dfrac{a}{b}$

33. 3

35. $\dfrac{x-6}{x-3}$

37. $\dfrac{1}{x+1}$

39. $x+1$

41. y

43. $\dfrac{x}{y}$

45. $-2x$

47. 40

49. $\dfrac{x+4}{x-2}$

2.6 CHAPTER REVIEW

1. (a) 0 (b) 4 (c) $2y^2-13y+15$
2. $4x^2-x-9$

3. $3x-4$

4. $-3x^3y+6x^2y^2+3xy^3$

5. $5x^4+28x^2-12$

6. $x^2y^2 - 5x^2 + 4y^2 - 20$

7. $2x^2 + 3x + 5 + \dfrac{5}{3x - 4}$

8. $y^2 - 2y + 5$

9. $x^2 - 7x - 1$

10. $t^3 + t - 2$

11. $4x^2(3x - 4)^2$

12. $(x - 2y)(x + 2y)(x^2 + 4y^2)$

13. $(x - 1)(x + 1)(x - 5)(x + 5)$

14. $x^3(3x - 1)(9x^2 + 3x + 1)$

15. $(x^2 - 2 - 3y)(x^2 - 2 + 3y)$

16. $(x - 1)(6x + 5)(6x^2 + x + 5)$

17. $(x - 7)(x + 1)(x^2 - x + 1)$

18. $(x - 9)(x + 9)(2x - 1)$

19. $\dfrac{x - 5}{x - 2}$

20. $\dfrac{y + 3}{y - 3}$

21. $\dfrac{2}{y - 6}$

22. $\dfrac{-x}{7}$

23. $\dfrac{7}{12(x - 5)}$

24. $\dfrac{4}{y + 4}$

25. $\dfrac{1}{x(x + 3)}$

26. $\dfrac{2(x^2 + 2)}{(x - 1)(x + 1)^2}$

27. $\dfrac{x + 3}{3x + 1}$

28. $\dfrac{x^2(x - 2)}{4(x - 4)}$

29. $\dfrac{5x + 1}{x + 5}$

30. $xy(x + y)$

31. $\dfrac{y + 4}{4}$

32. $2(x + 1)$

33. -1

34. $y + 5$

35. $\dfrac{1}{5}$

36. $\dfrac{15}{7}$

EXERCISES 3.1

1. 4

3. 5

5. -9

7. $\dfrac{1}{2}$

9. -3

11. $y = 10 - 2x$

13. $y = 5x - 5$

15. $x = \dfrac{8y - 4}{3}$

17. $z = 5x - 2y - 10$

19. $y = 8 + 4x - 4x^2$

21. $b = \dfrac{P - 2a}{2}$

23. $h = \dfrac{3A}{b^2}$

25. $n = \dfrac{r - s}{st}$

27. $m = \dfrac{2K}{v^2 + 2g}$

29. $x = s - r$ for $r \neq -s$

31. $x > 4$

33. $x \leq -2$

$-2 \ 0$

35. $x \leq -6$
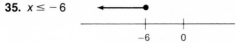
$-6 \qquad 0$

37. $x > 5$

$0 \qquad 5$

39. $x \leq 6$
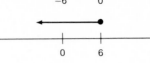
$0 \qquad 6$

41. $x < 4$
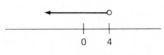
$0 \qquad 4$

43. $x \geq -8$
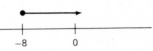
$-8 \qquad 0$

45. $x \geq 0$

0

47. $x > 1$
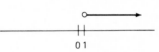
$0 \ 1$

49. $x < 3$

$0 \quad 3$

EXERCISES 3.2

1. $\{3, 7\}$ **3.** $\{-1, 5\}$

5. $\{-5, 0\}$ **7.** $\left\{-4, \dfrac{5}{2}\right\}$

9. $\{-5, 5\}$ **11.** $\{-2, -3, 3\}$

13. $\{0, 4, 6\}$ **15.** $\{-1, 1, -3, 3\}$

17. $\left\{0, \dfrac{1}{6}\right\}$ **19.** $\{2, 3, 4\}$

21. $\left\{\dfrac{1}{3}\right\}$ **23.** $\{15\}$

25. $\{7\}$ **27.** $\left\{\dfrac{3}{10}\right\}$

29. $\{5\}$ **31.** $\left\{\dfrac{1}{2}\right\}$

33. $\{10\}$ **35.** $\emptyset$ (Note that $x = 5$ is not in the domain)

37. $\{7\}$ **39.** $\left\{\dfrac{-2}{3}\right\}$

41. $\{2\}$ **43.** $\left\{5, \dfrac{8}{3}\right\}$

45. $\{4\}$ (Note that $x = 3$ is not in the domain)   **47.** $\{-9, 9\}$

49. $\{-5, -2\}$ **51.** $\{2a, 5a\}$

53. $\{a, b\}$

55. $x = \dfrac{ab}{a+b}$, $a \neq 0$, $b \neq 0$, $a \neq -b$

57. $x = \dfrac{2cd}{c+d}$, $c \neq 0$, $d \neq 0$, $c \neq -d$

59. $x = a + b$, $a \neq b$

61. $x = \dfrac{k}{10}$, $k \neq 0$

63. $x = \dfrac{a-b}{2}$, $b \neq -a$, $a \neq 3b$

EXERCISES 3.3

1. $\{-6, 6\}$

3. $\emptyset$

5. $\{-2, 10\}$

7. $\{8, 12\}$

9. $\{-3, 6\}$

11. $\left\{\dfrac{4}{5}, 4\right\}$

13. $\{-6, 2\}$

15. $\{0, 6\}$

17. $\{-1, 3\}$

19. $\emptyset$

21. $-7 < x < 7$

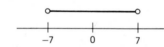

23. $x \leq -9$ or $x \geq 9$

25. $-2 \leq x \leq 10$

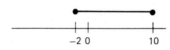

27. $x < -2$ or $x > 18$

29. $x \leq -4$ or $x \geq 1$

31. $-3 < x < 12$

33. $\dfrac{1}{3} \leq x \leq 5$

35. $x \leq -5$ or $x \geq 1$

37. $-1 < x < 3$

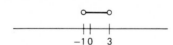

39. $x < -5$ or $x > -1$

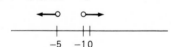

41. R

43. $\{x \mid x \leq 0\}$

45. R **47.** $\emptyset$
49. R

EXERCISES 3.4

1. $16; (x + 4)^2$

3. $36; (x - 6)^2$

5. $\frac{9}{4}; \left(y + \frac{3}{2}\right)^2$

7. $\frac{1}{4}; \left(t - \frac{1}{2}\right)^2$

9. $d^2; (x + d)^2$

11. $-2 \pm \sqrt{6}$

13. $5 \pm \sqrt{13}$

15. $3 \pm 3\sqrt{2}$

17. $-7 \pm 2\sqrt{5}$

19. $9 \pm 4\sqrt{2}$

21. $-50, 40$

23. $\frac{5 \pm \sqrt{3}}{2}, s = 5, p = \frac{11}{2}$

25. $\frac{-2 \pm i}{5}; s = \frac{-4}{5}, p = \frac{1}{5}$

27. $\frac{-1 \pm \sqrt{5}}{2}; s = -1, p = -1$

29. $\frac{1 \pm i\sqrt{14}}{3}; s = \frac{2}{3}, p = \frac{5}{3}$

31. $\frac{1 \pm 2\sqrt{2}}{2}; s = 1, p = \frac{-7}{4}$

33. $\frac{-3}{2}, \frac{2}{5}; s = \frac{-11}{10}, p = \frac{-3}{5}$

35. $\frac{2}{3}, \frac{-3}{4}; s = \frac{-1}{12}, p = \frac{-1}{2}$

37. $\frac{3 \pm i\sqrt{7}}{2}; s = 3, p = 4$

39. $\frac{5 \pm \sqrt{7}}{2}; s = 5, p = \frac{9}{2}$

41. Unequal, irrational

43. Unequal, rational

45. Unequal, imaginary

47. Unequal, irrational

49. Unequal, rational

51. Unequal, imaginary

53. Double, rational

55. $\{2, -2, 3, -3\}$

57. $\{5, -5\}$

59. $\{\sqrt{5}, -\sqrt{5}, 2i, -2i\}$

61. $\left\{1, 3, \frac{-1 \pm i\sqrt{3}}{2}, \frac{-3 \pm i\,3\sqrt{3}}{2}\right\}$

63. $\left\{4, -1, -2 \pm i\,2\sqrt{3}, \frac{1 \pm i\sqrt{3}}{2}\right\}$

65. $\left\{1, -1, \frac{1 \pm i\sqrt{3}}{2}, \frac{-1 \pm i\sqrt{3}}{2}\right\}$

67. $\{3, -3, 3i, -3i\}$

69. $\left\{\frac{1}{2}, \frac{-1}{2}, \frac{i}{3}, \frac{-i}{3}\right\}$

71. $\left\{-2, 3, \frac{1}{2}, \frac{-1}{3}\right\}$

73. $\{-b \pm \sqrt{b^2 - c}\}$

75. $\{a, -3a\}$

77. $\{a \pm a\sqrt{2}\}$

79. $\{ry - 1, 1 - ry\}$

81. -9

83. 13

85. 3

87. 4

91. 4 or -4

95. $x^2 - 3x - 28 = 0$

99. $x^2 + 10x + 34 = 0$

89. 20

93. (a) $k = 1$ (b) $k < 1$ (c) $k > 1$

97. $x^2 - 12x + 6 = 0$

EXERCISES 3.5

1. $3 < x < 6$

5. $x < \dfrac{1}{2}$ or $x > 3$

9. $-3 < x < 3$

13. $R - \{1\}$

17. $x < -1$ or $x > 1$

21. $1 \le x < 3$

25. $x < -8$ or $x > -4$

29. $-5 \le x < 0$ or $x \ge 5$

33. $\dfrac{-5 - \sqrt{73}}{6} \le x \le \dfrac{-5 + \sqrt{73}}{6}$

37. R

41. $\dfrac{1 - \sqrt{5}}{2} \le x \le \dfrac{1 + \sqrt{5}}{2}$

45. $x \le -2$ or $2 \le x \le 5$

47. $x \le -3$ or $-2 \le x \le 2$ or $x \ge 3$

49. $x < -7$ or $-\sqrt{5} < x < \sqrt{5}$

53. R

3. $x \le -2$ or $x \ge 7$

7. $2 \le x \le 4$

11. $y \le 0$ or $y \ge 9$

15. $\{2\}$

19. $0 < x < \dfrac{1}{5}$

23. $x < 0$

27. $x \le \dfrac{-1}{3}$ or $x > 3$

31. $x \le 6 - \sqrt{2}$ or $x \ge 6 + \sqrt{2}$

35. $\emptyset$

39. R

43. $-4 < x < 2$ or $x > 5$

51. $x \le -\sqrt{7}$ or $x \ge \sqrt{7}$

EXERCISES 3.6

A. **1.** $72''$, $90''$

5. (a) $40''$, $40''$, $34''$

    (b) $A = 1496$ sq. in.

9. 20 trees, 80 shrubs

11. 5 liters pure alcohol, 75 liters of 20%

13. Northbound 230 mph, southbound 270 mph

15. 17 mi.

17. 40 mph

19. 2 mph

21. 20 min.

3. 5 cm

7. $25,000 at 8%

    $15,000 at 10%

23. 2 hr.

25. 14 min.

B. **1.** $6\sqrt{2}$ ft.

3. 5 cm

5. 8 dm, 15 dm

7. 200 mph eastbound, 250 mph westbound

9. 30 mph

11. 10 hr, 15 hr

13. 10 days

15. 60 boxes

17. 40¢

19. $30 - \sqrt{275} \approx 13.42$ ft.

21. 0.00034

C. **1.** $3 \le x \le 30$

3. $-10 < c < 30$

5. $\dfrac{-1}{16} \le x \le \dfrac{1}{16}$

7. $x \ge 10$

9. $15 \le v \le 35$

3.7 CHAPTER REVIEW

---

1. 6

2. -5

3. $x = \dfrac{5y + 30}{6}$ and $y = \dfrac{6x - 30}{5}$

4. $f = \dfrac{mg - T}{m}$

5. $x \ge 2$

6. $x < -3$

7. $\{-1, 7\}$

8. $\{-4, 5\}$

9. $\{2, 7, -7\}$

10. $\{10, -10\}$

11. $\{11\}$

12. $\{-1\}$ (Note that 6 is not in domain)

13. $\left\{\dfrac{-5}{3}, 5\right\}$

14. $\{10, -7\}$

15. $-3 \le x \le \dfrac{-5}{3}$

16. $t < -1$ or $t > 7$

17. $x \le -7$ or $x \ge 3$

18. $-6 < x < 18$

19. $\emptyset$

20. R

21. $\{-6 \pm \sqrt{10}\}$

22. $\{4 \pm 2i\}$

23. $\left\{\dfrac{-1 \pm i\sqrt{5}}{3}\right\}$, $s = \dfrac{-2}{3}$, $p = \dfrac{2}{3}$

24. $\left\{\dfrac{4 \pm \sqrt{6}}{2}\right\}$, $s = 4$, $p = \dfrac{5}{2}$

25. $\{\sqrt{6}, -\sqrt{6}, 2i, -2i\}$

26. $\{2, -1 \pm i\sqrt{3}\}$

27. $-2 < x < 5$

28. $x \le -6$ or $x \ge -4$

29. $x \le -6$ or $x > -2$

30. $-5 < x < 0$ or $x > 2$

31. $-2 < x < 10$, $x \ne 4$

32. $x \le 1$

EXERCISES 4.1

1. Yes

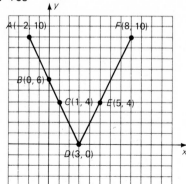

3. No

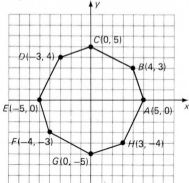

5. Yes

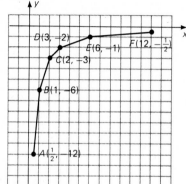

7. No

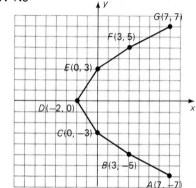

9. $d = \{5, 6, 7, 8\}$, $r = \{1, 2\}$, yes
11. $d = \{0, 1, 8\}$, $r = \{0, 1, 2, -1, -2\}$, no
13. $d = R$, $r = R$, yes
15. $d = R$, $r = \{y \mid y \geq -1\}$, yes
17. $d = \{x \mid x > 0\}$, $r = \{y \mid y \neq 0\}$, no
19. $d = \{x \mid x \leq 6\}$, $r = \{y \mid y \geq 0\}$, yes
21. $d = R$, $r = \{4\}$, yes
23. $d = \{0\}$, $r = R$, no

25. (a) $f(x) = 10 - 2x$
    (b) $f(5) = 0$
    (c) $f(-3) = 16$
    (d) $f(0) = 10$
    (e) $f(t + 1) = 8 - 2t$
    (f) $f(2c) = 10 - 4c$

27. (a) $f(x) = |2 - x|$
    (b) $f(5) = 3$
    (c) $f(-3) = 5$
    (d) $f(0) = 2$
    (e) $f(t + 1) = |1 - t|$
    (f) $f(2c) = |2 - 2c|$

29. (a) $f(x) = \dfrac{x}{x + 5}$

31. (a) $f(x) = 9 - x^2$

(b) $f(5) = \dfrac{1}{2}$

(c) $f(-3) = \dfrac{-3}{2}$

(d) $f(0) = 0$

(e) $f(t+1) = \dfrac{t+1}{t+6}$

(f) $f(2c) = \dfrac{2c}{2c+5}$

(b) $f(5) = -16$

(c) $f(-3) = 0$

(d) $f(0) = 9$

(e) $f(t+1) = 8 - t^2 - 2t$

(f) $f(2c) = 9 - 4c^2$

33. (a) $f(x) = 2\sqrt{x+4}$
   (b) $f(5) = 6$
   (c) $f(-3) = 2$
   (d) $f(0) = 4$
   (e) $f(t+1) = 2\sqrt{t+5}$
   (f) $f(2c) = 2\sqrt{2c+4}$

35. (a) $f(x) = 3x - 12$
   (b) $f(5) = 3$
   (c) $f(-3) = -21$
   (d) $f(0) = -12$
   (e) $f(t+1) = 3t - 9$
   (f) $f(2c) = 6c - 12$

37. $(3, -2), (0, -4), (0, -4)$

39. $(5, 4), (16, 8), (4, 0)$

41. $(-2, 12); (2 + 2\sqrt{2}, 4), (2 - 2\sqrt{2}, 4); (2, -4)$

43. $(5, 2), (5, -2); (5, -2), (-5, -2); (-3, 0)$

45. $(-2, -4), (1, 8), \left(\dfrac{-1}{2}, -16\right)$

EXERCISES 4.2

1.

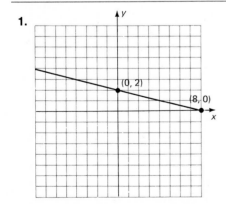

3.

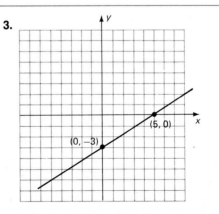

5.

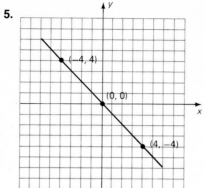

7.

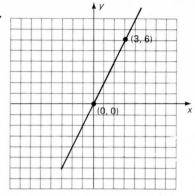

9.

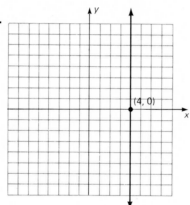

11.

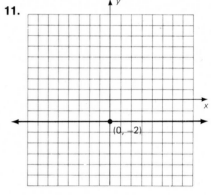

13. $\dfrac{-3}{4}$

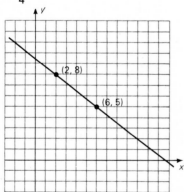

15. -3

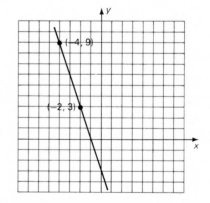

17. $\dfrac{4}{9}$

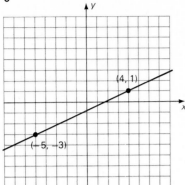

19. $\dfrac{5}{4}$

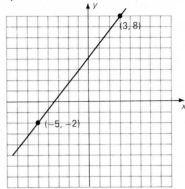

21. 0

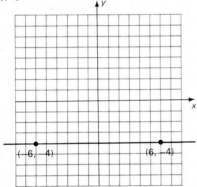

23. $x + y - 7 = 0$

25. $x + 2y + 2 = 0$

27. $y + 2 = 0$

29. $2x - y = 0$

31. $3x - y - 13 = 0$

33. $x + 2y + 2 = 0$

35. $5x - 4y = 0$

37. $y + 3 = 0$

39. $m = \dfrac{-3}{5}, b = 6$

41. $m = \dfrac{1}{4}, b = -4$

43. $m = \dfrac{-5}{2}, b = -10$

45. $m = 0, b = -8$

47. $m = 3, b = 0$

49. $m = -1, b = -1$

51. Intersect

53. Parallel

55. Intersect

57. Intersect

59. Parallel

61. Perpendicular

63. Not perpendicular

65. Perpendicular

67. Perpendicular

69. Not perpendicular

71. $2x - y - 8 = 0$

73. $x - y + 9 = 0$ **75.** $4x + y - 6 = 0$
77. $5x + 4y - 20 = 0$

79. Using points $(a, 0)$ and $(0, b)$, $m = \dfrac{b - 0}{0 - a} = \dfrac{b}{-a} = \dfrac{-b}{a}$

$y = \dfrac{-b}{a}x + b, \dfrac{b}{a}x + y = b, \dfrac{bx}{ab} + \dfrac{y}{b} = \dfrac{b}{b}$

$\dfrac{bx}{ab} + \dfrac{ay}{ab} = \dfrac{ab}{ab}$ and $\dfrac{x}{a} + \dfrac{y}{b} = 1$

81. $m_1 = \dfrac{-a_1}{b_1}$ and $m_2 = \dfrac{-a_2}{b_2}$

If $m_1 = m_2$, then $\dfrac{-a_1}{b_1} = \dfrac{-a_2}{b_2}$ and $a_1 b_2 = a_2 b_1$

Conversely, if $a_1 b_2 = a_2 b_1$, then $\dfrac{a_1}{b_1} = \dfrac{a_2}{b_2}, \dfrac{-a_1}{b_1} = \dfrac{-a_2}{b_2}$, and $m_1 = m_2$

EXERCISES 4.3

1. Vertex $(0, 0)$, axis $x = 0$

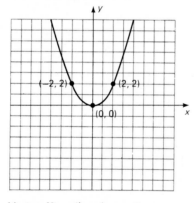

3. Vertex $(0, 0)$, axis $x = 0$

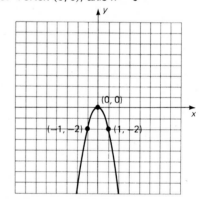

5. Vertex $(0, -4)$, axis $x = 0$

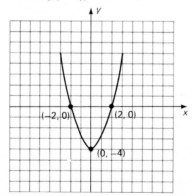

7. Vertex $(1, -1)$, axis $x = 1$

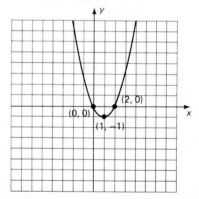

ANSWERS

9. Vertex $(4, 16)$, axis $x = 4$

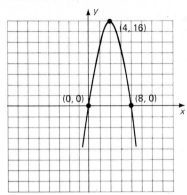

11. Vertex $(3, -4)$, axis $x = 3$

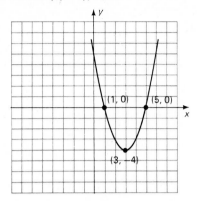

13. Vertex $(-1, 5)$, axis $x = -1$

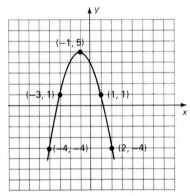

15. Vertex $(-2, -11)$, axis $x = -2$

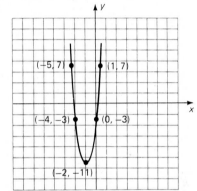

17. Vertex $\left(1, \dfrac{9}{2}\right)$, axis $x = 1$

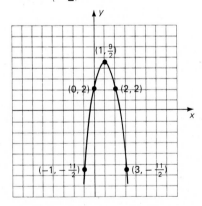

19. Vertex $(2, 0)$, axis $x = 2$

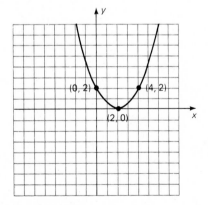

21. $-1 + \sqrt{5} \approx 1.2$

$-1 - \sqrt{5} \approx -3.2$

23. $-2 + \dfrac{1}{2}\sqrt{22} \approx 0.3$

$-2 - \dfrac{1}{2}\sqrt{22} \approx -4.3$

25. $1 + \dfrac{3}{5}\sqrt{5} \approx 2.3$

$1 - \dfrac{3}{5}\sqrt{5} \approx -0.3$

27. 120 ft., 240 ft.

29. 10,000 ft.

31. 400

33. $\dfrac{3}{5}$ hr = 36 min.

$d = 20$ mi.

35. 6, 18

EXERCISES 4.4

1.

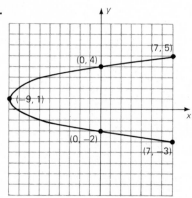

3.

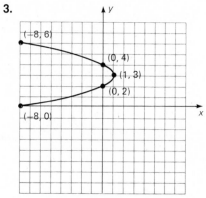

5.

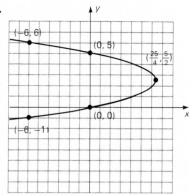

7.

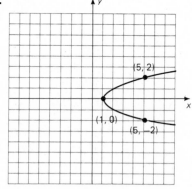

9.

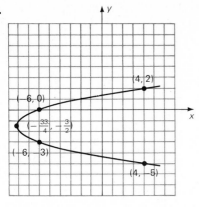

11.

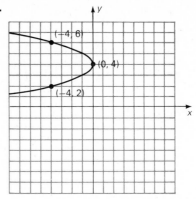

13.

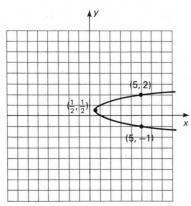

15.

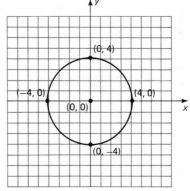

17.

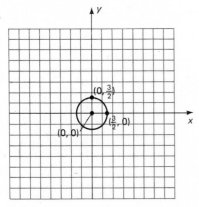

19.

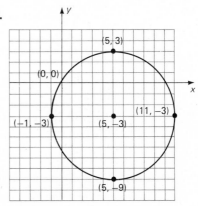

21.

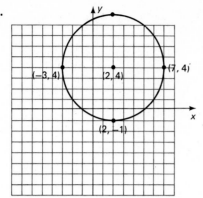

23.

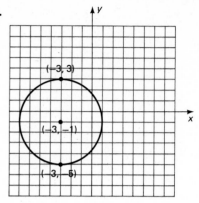

25.

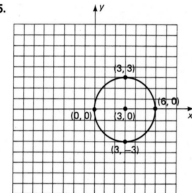

27. Ellipse

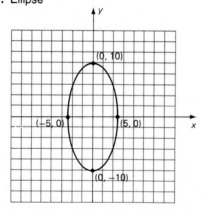

29. Hyperbola; $y = x$, $y = -x$

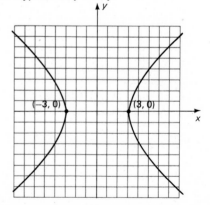

31. Hyperbola; $x = 0$, $y = 0$

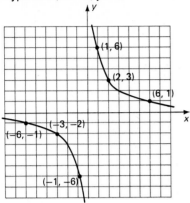

33. Ellipse

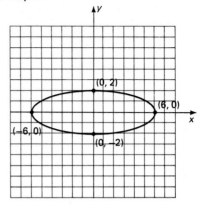

35. Hyperbola; $2y = 3x$, $2y = -3x$

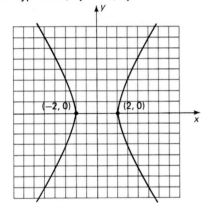

37. Hyperbola; $x = 0$, $y = 0$

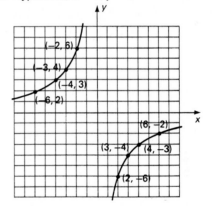

39. Hyperbola; $x = 4y$, $x = -4y$

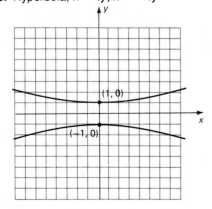

41. Ellipse

43. Circle

45. Parabola

47. Hyperbola

49. Hyperbola

EXERCISES 4.5

1.

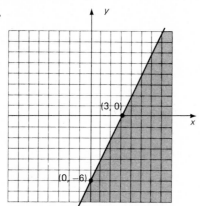

3.

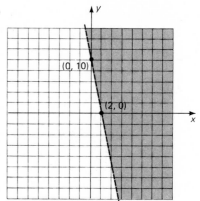

5.

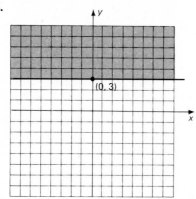

7.

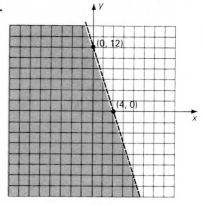

ANSWERS

9.

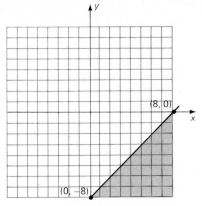

11.

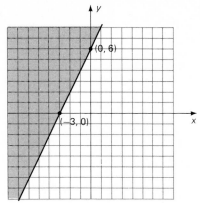

13.

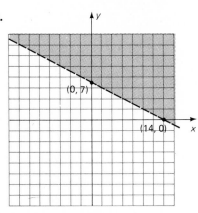

15.

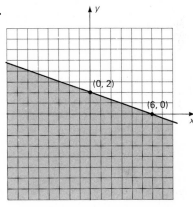

17.

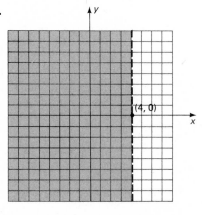

19.

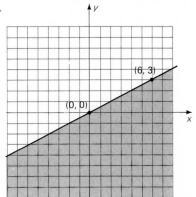

21.

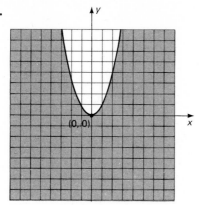

23.

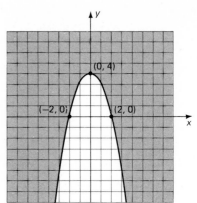

25.

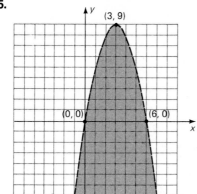

27.

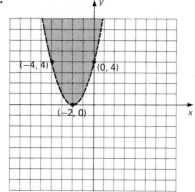

29.

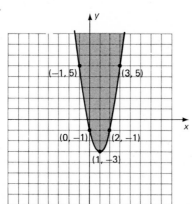

31.

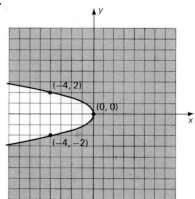

33.

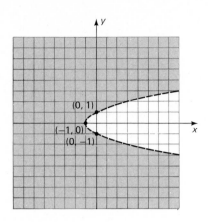

35.

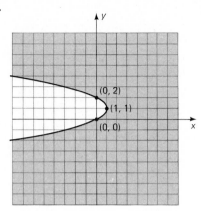

37.

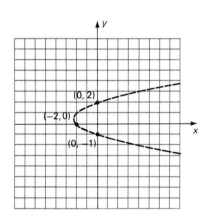

39.

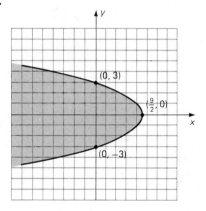

41.

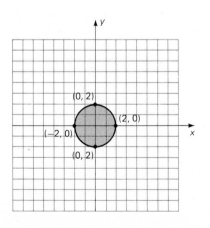

43.

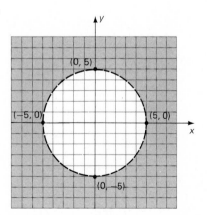

45.

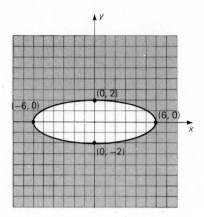

47.

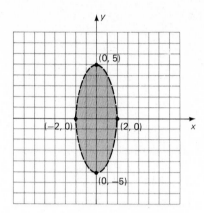

49.

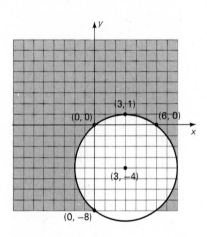

51.

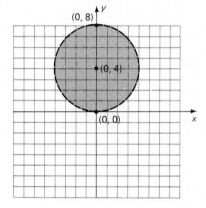

53.

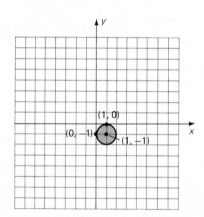

55. (a) Yes   (b) Yes   (c) No
59. (a) Yes   (b) No   (c) Yes
63. $2x + 3y > 12$
67. $9x^2 + y^2 \geq 9$

57. (a) No   (b) Yes   (c) No
61. $x^2 + y^2 < 1$
65. $y \leq 4 - x^2$
69. $x > y^2 - 4y$

EXERCISES 4.6

1. 105 lb.

3. $4.90

5. $42\frac{2}{3}$ ft.

7. $16\frac{2}{3}$ gal.

9. $d = 45t$; 180 mi.

11. $xy = 120$; 24 workers

13. $xy = 1,080$; 60 rpm

15. $A = \frac{1}{2} bh$; 150 sq. m.

17. $PV = \frac{8}{9} T$; 12.8 lb./sq. in.

19. $R = \dfrac{1.042 \times 10^{-5} L}{d^2}$; approximately 0.16 ohm

21. $r = \dfrac{15}{2} \sqrt{d}$; 22.5 gal./min.; 4 ft.

EXERCISES 4.7

1. Both are functions

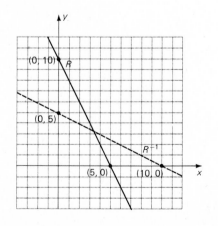

3. Both are functions

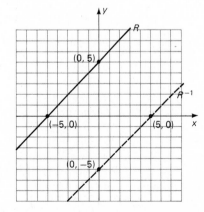

5. Both are functions

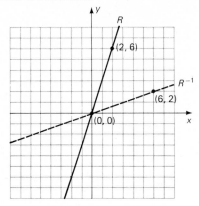

7. R is a function, R^{-1} is not

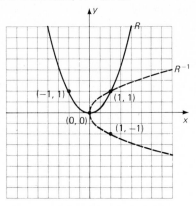

9. R is a function, R^{-1} is not

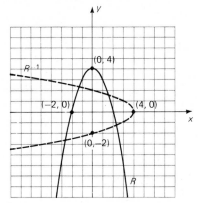

11. Both are functions

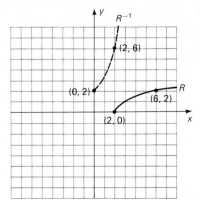

13. Neither R nor R^{-1} are functions

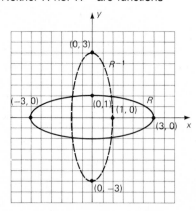

ANSWERS

15. R is a function, R^{-1} is not

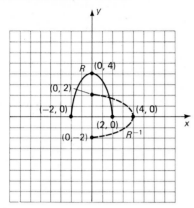

17. R is a function, R^{-1} is not

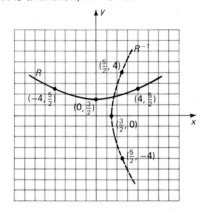

19. Neither R nor R^{-1} are functions

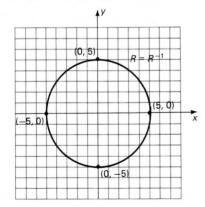

21. (a) $f(x) = 2\sqrt{x + 4},\ x \geq -4,\ y \geq 0$

(b) $f^{-1}(x) = \dfrac{x^2 - 16}{4},\ x \geq 0,\ y \geq -4$

(c) $f^{-1}(f(x)) = \dfrac{(2\sqrt{x + 4})^2 - 16}{4} = x$

$f(f^{-1}(x)) = 2\sqrt{\dfrac{x^2 - 16}{4} + 4} = x$ since $x \geq 0$

23. (a) $f(x) = x^3$, all x, all y

(b) $f^{-1}(x) = \sqrt[3]{x}$, all x, all y

(c) $f^{-1}(f(x)) = \sqrt[3]{x^3} = x$

$f(f^{-1}(x)) = (\sqrt[3]{x})^3 = x$

25. (a) $f(x) = \dfrac{10}{x},\ x \neq 0,\ y \neq 0$

(b) $f^{-1}(x) = \dfrac{10}{x},\ x \neq 0,\ y \neq 0$

(c) $f^{-1}(f(x)) = \dfrac{10}{\dfrac{10}{x}} = x$ and $f(f^{-1}(x)) = \dfrac{10}{\dfrac{10}{x}} = x$

27. (a) $f(x) = x^2,\ x \geq 0,\ y \geq 0$

(b) $f^{-1}(x) = \sqrt{x},\ x \geq 0,\ y \geq 0$

(c) $f^{-1}(f(x)) = \sqrt{x^2} = x$ since $x \geq 0$

$f(f^{-1}(x)) = (\sqrt{x})^2 = x$

29. (a) $f(x) = \sqrt{16 - 4x^2} = 2\sqrt{4 - x^2},\ 0 \leq x \leq 2,\ 0 \leq y \leq 4$

(b) $f^{-1}(x) = \dfrac{1}{2}\sqrt{16 - x^2},\ 0 \leq x \leq 4,\ 0 \leq y \leq 2$

(c) $f^{-1}(f(x)) = \dfrac{1}{2}\sqrt{16 - (\sqrt{16 - 4x^2})^2} = x$ since $x \geq 0$

$f(f^{-1}(x)) = \sqrt{16 - 4\left(\dfrac{1}{2}\sqrt{16 - x^2}\right)^2} = x$ since $x \geq 0$

31. (a) $f(x) = \dfrac{x}{x - 2},\ x \neq 2,\ y \neq 1$

(b) $f^{-1}(x) = \dfrac{2x}{x - 1},\ x \neq 1,\ y \neq 2$

(c) $f^{-1}(f(x)) = 2\left(\dfrac{x}{x - 2}\right) \Big/ \left(\dfrac{x}{x - 2} - 1\right) = x$ since $x \neq 2$

$f(f^{-1}(x)) = \dfrac{2x}{x - 1} \Big/ \left(\dfrac{2x}{x - 1} - 2\right) = x$ since $x \neq 1$

4.8 CHAPTER REVIEW

1. Yes, $x \geq 4,\ y \geq 0$

2. Yes, all $x,\ y \geq -9$

3. (a) $f(x) = x^2 - 4x - 12$
  (b) $f(0) = -12$
  (c) $f(4) = -12$
  (d) $f(-2) = 0$
  (e) $f(t-3) = t^2 - 10t + 9$

4. (a) $f(x) = -1 + \sqrt{4x+9}$
  (b) $f(0) = 2$
  (c) $f(4) = 4$
  (d) $f(-2) = 0$
  (e) $f(t-3) - -1 + \sqrt{4t-3}$

5. (a) $(0, 10), (0, -10)$ (b) $(5, 0), (-5, 0)$
  (c) $(-4, 6), (-4, -6)$ (d) $(3, 8), (-3, 8)$

6.

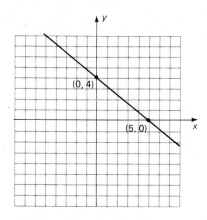

7.

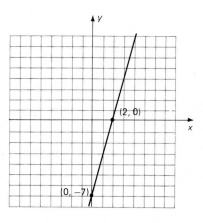

8.

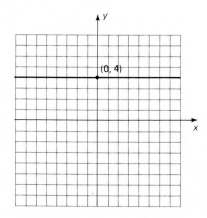

9.

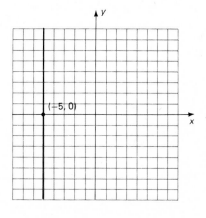

10. $\dfrac{-7}{4}$

11. $6x - 5y - 30 = 0$

12. $3x + y + 10 = 0$

13. $m = \dfrac{4}{3}, b = -8$

14. Intersect

15. Parallel

16. $(2, -4)$, $x = 2$

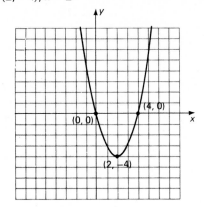

17. $(0, 8)$, $x = 0$

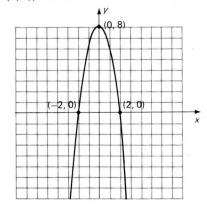

18. $\left(3, \dfrac{-17}{2}\right)$, $x = 3$

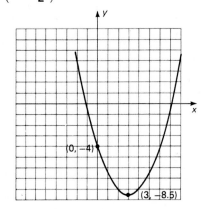

19. $\left(\dfrac{-3}{2}, \dfrac{25}{8}\right)$, $x = \dfrac{-3}{2}$

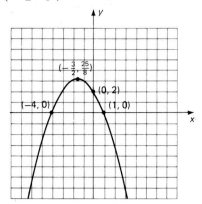

20. 30

21. Circle

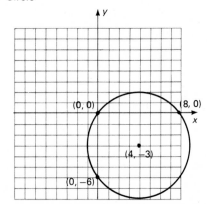

22. Parabola

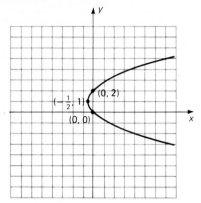

23. Hyperbola

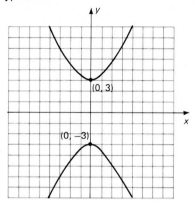

24. Circle

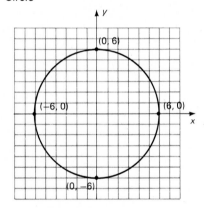

25. Ellipse

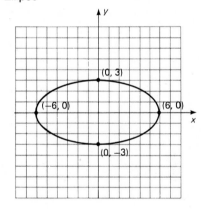

26. Hyperbola

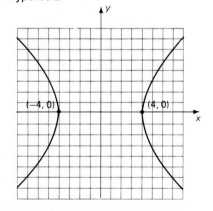

27. Parabola

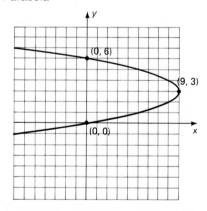

28. Hyperbola

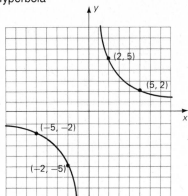

29.

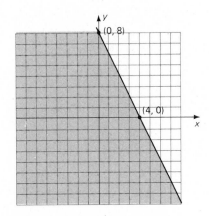

30.

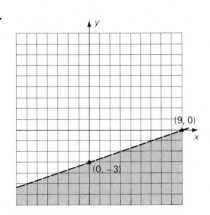

31.

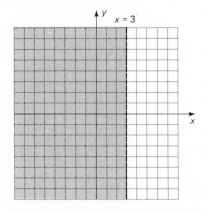

32.

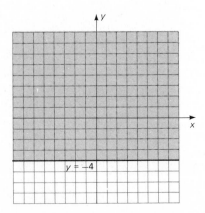

33.

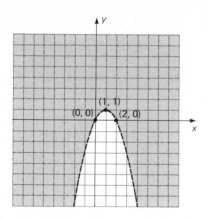

34.

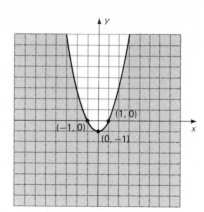

35.

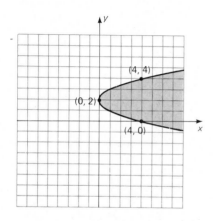

36.

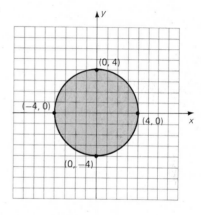

37.

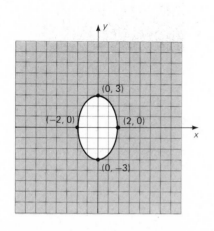

38.

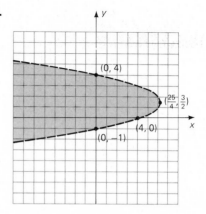

39. 8 liters

41. 6 days

43. R is a function

    R^{-1} is not

40. 15,600 lb./sq. ft.

42. $32

44. Both are functions

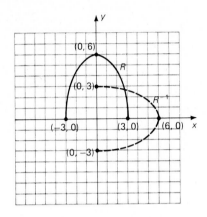

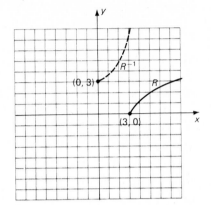

45. (a) $f(x) = \dfrac{1}{2}\sqrt{16 - 2x}$

    $x \le 8,\ y \ge 0$

  (b) $f^{-1}(x) = 8 - 2x^2$

    $x \ge 0,\ y \le 8$

  (c) $f^{-1}(f(x)) = 8 - 2\left(\dfrac{\sqrt{16 - 2x}}{2}\right)^2$

          $= x$ since $x \le 8$

    $f(f^{-1}(x)) = \dfrac{1}{2}\sqrt{16 - 2(8 - 2x^2)}$

          $= x$ since $x \ge 0$

46. (a) $f(x) = \dfrac{12}{x - 2}$

    $x \ne 2,\ y \ne 0$

  (b) $f^{-1}(x) = \dfrac{2x + 12}{x}$

    $x \ne 0,\ y \ne 2$

  (c) $f^{-1}(f(x)) = \dfrac{2\left(\dfrac{12}{x - 2}\right) + 12}{\dfrac{12}{x - 2}}$

          $= x$ since $x \ne 2$

    $f(f^{-1}(x)) = \dfrac{12}{\dfrac{2x + 12}{x} - 2}$

          $= x$ since $x \ne 0$

EXERCISES 5.1

1. $(6, 4)$

5. $(6, 0)$

3. $(20, 12)$

7. $\emptyset$

9. $(7, -4)$

11. $(x, 2x - 5)$ or $\left(\dfrac{y + 5}{2}, y\right)$

13. $(-2, 3)$

15. $(3 - 2y, y)$ or $\left(x, \dfrac{3 - x}{2}\right)$

17. $(20, 10)$

19. $(15, 8)$

21. $(3, 6)$

23. $(4, 10)$

25. $\left(\dfrac{1}{5}, \dfrac{1}{2}\right)$

27. $\left(\dfrac{-1}{4}, \dfrac{1}{3}\right)$

29. $(a + b, a - b)$

31. $\left(\dfrac{ab}{a + b}, \dfrac{ab}{a + b}\right)$ where $b \neq \pm a$

33. $(1, 0)$
$ad - bc \neq 0$

35. (a, b)
$a^2 + b^2 \neq 0$

37. gas, 75¢/gal.
oil, $1/qt.

39. 595 mph, airplane
35 mph, wind

41. 10%, mortgages
8%, bonds

43. 6 hr, 3 hr

45. 500 g, Food A
200 g, Food B

EXERCISES 5.2

1. $(5, 7, 4)$

3. $(5, -2, 6)$

5. $\left(2, \dfrac{5}{2}, \dfrac{3}{2}\right)$

7. $(5, 3, -5)$

9. $\emptyset$

11. $\left(\dfrac{1}{2}, \dfrac{1}{4}, \dfrac{1}{8}\right)$

13. $(0, 0, 0)$

15. $(x, 3x, 2x)$

17. $(z - 4, 3z + 2, z)$

19. $(2z + 1, 5 - z, z)$

21. $(4, 5, 3)$

23. $\left(\dfrac{1}{2}, 1, \dfrac{3}{2}\right)$

25. $\emptyset$

27. $(5y + 6, y, 3y - 2)$

29. 40, product P
25, product Q
6, product R

31. 90 min., machine 1
45 min., machine 2
60 min., machine 3

33. $(3, 4, 2)$

35. 12 ohms, resistor A
18 ohms, resistor B
36 ohms, resistor C

EXERCISES 5.3

1. $\{(1, 8), (-2, 5)\}$

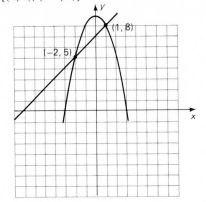

3. $\{(3, 0), (-5, -4)\}$

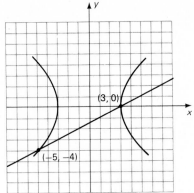

5. $\left\{\left(4, \frac{3}{2}\right), \left(-4, \frac{-3}{2}\right), (3, 2), (-3, -2)\right\}$

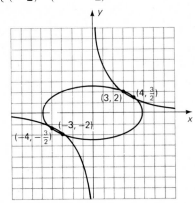

7. $\left\{\left(\frac{4}{5}, \frac{-9}{5}\right)\right\}$

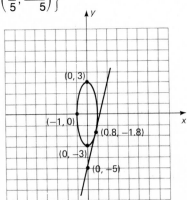

9. $\{(5, 0), (-3, 4), (-3, -4)\}$

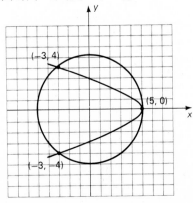

11. $\{(3 + i, -1 - 3i), (3 - i, -1 + 3i)\}$

13. $\left\{(0, 4), \left(\dfrac{-7}{4}, \dfrac{57}{16}\right)\right\}$

15. $\left\{(2, 3), \left(\dfrac{-3}{2}, -4\right)\right\}$

17. $\{(6, -2), (-3, 4)\}$

19. $\{(1, 2), (1, -2), (3, 2\sqrt{3}), (3, -2\sqrt{3})\}$

EXERCISES 5.4

1. $\{(4, 3), (4, -3), (-4, 3), (-4, -3)\}$

3. $\{(2, \sqrt{5}), (2, -\sqrt{5}), (-2, \sqrt{5}), (-2, -\sqrt{5})\}$

5. $\{(\sqrt{2}, 3), (\sqrt{2}, -3), (-\sqrt{2}, 3), (-\sqrt{2}, -3)\}$

7. $\left\{\left(\dfrac{3}{2}, \dfrac{4}{3}\right), \left(\dfrac{-3}{2}, \dfrac{-4}{3}\right)\right\}$

9. $\{(3\sqrt{5}, 2\sqrt{5}), (-3\sqrt{5}, -2\sqrt{5})\}$

11. $\{(4, 6), (4, 2), (-4, -6), (-4, -2)\}$

13. $\left\{(3, 2), (-3, -2), \left(\sqrt{2}, \dfrac{-\sqrt{2}}{2}\right), \left(-\sqrt{2}, \dfrac{\sqrt{2}}{2}\right)\right\}$

15. $\{(6, 4), (-8, -3)\}$

17. $\left\{(0, 0), \left(\dfrac{9}{5}, \dfrac{3}{5}\right)\right\}$

19. $4
45 towels

21. 5 units
$45

23. 5 ohms
20 ohms

25. 35 mph, 4 hr

EXERCISES 5.5

1.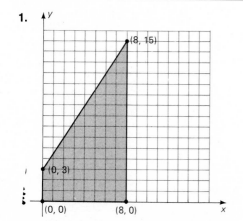

3.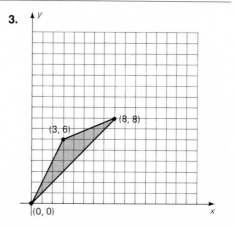

5.

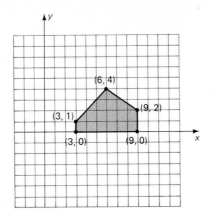

7.

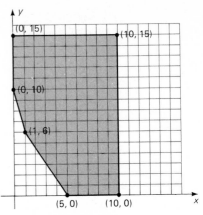

9. 6 of *A*, 8 of *B*, $62

13. 200 boxes oranges
500 boxes lemons
100 boxes limes

11. 30 oz. of *A*, 20 oz. of *B*

15. 20 living room sets
10 bedroom sets
10 dining sets

5.6 CHAPTER REVIEW

1. $(10, 7)$

2. $(-6, 9)$

3. $\emptyset$

4. $(4 - 3y, y)$

5. $(8, 16)$

6. $(8, 6, 7)$

7. $(7, -3, 5)$

8. $(x, x - 4, x - 2)$

9. $(6, 5, 4)$

10. $\{(-2, 3), (4, 6)\}$

11. $\left\{ \left(2\sqrt{3}, \dfrac{2\sqrt{3}}{3}\right), \left(-2\sqrt{3}, \dfrac{-2\sqrt{3}}{3}\right), (2i, -2i), (-2i, 2i)\right\}$

12. $\{(\sqrt{10}, 5), (\sqrt{10}, -5), (-\sqrt{10}, 5), (-\sqrt{10}, -5)\}$

13. $\left\{ (2, -3), (-2, 3), \left(\dfrac{\sqrt{2}}{2}, \sqrt{2}\right), \left(\dfrac{-\sqrt{2}}{2}, -\sqrt{2}\right)\right\}$

14. $\{(-30, -5), (15, 10)\}$

15. Vertices are $\{(0, 1), (0, 7), (6, 10), (12, 4), (12\ 0)\}$
Maximum = 62 at (6, 10)

16. Vertices are $\{(0, 8), (1, 4), (3, 2), (10, 0)\}$
Minimum cost is $14,000 operating *A* 3 days and *B* 2 days

EXERCISES 6.1

1. $(5, 2)$

3. $(3, -4)$

ANSWERS

5. $(4, 3)$

9. $(-5, 10, 5)$

13. $(6, -3, -2)$

17. $(-1, -3, 5)$

21. $(0.3, 0.5, 0.7)$

7. $(325, 150)$

11. $(3, 1, 2)$

15. $\left(2, \dfrac{-1}{3}, -1\right)$

19. $(20, 15, 10)$

23. $(1, -3, 3)$

EXERCISES 6.2

1. $(3, 2, 4)$

5. $(10, -5, 20)$

9. $(5, -1, -2)$

13. $\varnothing$

17. $(8, 2z, z)$

21. $(3, 0, -1, 1)$

3. $(2, 3, -2)$

7. $\left(\dfrac{1}{3}, \dfrac{2}{3}, \dfrac{1}{9}\right)$

11. $(3z + 3, 2z + 1, z)$

15. $(v + 5, v, 2v)$

19. $(3, 5, 4, 6)$

EXERCISES 6.3

1. 0

5. 288

9. $(-6, 7, -5, 4)$

13. $(x, 11x, 7x)$

17. $\left(x, \dfrac{7}{8}x, \dfrac{-3}{2}x, \dfrac{-3}{8}x\right)$

3. 140

7. 240

11. $(-2, 3, 5, -4)$

15. $(0, 0, 0)$

EXERCISES 6.4

1. $\begin{bmatrix} 4 & 5 \\ -7 & -1 \end{bmatrix}$

3. Undefined, dimensions are different

5. $\begin{bmatrix} 5 & 3 & -4 \\ 2 & -3 & 0 \\ 4 & -2 & -1 \end{bmatrix}$

9. $\begin{bmatrix} 6 & -9 & 12 \end{bmatrix}$

13. Undefined, dimensions not compatible

7. $\begin{bmatrix} 6 & 15 & 14 \end{bmatrix}$

11. $\begin{bmatrix} -2 & 10 \\ 16 & 11 \end{bmatrix}$

15. $\begin{bmatrix} a - 2b + & c + 3d & 7 \\ 2a & + 5c + 4d & -1 \end{bmatrix}$

17. $\begin{bmatrix} a & b & c \\ d & e & f \\ g & h & i \end{bmatrix}$

19. Undefined, different dimensions

21. $\begin{bmatrix} 6 & x + 2 & 4 & 0 \\ y + 3 & 0 & -1 & 2 \end{bmatrix}$

23. $w = -2, x = 3, y = 1, z = 5$

25. $x = -3, y = \dfrac{12}{5}$

27. $x = \dfrac{9}{2}, y = -2, z = -\dfrac{15}{4}$

29. $\begin{bmatrix} -1 & 1 \\ -2 & -1 \end{bmatrix}$

31. $\begin{bmatrix} 5 & 1 \\ 1 & 2 \end{bmatrix}$

33. $\begin{bmatrix} -5 & -1 \\ -1 & -2 \end{bmatrix}$

$\begin{bmatrix} \dfrac{2}{9} & -\dfrac{1}{9} \\ -\dfrac{1}{9} & \dfrac{5}{9} \end{bmatrix}$

37. $\begin{bmatrix} 1 & -1 \\ 1 & 0 \end{bmatrix}$

39. Unequal

41. $AB = \begin{bmatrix} 0 & 0 \\ 0 & 0 \end{bmatrix}$, no

43. $A + B = \begin{bmatrix} a + 1 & b + 1 \\ c - 1 & d + 1 \end{bmatrix} = \begin{bmatrix} 1 + a & 1 + b \\ -1 + c & 1 + d \end{bmatrix} = B + A$

45. $(A + B) + C = \begin{bmatrix} a + 2 & b + 2 \\ c & d \end{bmatrix} = A + (B + C)$

$(AB)C = \begin{bmatrix} 2a & -2b \\ 2c & -2d \end{bmatrix} = A(BC)$

47. $I = \begin{bmatrix} 1 & 0 \\ 0 & 1 \end{bmatrix}$

49. $A^{-1} = \begin{bmatrix} \dfrac{d}{ad - bc} & \dfrac{-b}{ad - bc} \\ \dfrac{-c}{ad - bc} & \dfrac{a}{ad - bc} \end{bmatrix}$; yes, $A^{-1}A = AA^{-1}$

51. $Y = A^{-1}B$ if $ad - bc \neq 0$

$\dfrac{B}{A} = A^{-1}B$ if $ad - bc \neq 0$

EXERCISES 6.5

1. $(5, 8)$

3. $(6, -4)$

5. $(0.5, 0.2)$

7. $(4, 6, 3)$

9. $(6, -2, -3)$

11. $\left(\dfrac{1}{2}, \dfrac{1}{4}, \dfrac{3}{4} \right)$

6.6 CHAPTER REVIEW

1. $(5, 8, 6)$

2. $\left(\dfrac{1}{4}, \dfrac{3}{8}, \dfrac{-7}{8}\right)$

3. $(4, 5, 9)$

4. $(6, 9, 5)$

5. $\varnothing$

6. $(20 - 5z, 6 - z, z)$

7. $(3, 5, 4, 2)$

8. $(x, 18x, 13x)$

9. $(0, 0, 0)$

10. $\begin{bmatrix} 3 & -2 & 4 \\ 1 & 5 & -3 \\ 6 & 0 & -1 \end{bmatrix}$

11. $\begin{bmatrix} 4 & 3 & 4 \\ 5 & 8 & 0 \end{bmatrix}$

12. $\begin{bmatrix} 3 & 0 \\ -9 & 28 \end{bmatrix}$

13. $\begin{bmatrix} -1 & -3 \\ -2 & -4 \end{bmatrix}$

14. $\dfrac{-1}{2} \begin{bmatrix} 4 & -3 \\ -2 & 1 \end{bmatrix}$

15. $\begin{bmatrix} 7 & 15 \\ 10 & 22 \end{bmatrix}$

16. $\left(\dfrac{3}{2}, \dfrac{5}{2}\right)$

17. $(5, 2, 1)$

EXERCISES 7.1

1. $2^7 = 128$

3. x^{k+n}

5. x^{n+1}

7. $5^6 = 15{,}625$

9. x^6

11. x^{3n}

13. $8x^3$

15. $x^5 y^5$

17. $x^n y^n$

19. $\dfrac{216}{x^3}$

21. $\dfrac{x^5}{y^5}$

23. $\dfrac{x^k}{5^k}$

25. 49

27. 1

29. x^8

31. $\dfrac{1}{9^4} = \dfrac{1}{6{,}561}$

33. $\dfrac{1}{x^5}$

35. 1

37. $\dfrac{1}{x^2}$

39. $\dfrac{1}{x^{k-n}}$

41. x^7

43. $4y^6$

45. $81x^4 y^{12}$

47. $x^5 y^8$

49. $\dfrac{x^{12}}{10{,}000}$

51. $\dfrac{9}{x^{10}}$

53. $\dfrac{5}{x^2}$

55. $\dfrac{8y}{x^6}$

57. $\dfrac{625x^4}{y^4}$

59. x^3

61. 5^{4x}

63. $(x + 6)^4$

65. x^n

67. $\dfrac{1}{x^8}$

69. y^2

71. 6^x

73. $10^{x+3} = 1{,}000(10^x)$

EXERCISES 7.2

1. 1

3. 1

5. 1

7. $\frac{1}{2}$

9. 4

11. $\frac{12}{7}$

13. 10

15. $\frac{1}{32}$

17. 25

19. $\frac{25}{4}$

21. $\frac{49}{100}$

23. $\frac{y}{x}$

25. 4

27. 1

29. $\frac{1}{32}$

31. 1

33. 11

35. 150

37. 16

39. $\frac{1}{100,000}$

41. 100

43. $\frac{1}{125}$

45. $\frac{2}{x^3}$

47. $\frac{x^8}{81}$

49. $\frac{16}{x^2}$

51. $\frac{4}{25}$

53. $\frac{1}{x^5}$

55. $\frac{125}{xy^2}$

57. $\frac{1}{x+y}$

59. $\frac{x+y}{x}$

61. $\frac{y^6}{8x^9}$

63. $\frac{a^2+2ab+b^2}{ab}$

65. $\frac{4a^8b^2}{25}$

67. $\frac{1}{16y^{12}}$

69. $\frac{b^{n+1}}{a^{n+1}}$

71. $\frac{y^{4n}}{x^{2n}}$

73. $a^{2n}b^{2n-2}$

75. $\frac{b^{2n}-a^{2n}}{a^{2n}b^{2n}}$

EXERCISES 7.3

1. 2

3. 5

5. 5

7. 2

9. $\frac{1}{5}$

11. $\frac{3}{5}$

13. $\frac{1}{3}$

15. 7

17. 16

19. $\frac{1}{27}$

21. $\frac{1}{32}$

23. $\frac{36}{49}$

25. $\frac{1,000}{729}$

27. $\frac{1}{27}$

29. $\sqrt{10}$

31. $10\sqrt{10}$

33. $\sqrt[5]{1,000}$

35. $\frac{1}{5}$

37. 2

39. $\frac{7}{8}$

41. $\sqrt[3]{x^2}$

43. $\sqrt{x}$

45. $\frac{1}{\sqrt{x}} = \frac{\sqrt{x}}{x}$

47. $\frac{x^2}{y}$

49. x

51. $\dfrac{y^2}{x}$

53. $\sqrt[3]{y}$

55. $x + \sqrt[3]{x}$

57. $x - y$

59. $x - 1$

61. $\dfrac{y^2}{2x}$

EXERCISES 7.4

1. $6x\sqrt[3]{x^2}$

3. $2y^2\sqrt[5]{7y^2}$

5. $\sqrt{3x}$

7. $\sqrt[3]{12xy}$

9. $b\sqrt{5ab}$

11. $\sqrt{3t}$

13. $2x$

15. $\dfrac{\sqrt{10t}}{10t}$

17. $\dfrac{\sqrt[3]{6xy}}{6xy}$

19. $\dfrac{\sqrt[3]{49xy^2}}{7xy}$

21. $\dfrac{\sqrt{5x}}{5x}$

23. $\sqrt[4]{216t^3}$

25. $\sqrt[3]{3x}$

27. 6

29. $\sqrt[4]{1,000}$

31. $\sqrt{2t}$

33. $\sqrt[3]{2x^2}$

EXERCISES 7.5

1. $\{6\}$

3. $\emptyset$

5. $\{4\}$

7. $\{8, -2\}$

9. $\{0\}$

11. $\{8\}$

13. $\{0\}$

15. $\{2\}$

17. $\{9\}$

19. $\{6\}$

21. $\left\{\dfrac{5}{8}\right\}$

23. $\{-64\}$

25. $\{1, -4\}$

27. $\{100\}$

29. $\{14\}$

31. $y = x + a - 2\sqrt{ax}$

7.6 CHAPTER REVIEW

1. $125x^6$

2. $-4x^5$

3. $2,025x^{10}y^{10}$

4. $\dfrac{16x^8}{625y^4}$

5. x^8

6. $\dfrac{243}{x^5y^{10}}$

7. x

8. $\dfrac{x^{n-1}}{y^{2n}}$

9. 10^n

10. $\dfrac{10}{x^n}$

11. $\dfrac{25}{9}$

12. 500

13. $\dfrac{x^2}{36}$

14. $\dfrac{-1}{27y^6}$

15. $\dfrac{625}{y^4}$

16. x

17. $\dfrac{1}{16}$

18. $\dfrac{65}{64}$

19. $x^n y$

20. $\dfrac{xy}{x^2 + 2xy + y^2}$

21. $10\sqrt{10}$

22. $\dfrac{x^4 y^2}{4}$

23. $\dfrac{125}{x^3}$

24. $\dfrac{6\sqrt{6x}}{x^2}$

25. $\sqrt{5}$

26. $x^2\sqrt{y}$

27. $\dfrac{x^2 - 2x + 1}{x}$

28. $x\sqrt{x}$

29. $7x\sqrt[3]{2x}$

30. $2y\sqrt[5]{4y^2}$

31. $\dfrac{\sqrt[3]{12x^2 y}}{6xy}$

32. $\dfrac{\sqrt[4]{250t}}{5t}$

33. $\sqrt{7n}$

34. $2\sqrt{x}$

35. $\sqrt[3]{4x^2}$

36. $\dfrac{\sqrt[3]{100y^2}}{10y}$

37. $\{8\}$

38. $\varnothing$

39. $\{7, -5\}$

40. $\{9\}$

41. $\{6\}$

42. $\{26\}$

EXERCISES 8.1

1.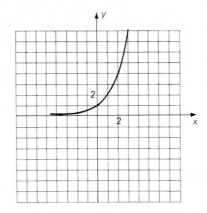

3.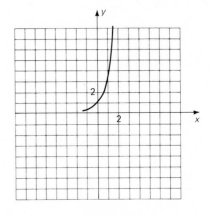

5.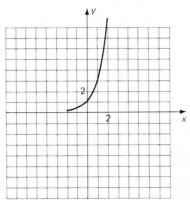

7.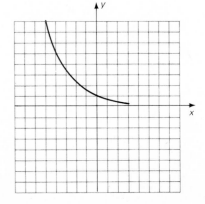

9. 4

11. -5

13. -2

15. 2.5

17. -2

19. $\dfrac{1}{2}$

21. 6

23. $\dfrac{1}{2}$

25. -1

27. 2^{10}

29. $5\sqrt{6}$

31. 9

33. $10^{4\sqrt{2}}$

35. $10^{\sqrt{6}}$

37. $2^{4\sqrt{5}}$

EXERCISES 8.2

1. $\log_3 81 = 4$

3. $\log_{10} 0.001 = -3$

5. $\log_9 3 = \dfrac{1}{2}$

7. $\log_{10} 1 = 0$

9. $\log_8 \dfrac{1}{4} = \dfrac{-2}{3}$

11. $\log_5 5 = 1$

13. $\log_{10} 10,000 = 4$

15. $\log_{10} 10\sqrt{10} = \dfrac{3}{2}$

17. $8^2 = 64$

19. $2^{-3} = 0.125$

21. $3^{3/2} = 3\sqrt{3}$

23. $27^{-2/3} = \dfrac{1}{9}$

25. $25^{-0.5} = 0.2$

27. $5^0 = 1$

29. $10^1 = 10$

31. 2

33. 5

35. -2

37. -4

39. 1.5

41. $\dfrac{2}{3}$

43. 1

45. 0

47. $\dfrac{-2}{3}$

49. $\dfrac{-2}{3}$

51. $\log_2 40$

53. $\log_{10} 8$

55. $\log_5 (31)^2 = \log_5 961$

57. $\log_{10} 6$

59. $\log_2 (120)^3$

61. $\log_3 \sqrt[5]{\dfrac{24}{35}}$

63. $\log_5 \dfrac{49}{6}$

65. $\log_2 52 \sqrt[3]{19}$

67. $\log_{10} \left(\dfrac{a}{b}\right)^3$

69. $\log_5 \sqrt{ab}$

71. $\log_2 \dfrac{a^2}{\sqrt{b}}$

73. 1.1761

75. 0.5229

77. 0.9030

79. 0.1505

81. 1.5562

85. 0.2594

89. $\dfrac{0.4771}{0.3010} \approx 1.585$

93. -0.05

97. 3, -3

83. 1.9542

87. -0.4515

91. 2.4

95. 30

99. -3.2

EXERCISES 8.3

1. 0.7657

5. 2.8451

9. -0.00656

13. -3.0969

17. 863.0

21. 0.592

25. 30.2

29. 0.0207

33. 1.5539

37. 58,695

3. 1.5416

7. 5.7042

11. -2.3372

15. 7.589

19. 4910.

23. 0.00255

27. 811.0

31. 0.000952

35. -2.1290

39. 0.6966

EXERCISES 8.4

1. 8430.

5. 24.3

9. 1922

13. 2.36

17. 3.227 sec.

21. 7.9

25. 3.2×10^{-8}

29. {6}

33. {7}

37. {12}

3. 0.157

7. 1.11

11. 0.0412

15. 2.39

19. 5.2

23. 2.0×10^{-5}

27. 3.98 decibels

31. {3, 6}

35. {4}

39. {4}

EXERCISES 8.5

1. $\log_3 20 = \dfrac{\log 20}{\log 3} \approx 2.73$

3. $\dfrac{\log 25}{3 \log 2} \approx 1.55$

ANSWERS

5. $3 + \dfrac{\log 4.5}{\log 6} \approx 3.84$

7. $\dfrac{\log 5}{\log 5 - \log 3} = \dfrac{\log 5}{\log \frac{5}{3}} \approx 3.15$

9. $\dfrac{2 \log 9 - \log 2}{2 \log 2 - \log 9} = \dfrac{\log 40.5}{\log \frac{4}{9}} \approx -4.56$

11. $\dfrac{\log 2}{\log 1.05} \approx 14.2$

13. 2.01

15. -0.432

17. 32.8

19. 0.0944

21. $\dfrac{\log 30}{\log 2} \approx 4.907$

23. $\dfrac{1}{\log 3} \approx 2.096$

25. 2.303

27. 3.532

29. -0.8347

31. 15.7 years

33. 361,608 bacteria

35. 8.04 years

37. 0.0595

8.6 CHAPTER REVIEW

1.

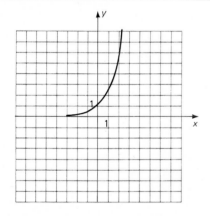

2.

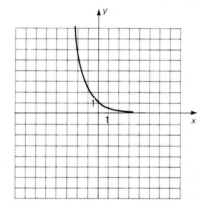

3. 3

4. $\dfrac{-1}{2}$

5. -3

6. $1 \pm \sqrt{7}$

7. 2

8. $\dfrac{-3}{2}$

9. $5^{3\sqrt{10}}$

10. 5^{20}

11. $6^{\sqrt{3}}$

12. 10^5

13. $\log_2 0.125 = -3$

14. $\log 91.2 = 1.96$

15. $\log_6 1 = 0$

16. $\log_8 8 = 1$

17. $10^{2.49} = 309$

18. $5^{-4} = 0.0016$

19. $(2.5)^1 = 2.5$

20. $6^0 = 1$

21. 5

22. $\dfrac{-3}{2}$

23. 8

24. 1

25. 0

26. -1

27. $\log_5 6.4$

28. $\log_3 \sqrt{35}$

29. $\log_b s^2 \sqrt[3]{r}$

30. $\log_b \left(\dfrac{4}{\sqrt{s}}\right)^3$

31. $\log_{10} \dfrac{(ab)^2}{\sqrt[5]{c}}$

32. $\log_2 \sqrt[4]{\dfrac{a}{bc^2}}$

33. 4.6830

34. -1.6289

35. -2.0144

36. 3.8761

37. 725.9

38. 0.07079

39. 0.4367

40. 24.55

41. 0.315

42. 1.104

43. 5.1

44. 2.5×10^{-9}

45. 4

46. 5, 6

47. $\dfrac{\log 15}{\log 2} \approx 3.91$

48. $\dfrac{\log 2}{2 \log 1.08} \approx 4.50$

49. $\dfrac{\ln 8.2}{3} \approx 0.701$

50. $\dfrac{-\ln 0.46}{2} \approx 0.388$

51. $\dfrac{\log 6}{\log 12} \approx 0.721$

52. $-\log 16 \approx -1.2041$

53. $e^{-2} \approx 0.1353$

54. $e^{1.75} \approx 5.755$

55. $\dfrac{\log 25}{\log 4} \approx 2.322$

56. $\dfrac{\log 1.24}{\log 3} \approx 0.1958$

57. 2.219

58. -1.139

EXERCISES 9.1

1. $R = 3 = P(-2)$

3. $R = 1 = P(1)$

5. $R = 57 = P(-3)$

7. $R = 0 = P\left(\dfrac{1}{2}\right)$

9. $R = 0 = P\left(\dfrac{-7}{4}\right)$

11. $R = 0 = P(\sqrt{5})$

13. $R = 0 = P(2i)$

15. $R = 32 = P(-2i)$

17. $1, x - 1$
$\quad -5, x + 5$

19. $3, x - 3$
$\quad -3, x + 3$

21. $\dfrac{5}{2}, x - \dfrac{5}{2}$

23. $-1, x + 1$

25. $1, x - 1; -1, x + 1; i, x - i; -i, x + i$

27. $\sqrt{2}, x - \sqrt{2}; 2i, x - 2i$

29. $(x - 1)(x^3 + x^2 + x + 4)$

31. $(x + 5)(x^2 - x + 5)$ **33.** $(2x - 1)(2x^3 + 3)$
35. $(x - 2 + \sqrt{3})(x^2 - \sqrt{3}\,x - 4 - 2\sqrt{3})$
37. $(x + 1 - i)(x^2 + ix - 1 - i)$ **39.** -2
41. 9   **43.** -21

45. 6   **47.** 0 or $\dfrac{1}{3}$

49. Yes   **51.** Yes   **53.** No
55. No   **57.** No   **59.** Yes

EXERCISES 9.2

1. $(x - 2)(x^2 - 6x + 4)$ **3.** $(x - 3)^2(x + 2)$
$\{2, 3 + \sqrt{5}, 3 - \sqrt{5}\}$ $\{3, -2\}$
5. $(x + 2)(x^2 - 2x + 4)$ **7.** $(x - \sqrt{3})(x + \sqrt{3})(x^2 + 3)$
$\{-2, 1 + i\sqrt{3}, 1 - i\sqrt{3}\}$ $\{\sqrt{3}, -\sqrt{3}, i\sqrt{3}, -i\sqrt{3}\}$
9. $(x + 1)(x^2 - 2x + 5)$ **11.** $\{3, 2i, -2i\}$
$\{-1, 1 + 2i, 1 - 2i\}$

13. $\left\{1, \dfrac{-1}{2}\right\}$ **15.** $\{-4, 1 + \sqrt{3}, 1 - \sqrt{3}\}$

17. $\{-1, 1 + i, 1 - i\}$ **19.** $\{\sqrt{2}, -\sqrt{2}, 3 + \sqrt{2}, 3 - \sqrt{2}\}$
21. $\{2 - \sqrt{5}, 2 + \sqrt{5}, -2 + \sqrt{5}, -2 - \sqrt{5}\}$
23. $\{\sqrt{2}, -\sqrt{2}, 1 - i, 1 + i\}$
25. $\{1 + \sqrt{2}, 1 - \sqrt{2}, -1 - 2i, -1 + 2i\}$

27. $\left\{2i, -2i, \dfrac{-1 + \sqrt{17}}{2}, \dfrac{-1 - \sqrt{17}}{2}\right\}$

29. $x^3 - 9x^2 + 33x - 25$ **31.** $x^3 - 3x^2 - 14x + 12$
33. $x^4 - 6x^3 + 7x^2 - 30x + 10$ **35.** $x^4 - 14x^3 + 67x^2 - 120x + 48$
37. $x^3 + 15x^2 + 75x + 125$ **39.** $x^4 - 4x^3 + 8x + 4$

41. $2 - \sqrt{3}, \dfrac{-1}{4}, c = 0, d = 1$

EXERCISES 9.3

1. $-1, 2$ **3.** $-1, 2, \dfrac{-3}{2}$

5. $2, \dfrac{-7}{4}$ **7.** None

9. 4   **11.** $5, -3, \dfrac{1}{2}$

13. None

15. $-2, \dfrac{1}{4}$

17. $\dfrac{1}{3}, -2$

19. $\left\{ -1, \dfrac{1}{2} \right\}$

21. $\left\{ \dfrac{3}{2}, \dfrac{-3}{2}, \dfrac{2i}{3}, \dfrac{-2i}{3} \right\}$

23. $\{1, -2, i\,2\sqrt{3}, -i\,2\sqrt{3}\}$

25. $\{5, 6, i\sqrt{2}, -i\sqrt{2}\}$

27. $\{3, 4, -1 + 2i, -1 - 2i\}$

29. $\left\{ \dfrac{-1}{3} \right\}$

31. Possible rational roots are $1, -1, 2, -2$
$P(1) = -1, P(-1) = -1, P(2) = 2, P(-2) = 2$

EXERCISES 9.4

1. One positive real root, zero or two negative real roots
3. One positive real root, zero or two negative real roots
5. No positive real roots, one negative real root, four imaginary roots
7. No positive real roots, no negative real roots, six imaginary roots
9. One or three positive real roots, one negative real root
11. $P(2) = -5, P(3) = 15$

13. $P(-4) = -5, P(-3) = 7$

15. $P(0) = 6, P(1) = -10$

17. $-2 < r < -1$

19. $-1 < r < 0, 0 < r < 1, -5, < r < -4$
21. $-2 < r < -1, 1 < r < 2$
23. 2.65

25. 3.25

27. -1.61

29. 0.33

31. 1.71
33. (1) By Descartes' rule of signs, there is exactly one real positive root and exactly one real negative root.
    (2) Since $P(x) \neq 0$ for x in $\{1, -1, 2, -2, 4, -4\}$, there are no rational roots.
35. (a) n odd.
    By Descartes' rule of signs, there is exactly one real positive root and no real negative roots.
    (b) n even.
    By Descartes' rule of signs, there is exactly one real positive root and exactly one real negative root.

9.5 CHAPTER REVIEW

1. -1

2. 46

3. $-2, x + 2; 4, x - 4$

4. 12

5. $R = 6, Q(x) = 2x^3 + 6x^2 + 3x + 3$

ANSWERS

6. $R = 8$, $Q(x) = x^2 + 2x - 3$

7. $\{4, 2 + \sqrt{5}, 2 - \sqrt{5}\}$

8. $\{-2, 3i, -3i\}$

9. $\{2, -1 + 2i, -1 - 2i\}$

10. $\{-3, 2\}$

11. $\{\sqrt{5}, -\sqrt{5}, 4 + \sqrt{2}, 4 - \sqrt{2}\}$

12. $\{1, -1, 3 - 2i, 3 + 2i\}$

13. $x^4 - 6x^3 + 12x^2 - 18x - 5$

14. $x^3 - 13x^2 + 56x - 80$

15. $\{3, -4, 1 + \sqrt{2}, 1 - \sqrt{2}\}$

16. $\left\{\dfrac{1}{3}, -3 + \sqrt{7}, -3 - \sqrt{7}\right\}$

17. $\left\{\dfrac{5}{2}, \dfrac{-5 + \sqrt{17}}{2}, \dfrac{-5 - \sqrt{17}}{2}\right\}$

18. $\{2, 3, -3, -1 + i\sqrt{3}, -1 - i\sqrt{3}\}$

19. One positive real root, two imaginary roots.

20. Either (3 positive real roots and 1 negative real root) or (1 positive real root, 1 negative real root, and 2 imaginary roots)

21. One positive real root, one negative real root, and two imaginary roots

22. $3 < r < 4$

23. $1 < r < 2$, $-3 < r < -2$

24. 1.37

25. 0.62, -1.62

EXERCISES 10.1

1. $-1, 1, 3, 5, 7$

3. $\dfrac{1}{3}, \dfrac{1}{9}, \dfrac{1}{27}, \dfrac{1}{81}, \dfrac{1}{243}$

5. $\dfrac{1}{3}, \dfrac{-1}{5}, \dfrac{1}{7}, \dfrac{-1}{9}, \dfrac{1}{11}$

7. $-6, 12, -24, 48, -96$

9. $0, \dfrac{1}{3}, \dfrac{1}{2}, \dfrac{3}{5}, \dfrac{2}{3}$

11. $0.43, 0.0043, 0.000043, 0.00000043, 0.0000000043$

13. $1, \dfrac{-1}{8}, \dfrac{1}{27}, \dfrac{-1}{64}, \dfrac{1}{125}$

15. $2, 8, 20, 40, 70$

17. $\dfrac{1}{3}, \dfrac{-1}{9}, \dfrac{1}{27}, \dfrac{-1}{81}, \dfrac{1}{243}$

19. $-x, \dfrac{x^2}{2}, \dfrac{-x^3}{3}, \dfrac{x^4}{4}, \dfrac{-x^5}{5}$

21. $-1 + 2 + 5 + 8 + 11$

23. $2 + 6 + 12 + 20$

25. $\dfrac{1}{9} + \dfrac{1}{27} + \dfrac{1}{81} + \dfrac{1}{243} + \dfrac{1}{729}$

27. $\dfrac{4}{5} + \dfrac{4}{25} + \dfrac{4}{125} + \cdots$

29. $0.35 + 0.0035 + 0.000035 + \cdots$

31. $\dfrac{-1}{6} + \dfrac{1}{12} - \dfrac{1}{20} + \dfrac{1}{30} - \dfrac{1}{42} + \cdots$

33. $-2\sqrt{2} + 3\sqrt{3} - 8 + 5\sqrt{5} - 6\sqrt{6}$

35. $\dfrac{x^2}{2} + \dfrac{x^4}{4} + \dfrac{x^6}{6} + \dfrac{x^8}{8} + \dfrac{x^{10}}{10} + \dfrac{x^{12}}{12}$

37. $x - \dfrac{x^2}{4} + \dfrac{x^3}{9} - \dfrac{x^4}{16} + \cdots$

39. $\dfrac{x}{5} + \dfrac{2x^2}{25} + \dfrac{3x^3}{125} + \dfrac{4x^4}{625} + \cdots$

41. $\sum_{k=1}^{5} 6k$

43. $\sum_{k=1}^{\infty} 2k$

45. $\sum_{k=1}^{\infty} \frac{1}{k}$

47. $\sum_{k=1}^{\infty} \frac{(-1)^{k+1}}{2^{k-1}}$

49. $\sum_{k=1}^{\infty} (-1)^{k+1} 6^k$

51. $\sum_{k=1}^{5} 5^{1-k}$

53. $\sum_{n=1}^{6} (-1)^{n+1}(1 - 2^{1-n})$

55. $1 + \sum_{k=1}^{5} (-1)^n \frac{2^n - 1}{2^n}$

EXERCISES 10.2

1. (a) If $n = 1$, then
$(2 \cdot 1 - 1) = 1^2,$
which is true.
(b) Assume true for $n = k$,
$1 + 3 + 5 + \cdots + (2k - 1) = k^2.$
Then for $n = k + 1$,
$1 + 3 + \cdots + (2k - 1) + (2k + 1) = k^2 + (2k + 1) = (k + 1)^2$
Therefore true for all positive integers n.

3. (a) If $n = 1$, then
$$1^2 = \frac{1(1 + 1)(2 + 1)}{6} = 1$$
Assume true for $n = k$,
$$1^2 + 2^2 + \cdots + k^2 = \frac{k(k + 1)(2k + 1)}{6}$$
Then for $n = k + 1$,
(b) $1^2 + 2^2 + \cdots + k^2 + (k + 1)^2 = \frac{k(k + 1)(2k + 1)}{6} + (k + 1)^2$
$$= \frac{(k + 1)(k + 2)(2k + 3)}{6}$$
Therefore true for all positive integers n.

5. (a) If $n = 1$, $2^{1-1} = 2^0 = 1$ and $2^1 - 1 = 1$,
which is true.
(b) Assume true for $n = k$,
$1 + 2 + \cdots + 2^{k-1} = 2^k - 1$
Then for $n = k + 1$,
$1 + 2 + \cdots + 2^{k-1} + 2^{(k+1)-1} = 2^k - 1 + 2^k$
$$= 2(2^k) - 1 = 2^{k+1} - 1$$
Therefore true for all positive integers n.

7. (a) If $n = 1$, then
$$\frac{1}{1 \cdot 3} = \frac{1}{3} = \frac{1}{2 + 1},$$
which is true.

(b) Assume true for $n = k$,

$$\frac{1}{1 \cdot 3} + \frac{1}{3 \cdot 5} + \cdots + \frac{1}{(2k-1)(2k+1)} = \frac{k}{2k+1}$$

Then for $n = k + 1$,

$$\frac{1}{1 \cdot 3} + \frac{1}{3 \cdot 5} + \cdots + \frac{1}{(2k-1)(2k+1)} + \frac{1}{(2k+1)(2k+3)}$$

$$= \frac{k}{2k+1} + \frac{1}{(2k+1)(2k+3)} = \frac{k+1}{2k+3} = \frac{(k+1)}{2(k+1)+1}$$

Therefore true for all positive integers n.

9. (a) If $n = 1$, then $1 \cdot 2 = 2 = \dfrac{1(1+1)(1+2)}{3}$,

which is true.

(b) Assume true for $n = k$,

$$1 \cdot 2 + \cdots + k(k+1) = \frac{k(k+1)(k+2)}{3}$$

Then for $n = k + 1$,

$$1 \cdot 2 + \cdots + k(k+1) + (k+1)(k+2)$$

$$= \frac{k(k+1)(k+2)}{3} + (k+1)(k+2)$$

$$= \frac{(k+1)(k+2)(k+3)}{3}$$

Therefore true for all positive integers n.

11. (a) If $n = 1$, then

$$3 = \frac{3(1)(1+1)}{2} = 3,$$

which is true.

(b) Assume true for $n = k$,

$$3 + 6 + \cdots + 3k = \frac{3k(k+1)}{2}$$

Then for $n = k + 1$,

$$3 + 6 + \cdots + 3k + 3(k+1) = \frac{3k(k+1)}{2} + 3(k+1)$$

$$= 3(k+1)\left(\frac{k}{2} + \frac{2}{2}\right)$$

$$= \frac{3(k+1)([k+1]+1)}{2}$$

Therefore true for all positive integers n.

13. (a) If $n = 1$, then

$$2^2 = 4 = \frac{2(1)(1+1)(2+1)}{3},$$

which is true.

(b) Assume true for $n = k$,

$$2^2 + 4^2 + \cdots + (2k)^2 = \frac{2k(k+1)(2k+1)}{3}$$

Then for $n = k + 1$,

$$2^2 + 4^2 + \cdots + (2k)^2 + [2(k + 1)]^2 = \frac{2k(k + 1)(2k + 1)}{3} + [2(k + 1)]^2$$

$$= \frac{2(k + 1)(k + 2)(2k + 3)}{3}$$

$$= \frac{2(k + 1)([k + 1] + 1)(2[k + 1] + 1)}{3}$$

Therefore true for all positive integers n.

15. (a) If $n = 1$, then

$$5^{-1} = \frac{1}{5} = \frac{5 - 1}{4(5)},$$

which is true.

(b) Assume true for $n = k$,

$$\frac{1}{5} + \cdots + \frac{1}{5^k} = \frac{5^k - 1}{4(5^k)}$$

Then for $n = k + 1$,

$$\frac{1}{5} + \cdots + \frac{1}{5^k} + \frac{1}{5^{k+1}} = \frac{5^k - 1}{4(5^k)} + \frac{1}{5^{k+1}}$$

$$= \frac{5^{k+1} - 5 + 4}{4(5^{k+1})} = \frac{5^{k+1} - 1}{4(5^{k+1})}$$

Therefore true for all positive integers n.

17. (a) If $n = 1$, then

$$ar^{1-1} = a = \frac{a(1 - r^1)}{1 - r},$$

which is true.

(b) Assume true for $n = k$,

$$a + ar + \cdots + ar^{k-1} = \frac{a(1 - r^k)}{1 - r}$$

Then for $n = k + 1$,

$$a + ar + \cdots + ar^{k-1} + ar^k = \frac{a(1 - r^k)}{1 - r} + ar^k$$

$$= \frac{a - ar^k + ar^k - ar^{k+1}}{1 - r}$$

$$= \frac{a(1 - r^{k+1})}{1 - r}$$

Therefore true for all positive integers n.

19. Induction on n

(a) If $n = 1$, then $x^m x^1 = xx^{(m+1)-1} = x^{m+1}$ by definition.

(b) Assume true for $n = k$,

$$x^m x^k = x^{m+k}.$$

Then for $n = k + 1$,

$$x^m x^{k+1} = x^m(x^k x) = x^{m+k} x^1 = x^{m+(k+1)}$$

Therefore true for all positive integers n.

21. (a) If $n = 1$, then $(xy)^1 = xy = x^1 y^1$ by definition.

(b) Assume true for $n = k$,
$$(xy)^k = x^k y^k.$$
Then for $n = k + 1$,
$$(xy)^{k+1} = (xy)^k (xy)^1 = x^k y^k x^1 y^1 = x^{k+1} y^{k+1}$$
Therefore true for all positive integers n.

23. (a) If $n = 1$, then $\left(\dfrac{x}{y}\right)^1 = \dfrac{x}{y} = \dfrac{x^1}{y^1}$ by definition.

(b) Assume true for $n = k$,
$$\left(\frac{x}{y}\right)^k = \frac{x^k}{y^k}.$$
Then for $n = k + 1$,
$$\left(\frac{x}{y}\right)^{k+1} = \left(\frac{x}{y}\right)^k \left(\frac{x}{y}\right)^1 = \frac{x^k}{y^k} \cdot \frac{x^1}{y^1} = \frac{x^{k+1}}{y^{k+1}}$$
Therefore true for all positive integers n.

25. (a) If $n = 1$, then
$$1 = \frac{1(1 + 1)}{2} = 1,$$
which is true.

(b) Assume true for $n = k$,
$$\sum_{i=1}^{k} i = \frac{k(k + 1)}{2}$$
Then for $n = k + 1$,
$$\sum_{i=1}^{k+1} i = k + 1 + \frac{k(k + 1)}{2} = \frac{2k + 2 + k^2 + k}{2} = \frac{(k + 1)(k + 2)}{2}$$
Therefore true for all positive integers n.

27. (a) If $n = 1$, then
$$1^3 = 1 = \frac{1(2)^2}{4},$$
which is true.

(b) Assume true for $n = k$,
$$\sum_{i=1}^{k} i^3 = \frac{k^2(k + 1)^2}{4}$$
Then for $n = k + 1$,
$$\sum_{i=1}^{k+1} i^3 = (k + 1)^3 + \frac{k^2(k + 1)^2}{4} = \frac{4(k + 1)^3 + k^2(k + 1)^2}{4}$$
$$= \frac{(k + 1)^2(4[k + 1] + k^2)}{4} = \frac{(k + 1)^2(k + 2)^2}{4}$$
$$= \frac{(k + 1)^2([k + 1] + 1)}{4}$$
Therefore true for all positive integers n.

29. $\dfrac{2n^3 + 3n^2 - 7n}{2}$

31. $\dfrac{4n^3 + 11n^2 - n}{2}$

33. $\dfrac{n(n-1)(n+1)(n+2)}{4}$

35. $\dfrac{n(n+1)(3n^2+7n+8)}{12}$

EXERCISES 10.3

1. 3, 8, 13, 18

3. $-10, -4, 2, 8, 14, 20$

5. 12, 10, 8, 6, 4

7. 19, 15, 11, 7, 3

9. 79

11. -213

13. 21

15. $10 - 15\sqrt{2}$

17. $x + 13k$

19. 79

21. 5, 0, -5

23. 48, 51

25. 1.2, 1.4, 1.6, 1.8

27. 0.125, 0.250, 0.375, 0.500, 0.625, 0.750, 0.875

29. 2,550

31. -36

33. $\dfrac{3n(n+1)}{2}$

35. $2n(n+2)$

37. 662.5

39. 400

41. 30,100

43. $\dfrac{n(1-3n)}{2}$

45. $\dfrac{n(n+1)}{2}$

47. 13

49. \$12,250

51. \$12.30

53. Job 1, \$13,000
Job 2, \$13,200

55. \$1,600

57. 48

59. \$9,031.25, 101 ft.

EXERCISES 10.4

1. 5, 10, 20, 40, 80, 160

3. $6, 2, \dfrac{2}{3}, \dfrac{2}{9}, \dfrac{2}{27}$

5. 1, -4, 16, -64

7. $3, \dfrac{-3}{2}, \dfrac{3}{4}, \dfrac{-3}{8}, \dfrac{3}{16}, \dfrac{-3}{32}$

9. 2,401

11. 512

13. 9×10^{-14}

15. $\dfrac{256}{243}$

17. $\dfrac{\sqrt{2}}{2}$

19. 36, 24

21. 6, 18, 54 or -6, 18, -54

23. $4\sqrt[3]{5}, 4\sqrt[3]{25}$

25. 15

27. -20

29. $10\sqrt{6}$

31. $-4\sqrt{6}$

33. 19,608

35. 189

37. 62.5008

39. 9,841

41. $\dfrac{5,320}{3,645}$

43. 515.20151

45. 2

47. $\dfrac{3}{5}$

49. $\dfrac{25}{2}$

51. $\dfrac{2}{275}$

53. 300

55. 25

57. $\dfrac{4}{9}$

59. $\dfrac{28}{11}$

61. $\dfrac{80}{99}$

63. $\dfrac{113}{90}$

65. $\dfrac{6,126}{999}$

67. 360 in.

69. 8,190

71. 5

73. (a) 2.56%   (b) 6

75. $128

10.5 CHAPTER REVIEW

1. 0, 2, 6, 12, 20

2. $\dfrac{3}{2}, \dfrac{-9}{4}, \dfrac{27}{8}, \dfrac{-81}{16}, \dfrac{243}{32}$

3. $2, \dfrac{\sqrt{2}}{2}, \dfrac{2\sqrt{3}}{9}, \dfrac{1}{4}, \dfrac{2\sqrt{5}}{25}$

4. 1, 3, 6, 10, 15

5. $3 + \dfrac{3}{7} + \dfrac{3}{17} + \dfrac{3}{31} + \dfrac{3}{49} + \dfrac{3}{71}$

6. $\dfrac{x^2}{4} - \dfrac{x^4}{16} + \dfrac{x^6}{36} - \dfrac{x^8}{64} + \cdots$

7. (a) If $n = 1$, then
$$1 = \frac{1}{2}(3 - 1) = 1,$$
which is true.

(b) Assume true for $n = k$,
$$1 + \cdots + (3k - 2) = \frac{k(3k - 1)}{2}.$$
Then for $n = k + 1$,
$$1 + \cdots + (3k - 2) + (3k + 3 - 2) = \frac{k(3k - 1)}{2} + 3k + 1$$
$$= \frac{3k^2 - k + 6k + 2}{2}$$

$$= \frac{(k+1)(3k+2)}{2} = \frac{(k+1)(3[k+1]-1)}{2}$$

Therefore true for all positive integers n.

8. (a) If $n = 1$,

$$1 - 1 = 0 = \frac{1(0)}{3},$$

which is true.

(b) Assume true for $n = k$,

$$1 + \cdots + (k^2 - k) = \frac{k(k^2 - 1)}{3}.$$

Then for $n = k + 1$,

$$1 + \cdots + (k^2 - k) + (k^2 + 2k + 1 - k - 1) = \frac{k(k^2 - 1)}{3} + k^2 + k$$

$$= \frac{k(k+1)}{3}(k - 1 + 3) = \frac{(k+1)k(k+2)}{3}$$

$$= \frac{(k+1)[(k+1)^2 - 1]}{3}$$

Therefore true for all positive integers n.

9. 79

10. 5.125

11. 91

12. $\frac{1}{3}, \frac{2}{3}$

13. $-3.5, -1, 1.5$

14. 975

15. $7n - n^2$

16. $\frac{729}{4}$

17. $\frac{-512}{243}$

18. 100, 20

19. 40, 100, 250, or $-40, 100, -250$

20. $5\sqrt{6}$

21. -28

22. 671,846

23. $15 + 7\sqrt{2}$

24. 4

25. $\frac{27}{5}$

26. $\frac{175}{33}$

EXERCISES 11.1

1. 24	3. 720	5. 5,040
7. 6,720	9. 720	11. 30
13. 72	15. 870	17. 40
19. 60	21. 360	23. 1,440
25. 729	27. 144	29. 720
31. 5,040	33. 5,040	

ANSWERS

EXERCISES 11.2

1. 84

3. 210

5. 1

7. 8

9. 21

11. 105

13. 120

15. 1,330

17. 56

19. 2,520

21. 4,410

23. 4,512

25. 4,400

EXERCISES 11.3

1. $x^3 + 12x^2 + 48x + 64$

3. $16x^4 - 96x^3y + 216x^2y^2 - 216xy^3 + 81y^4$

5. $t^{10} + 10t^8 + 40t^6 + 80t^4 + 80t^2 + 32$

7. $y^7 + 7y^6 + 21y^5 + 35y^4 + 35y^3 + 21y^2 + 7y + 1$

9. $x^8 - 8x^7 + 28x^6 - 56x^5 + 70x^4 - 56x^3 + 28x^2 - 8x + 1$

11. $x^6 + 6x^4 + 15x^2 + 20 + \dfrac{15}{x^2} + \dfrac{6}{x^4} + \dfrac{1}{x^6}$

13. $x^9 - 9x^8y + 36x^7y^2 - 84x^6y^3 + 126x^5y^4 - 126x^4y^5 + 84x^3y^6 - 36x^2y^7 + 9xy^8 - 1$

15. $17{,}010x^6$

17. $14{,}080y^9$

19. $\dfrac{126}{x}$

21. $860{,}160x^{13}y^2$

23. 1.268

25. 1.71

27. 0.737

29. 107.21

31. 16

EXERCISES 11.4

1. $1 + \dfrac{x}{3} - \dfrac{x^2}{9} + \dfrac{5x^3}{81} - \dfrac{10x^4}{243}$

3. $1 - \dfrac{x}{2} - \dfrac{x^2}{8} - \dfrac{x^3}{16} - \dfrac{5x^4}{128}$

5. $1 - 2x^2 + 3x^4 - 4x^6 + 5x^8$

7. $\dfrac{\sqrt{2}}{2}\left(1 - \dfrac{x}{4} + \dfrac{3x^2}{32} - \dfrac{5x^4}{128} + \dfrac{35x^6}{2{,}048}\right)$

9. $x^{1/4}\left(1 - \dfrac{1}{4x^2} - \dfrac{3}{32x^4} - \dfrac{7}{128x^6} - \dfrac{77}{2{,}048x^8}\right)$

11. 1.007

13. 0.9980

15. 0.2582

17. 2.024

19. 0.8203

EXERCISES 11.5

1. $\frac{1}{3}$ 3. $\frac{2}{3}$ 5. 0

7. $\frac{1}{4}$ 9. $\frac{3}{4}$ 11. 1

13. $\frac{1}{13}$ 15. $\frac{1}{2}$ 17. $\frac{3}{4}$

19. $\frac{2}{13}$ 21. $\frac{1}{6}$ 23. $\frac{3}{4}$

25. $\frac{1}{2}$ 27. $\frac{1}{12}$ 29. $\frac{5}{36}$

31. $\frac{1}{18}$ 33. $\frac{2}{9}$ 35. $\frac{5}{33}$

37. $\frac{28}{33}$ 39. $\frac{11}{188}$ 41. $\frac{177}{188}$

43. $\frac{1}{5,525}$ 45. $\frac{5,524}{5,525}$ 47. 68%, approx.

49. 98%, approx.   51. 18%, approx.   53. $\frac{c(20,5)}{c(80,5)} = \frac{51}{79,079}$

55. (a) 125   (b) Yes   57. About 210

EXERCISES 11.6

1. (a) $\frac{1}{13}$ (b) $\frac{1}{17}$ (c) $\frac{1}{12}$ 3. $\frac{3}{10}$ 5. $\frac{3}{14}$

7. $\frac{7}{13}$ 9. $\frac{1}{2}$ 11. $\frac{1}{169}$

13. $\frac{1}{13}$ 15. $\frac{1}{17}$ 17. $\frac{25}{204}$

19. $\frac{7}{30}$ 21. $\frac{4}{5}$ 23. $\frac{1}{5}$

25. $\frac{1}{2}$ 27. $\frac{3}{8}$ 29. $\frac{7}{8}$

31. $\frac{13}{20}$ 33. $\frac{7}{20}$ 35. $\frac{9}{16}$

37. 23%   39. 50%   41. 14%

43. $\frac{5}{18}$ 45. $\frac{5}{216}$ 47. (a) $\frac{1}{25}$ (b) $\frac{8}{25}$

49. $\dfrac{1,024}{10,000,000,000} \approx \dfrac{1}{10,000,000}$

51. $\dfrac{4}{25}$ **53.** $\dfrac{8}{125}$ **55.** $\dfrac{5}{8}$

57. $\dfrac{1}{8}$ **59.** $\dfrac{14}{33}$ **61.** $\dfrac{16}{33}$

63. $\dfrac{5}{28}$ **65.** $\dfrac{15}{56}$ **67.** $\dfrac{19}{396}$

69. $\dfrac{97}{990}$ **71.** $\dfrac{13}{24}$

11.7 CHAPTER REVIEW

1. 40,320   **2.** 43,680   **3.** 720
4. 32,760   **5.** 362,880   **6.** 1,320
7. 5,040   **8.** 56   **9.** 210
10. 1   **11.** 21   **12.** 44,352
13. $729x^6 - 1458x^5 + 1215x^4 - 540x^3 + 135x^2 - 18x + 1$
14. $32x^5 + 40x^4 + 20x^3 + 5x^2 + 0.625x + 0.03125$

15. $448y^5$ **16.** $\dfrac{210}{x^2}$ **17.** 5,194

18. 0.598   **19.** 1.012   **20.** 1.504

21. (a) $\dfrac{21}{26}$ **22.** (a) $\dfrac{5}{36}$ **23.** (a) $\dfrac{1}{38}$

     (b) $\dfrac{25}{676}$ (b) $\dfrac{1}{9}$ (b) $\dfrac{9}{19}$

     (c) $\dfrac{651}{676}$ (c) $\dfrac{5}{6}$ (c) $\dfrac{6}{19}$

     (d) 1   (d) 0   (d) $\dfrac{1}{19}$

24. $\dfrac{2}{15}$ **25.** (a) $\dfrac{5}{6}$ (b) $\dfrac{1}{2}$

26. (a) $\dfrac{7}{8}$ **27.** (a) $\dfrac{1}{54}$ **28.** (a) $\dfrac{1}{28}$

     (b) $\dfrac{1}{4}$ (b) $\dfrac{1}{81}$ (b) $\dfrac{13}{28}$

     (c) $\dfrac{1}{2}$ (c) $\dfrac{15}{28}$

     (d) $\dfrac{1}{8}$ (d) $\dfrac{3}{14}$

INDEX